PUTNAM'S HISTORY OF AIRCRAFT

Faster, Further, Higher

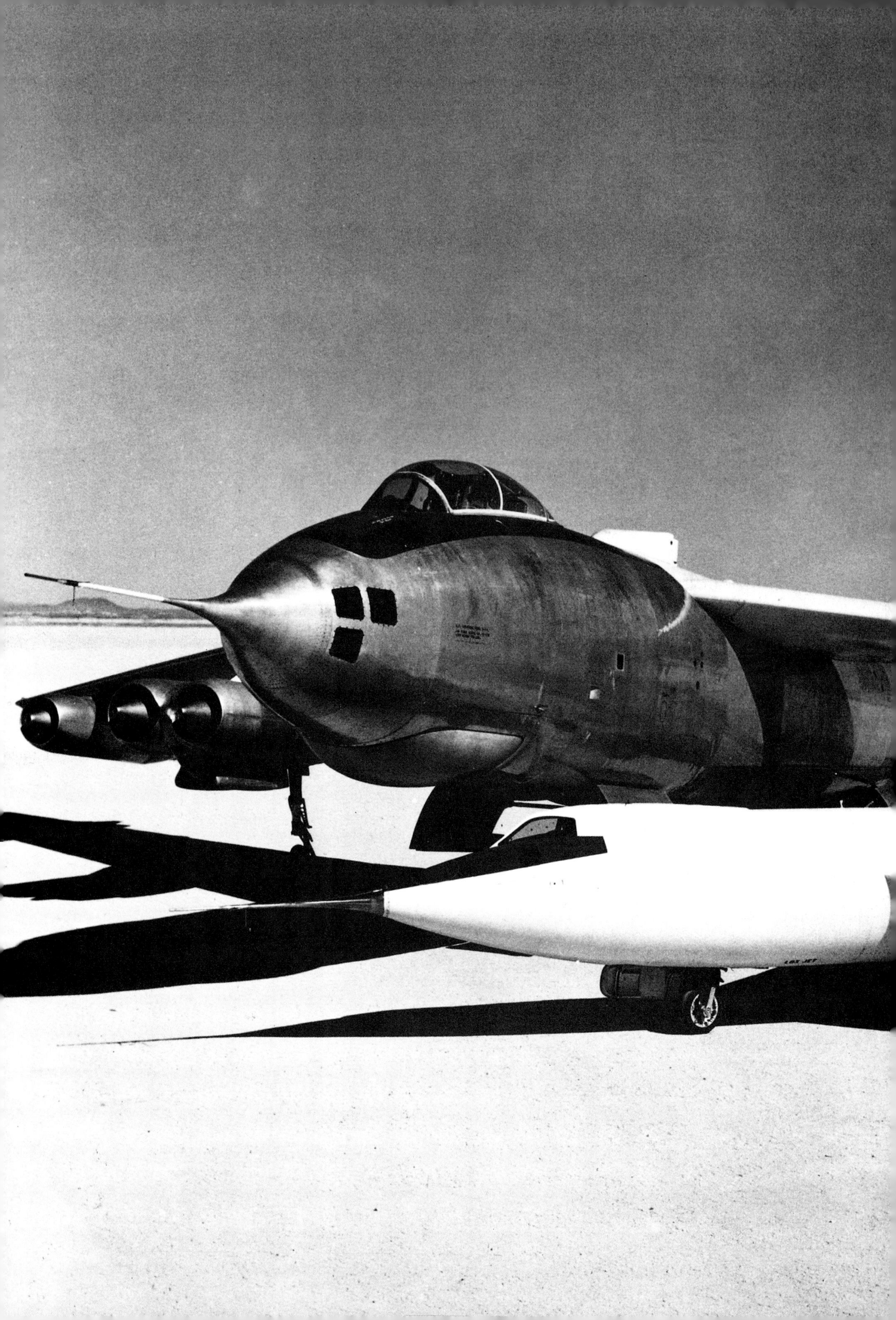

PUTNAM'S HISTORY OF AIRCRAFT

Faster, Further, Higher

Leading-edge Aviation Technology since 1945

Series Editor: Philip Jarrett

PUTNAM AERONAUTICAL BOOKS

Title page photograph: Two leading-edge aeroplanes. The first production Boeing B-47A Stratojet bomber and a Douglas D-558-2 Skyrocket No.2 at NACA's High Speed Flight Research Station at Edwards AFB, California, in the early 1950s.

Faster, Further, Higher

Series Editor: Philip Jarrett
Philip Jarrett is a freelance author, editor, sub-editor and consultant specialising in aviation. He began writing on aviation history in 1967, and in 1971 became assistant editor of *Aerospace*, the Royal Aeronautical Society's newspaper. He was assistant editor of *Aeroplane Monthly* from 1973 to 1980, and production editor of *Flight International* from 1980 to 1989.

Bruce Astridge
An engineer educated in New Zealand, Bruce Astridge worked for Rolls-Royce for thirty-eight years until his retirement in 1988. His early, and prime career was in gas turbine engine performance, in both UK and overseas posts. After a period in product support and a North American posting in engine and aircraft evaluation, his last position for the company was as Head of Aircraft Performance Evaluation.

Dr Norman Barfield
After joining Vickers-Armstrongs as an apprentice in 1947, Norman Barfield spent the whole of his working life there, concentrating on the design, engineering development, promotion and management of the company's postwar commercial and military aircraft family.

Les Coombs
After service in the RAF from 1939 to 1946, Les Coombs became a researcher and writer on avionics and ergonomics. He specialises in the history of air force technology and in the design and history of the aircraft cockpit. He is the author of *The Aircraft Cockpit* and *Fighting Cockpits 1914-2000*.

Colin Cruddas
In a design and flight test engineering career extending more than forty years, Colin Cruddas has worked for Fairey Aviation, Blackburn Aircraft and the British Aircraft Corporation in the UK, for Boeing and McDonnell Douglas in the USA and for Atlas Aircraft in South Africa. In 1979 he became engineering manager of the Aerospace Components Division of Flight Refuelling Ltd, then head of customer support. Leaving in 1992, he became the consultant archivist to Cobham plc and wrote his first book, *In Cobham's Company*.

Ian Goold
Ian Goold is an international aviation writer, editor and consultant. After an initial career in the British Aircraft Corporation structures design office, he spent twenty years on the editorial staff of *Flight International* before working freelance. Specialising in civil aviation, he contributes to publications in Europe, North America and the Middle East.

Dr Richard Hallion
Dr Richard P Hallion is the Air Force Historian at Headquarters, United States Air Force, Washington, D.C. He has been a curator and visiting professor at the National Air and Space Museum; has taught for the University of Maryland, Chapman College, and Golden Gate University; served as the historian of the Air Force Flight Test Center; and has been a visiting professor at the US Army War College.

Mike Hirst
A university lecturer, Mike Hirst has had an interest in aviation since childhood, and takes an especial interest in the technical development and operation of aircraft, both civil and military. He was technical editor of *Flight International*, and has served in design and education roles in the aviation industry. He has been published extensively in magazines and books.

Ian Moir
During twenty years as an RAF Engineering Officer, Ian Moir was involved with the first-line servicing, maintenance and rectification of fast jets and transport aircraft, and in systems development and project work relating to the entry into service of the Panavia Tornado. He then spent eighteen years at Smiths Industries on the introduction of avionics technology into aircraft utilities systems, selecting and developing new integrated systems for the Boeing AH-64C/D Longbow Apache attack helicopter and the Boeing 777. He is now an international aerospace consultant.

Stephen Ransom
After studying aeronautical engineering at the University of Salford, Stephen Ransom worked in the British Aircraft Corporation's advanced projects office. Moving to Germany in 1976, he was involved with the Spacelab programme at ERNO Raumfahrttechnik before working on advanced projects at VFW and, from 1979 to 1983, in MBB's Special Projects Division. He then returned to the space industry, working on spacecraft and planetary exploration projects. He is the co-author of *English Electric Aircraft and their Predecessors* and a two-volume history of the Me 163 rocket-powered interceptor.

Jeremy C Scutts
Jerry Scutts has been a freelance aviation author since 1970 and to has published numerous books, some of which have included his own profile and perspective artwork. A journalist and editor for over twenty years, he was a press officer with he UK Civil Aviation Authority in the early 1980s, before publishing his own aviation magazine. He specialises in combat aircraft operations and markings, with an emphasis on the US air forces and the Luftwaffe.

John Snow
A graduate of Southampton University and the Cranfield Institute of Technology, for the past seven years John Snow has been a senior lecturer with the Air Transport Group, Cranfield College of Aeronautics. He was previously Vice-President of Market Development and Engineering at Saab Aircraft, and before that he held a similar position at Airbus Industrie and worked with carriers in Scandinavia.

Ray Whitford
A senior lecturer at Cranfield University, Ray Whitford started his career with the British Aircraft Corporation in the 1960s and went on to write two books and more than forty papers on aircraft design. He has also been a professor at the United States Air Force Academy. In 1997 he won one of the Royal Aeronautical Society's 'Aerospace Journalist of the Year' awards.

© Putnam Aeronautical Books 2002

All rights reserved. No part of this book may be reproduced or transmitted in any form without prior written permission from the publisher.

First published in the United Kingdom in 2002 by Putnam Aeronautical Books.
Putnam Aeronautical Books is a division of Chrysalis Books plc
64 Brewery Road
London N7 9NT
www.putnamaeronautical.com

A member of Chrysalis Books plc

9 8 7 6 5 4 3 2 1

A CIP catalogue record for this book is available from the British Library.

ISBN 0 85177 876 3

Designed and typeset by Stephen Dent.
Printed and bound in Spain.

Contents

	Introduction	7
1	New-age Aerodynamics	10
2	Technology for the Supersonic Era	29
3	Extending the Reach	53
4	The Upper Reaches	73
5	Taking the Strain	92
6	Propulsion	111
7	New-age Systems	136
8	Exploring the Limits	170
9	Advanced Aircraft in Production	189
10	Demise of the Drawing Board	203
11	Simulation	212
12	The Future SST	228
13	Stepping into Space	241
	Index	254

Introduction
Philip Jarrett

It is appropriate that the aeronautical term 'leading edge' has become an adjective to describe anything at the forefront of its technology, as aerospace technology has usually led where others have followed. Since the beginning of the twentieth century, aircraft have stood at the forefront of human achievements, pointing the way ahead, and there is every indication that this will continue to be the case in the twenty-first century. The second half of the last century was especially exciting as far as aviation was concerned. The advent of the jet engine, supersonic flight, mass air transport and space travel made it one of the most outstanding eras of technical development in the whole of human history. Hand-in-hand with these advances came advances in materials and manufacturing, instrumentation, testing, medicine, chemistry, information technology and communications, and all the other allied disciplines upon which the world of aeronautics depended, and without which it would not have progressed. Jet engines required metals that could withstand extreme heat without weakening; high-speed flight brought problems of kinetic heating and the protection of systems and the crew and passengers. Barriers that had to be overcome ranged from the so-called 'sound barrier' that appeared to exist at Mach 1 to human physical limitations. The speed of sound was exceeded in level flight in 1947, with the help of rocket propulsion. Flight at the edge of space and at hypersonic speeds (greater than Mach 5) required new alloys of titanium, while jet flight at 85,000ft demanded complex engine-inlet control systems to enable turbojets to cope with the thinner air. In-flight refuelling technologies and the drive for fuel efficiencies extended the non-stop range of aircraft to thousands of miles.

It fell to designers, scientists, engineers, aerodynamicists and test pilots to tackle the numerous challenges that stood in the path of progress. That they have repeatedly done so with extraordinary success testifies to their skill, ingenuity, foresight, boldness and, often, tremendous courage. As more was learnt, so the methodology improved, and the ever-present dangers at the leading edge were rendered less of a threat to human life. Even so, the dangers are still there, but there seems to be no lack of people prepared to stare them in the face and explore the realms on the borders of our knowledge and understanding.

There is no dedication in this volume, but if there was, it might be to the test pilots who shoulder the responsibility of proving the soundness of the designers' and engineers' theories by taking aeroplanes aloft and systematically putting them through their paces. Even in this age of unmanned air vehicles there are frequent occasions when the ultimate test is sending the new machine aloft for its first piloted flights. And despite all that has been learnt, there are still many problems that have not been fully resolved. The F-117 fell victim to flutter, a phenomenon first encountered in the early decades of the twentieth century and still not eliminated from aeroplane design. Aeroplanes continue to succumb to natural forces as well as human fallibility, and as long as that is so, there will be a need for people with adventurous spirits, whether their skills be directed into computer-based design, advanced engineering or piloting itself.

This volume concentrates on leading-edge aviation technology since 1945; the key developments and the research aircraft and advanced demonstrators that prepared the way for today's sophisticated civil and military aeroplanes. Advancing technology has persistently extended the limits of aeroplane performance — the speed, range and altitude that give this book its title. The enhancement that occurred as part of national war efforts during the Second World War continued into the postwar world of confrontation between the Soviet Union and the USA. Aviation's future was explored in manufacturers' experimental programmes, laboratories and high-security governmental research establishments.

However, this volume is not limited to research flying alone. It also examines the crucial roles played by technology in equipping pilots and scientists with the means of coping with faster speeds, higher altitudes and greater distances than ever before. The volume concludes with the latest challenge to be attacked by technology; winged aircraft that can cross the boundary between the Earth's atmosphere and space. Even in this specialised field there is no shortage in the variety of ingenious solutions and methods devised by scientists and engineers worldwide. Who is to deny that their expertise will come in useful if humankind ever sets foot on another planet and is faced with new problems of mobility over its surface? Judging by their previous record, they will tackle that challenge with the same resourcefulness and ingenuity that enabled their forebears to conquer skies above the Earth.

In the opening chapter, Dr Norman Barfield looks at the way advances in transonic and supersonic aerodynamics have led aeroplane design, and spotlights significant aerodynamic developments. After observing the distinct 'phases and fashions' through which combat aircraft have passed, and the remarkable developments in the world of transport aircraft, he surveys the latest developments and future trends, often dictated by the essential need to keep costs in this most expensive of industries at a sustainable level. Lastly, he takes a look at the technological spin-off to other high-technology industries.

Dr Richard Hallion then examines the developments in investigations into high-speed flight, and the various approaches by aircraft manufacturers and research institutes worldwide. He pays particular attention to the initial use of research aeroplanes to probe the transonic region, when windtunnels failed to yield reliable information, and to the various configurations tested.

The development of practical in-flight refuelling systems to extend non-stop range is covered by Colin Cruddas, who first of all surveys the pioneering developments in the interwar years, starting with simple transfers of cans of fuel and gradually progressing to practical aircraft-to-aircraft fuel transfer using hose systems, pioneered in the UK by the Royal Aircraft Establishment and Sir Alan Cobham. Although air-to-air refuelling was little used during the Second World War, it was enthusiastically taken up by the military in the postwar years, and various systems were adopted. The author describes these, and recounts some of the outstanding flights achieved.

Jerry Scutts pinpoints the significant stages in the conquest of the upper atmosphere in his chapter on pressurisation and high-altitude flight. Although crude oxygen systems were used during the First World War, it was not until the 1930s that effective pressurised cockpits and cabins were developed. Although the first pressurised airliner flew in 1937, the onset of war delayed further developments in this direction, though pressurised crew compartments were used in a number of wartime aircraft on both sides. Postwar, the technologies evolved steadily, special clothing protecting the pilots and crew of small military aircraft, and high-altitude flight in comfortable pressurised cabins replacing bumpy low-altitude flights for the commercial passengers. The unceasing effort to go higher has now taken humans into space, and the author touches on the techniques and machines that have enabled us to travel beyond the confines of our planet.

None of the foregoing could have happened without the appropriate materials. The unprecedented demands made on airframe structures by high-speed and high-altitude flight brought needs for new materials and new structural techniques, and these form the subject of Ray Whitford's chapter. Likewise, newly-developed aircraft configurations also brought structural problems. The author describes integral construction, the challenges of aeroelasticity, and the development of suitable metals and, more recently, composites, and the means of fashioning them.

Next, Bruce Astridge delves into the challenges and solutions in the post-war development of jet engines, and looks at future prospects in the fascinating field of aircraft propulsion. From the ultra-high-powered piston engines of the late 1940s, he turns to the first early jets and rockets, and then surveys early high-speed projects. Second-generation turbojets brought great improvements in efficiency and hence, performance, and there then followed the development of supersonic turbojets. Ever-increasing demands of range and fuel efficiency for commercial aircraft led the development of the turbofan, and eventually the high-bypass turbofan, which began as a powerplant for large military transports. This chapter concludes with a look at future engine technologies and alternatives to oil-based fuels.

Also included in this chapter is Leonard Clow's masterly cutaway drawing of the Rolls-Royce Avon-powered Fairey F.D.2, the first aircraft to exceed 1,000mph in level flight. First published in *Flight* in 1958, this classic example of the technical draughtsman's art includes detail diagrams of the aircraft's nose-droop mechanism and air-conditioning system.

In 'New-age Systems', Ian Moir presents a study of the evolution of modern aircraft systems and cockpits, coming up to date with glass cockpits and fly-by-wire. In separate sections covering sensors, communication and navigation aids, navigation, displays, safety aids flight control and future air navigation systems, he describes their purpose and means of operation in terms that will help the interested layman grasp the basics of this complex subject.

In 'Exploring the Limits', the many different aspects of postwar research flying and the procedures adopted are covered by Ian Goold, who presents some significant examples of experimentation using both purpose-designed aircraft and specially modified test beds to obtain essential data throughout the speed range and under all regimes and conditions. Such machines have served to evaluate everything from aerodynamics to icing, and to allow new control systems and technology to be proved in flight and brought to a level of refinement that enables them to be put into widespread service.

Mike Hirst then describes the problems encountered in putting advanced aircraft into production, and the development of new manufacturing techniques to cope with the difficulties posed by exceptional airframe heating, exceptionally hard metals or materials such as carbon-fibre reinforced plastic. In the following chapter he looks at the momentous changes wrought in design and manufacture by the universal adoption of computer systems that have rendered the drawing board extinct, enabling aeroplanes to be created on screen. More than that, it is now possible to determine the necessary thickness of wing skins, the anticipated stresses in the structure and probable points where deformation or fatigue might develop. Even production processes and machine-hours can now be assessed in advance, enabling the needs of the production process to be predetermined.

Les Coombs tackles the increasingly important role

INTRODUCTION

played by simulators in crew training and risk management in new programmes. In a brief introductory historical survey he describes the very early ground trainers, the introduction of the Link trainer in the 1930s, and the devices adopted during the Second World War. The simulators of the 1960s and 1970s introduced increasingly complex mechanical systems and ever greater realism with regard to such effects as motion, noise, turbulence and the synthetically-created 'outside world'. Modern computer systems and the advent of virtual reality have brought even greater advances, and the future promises increasing integration of real aircraft with simulators.

After outlining the progress of human mobility, John Snow describes the dramatic changes brought by Concorde and then considers the future for supersonic transports, especially in the light of changing environmental requirements since Concorde made its debut. He also investigates the potential for ultra-high-capacity aeroplanes, and the advent of the high-subsonic airliner, which might offer an attractive alternative to the SST.

The volume concludes with Stephen Ransom's study of the development of rocket-powered lifting-body aircraft as a means of creating fully-recoverable space transportation systems. Over the past forty years there have been a surprising number of such projects, and the USA is by no means alone in its efforts to perfect the technology. The UK, Germany, Japan and Russia have also put considerable effort into research in this field, as the author's outline of the various programmes demonstrates. These are, perhaps, the ultimate faster, further, higher winged vehicles; at least for the foreseeable future.

Philip Jarrett

1
New-age Aerodynamics
Dr Norman Barfield

'Longer, larger, farther, faster, higher, quieter, smoother', the descriptive marketing legend stretching boldly along the fuselage of the Airbus A340-600, encapsulates all the competitive performance aspirations of today's commercial and military aircraft designers. Each objective implies a challenging limitation or demands a significant, if not dominating, aerodynamic ingredient. Each objective has to earn its place in the complex matrix of factors, beyond the technical ones, in all spheres of the aerospace business.

Humans ceaselessly contest the fundamental natural phenomena of gravitational force, terrain, climate, weather, wind shear, and pressure altitude. These are now exaggerated by the implications of compressibility, unsteady transonic and shock-wave flow, and kinetic heating as flight speeds meet the expansive supersonic- and hypersonic-flow regimes. Global environmental concerns have also conspired to demand minimum noise and pollution from civil aircraft, adding a further challenge. This applies especially in engine design, in the persistent drive for optimum propulsive efficiency at minimum operating cost.

The dramatic progress made in satellite-linked electronic warfare techniques and equipment has also resulted in the incorporation of a high degree of stealth technology in the latest high-agility combat aircraft, to give small radar signatures. This also relies heavily on the aerodynamic contribution. Moreover, as the limits of human tolerance of excessive gravitational pull are being reached, more attention is being paid to the conception of

The multi-virtuous Airbus A340-600, the world's longest airliner to date at 246ft 3in (75.1m), typifies the extensive fuselage stretching process that has characterised most large (and, more recently, small) jet airliners since the extreme stretching of the Douglas DC-8 in the 1960s. The limiting factor in this process tends to be the structural stability of the unsupported forward fuselage, rather than the aerodynamic considerations of slenderness ratio and tail-down angle on take-off. The smaller capacity of the following A340-500 sibling (313 seats and 222ft 6in (67.8m) long) is traded for extra range (8,500 miles; 13,680km), such that the first example carries the legend 'the longest-range aircraft in the world'. In can fly non-stop between key cities in southeast Asia and the USA.

a new generation of unmanned reconnaissance and combat aircraft.

Above all of these crucial factors is the enormous cost of the necessary technological solutions in a balanced and effective fashion, costs that have demanded either industrial alliances or corporate mergers to spread the financial risks. To the constraints of cost is added the problem of the long timescales now involved in completing major new aviation projects. Consequently, the number of wholly-new concepts is becoming so limited that some are being conceived without competition, other than that between the participating partners in bidding for their preferred share of the complete project. This means that primary aerodynamic expertise will in future rest only with the biggest contractors.

In research, this situation has also resulted in widespread sharing of the crucial but very expensive specialised resources, notably windtunnels and the directly-related extreme environmental testing facilities. Much more such testing is conducted within (and funded by) the industry itself, together with the contracting-in of the expertise and capabilities accumulated in universities and other higher educational institutions. The major part of this research effort must now be devoted to direct applications in the small number of next-generation mainstream aircraft programmes. Basic research into new aerodynamic inventions has suffered.

However, these many constricting influences are largely counterbalanced by the tremendous advances in computational power and programming, visualisation, modelling and simulation. This advantage is being expressed through the already very sophisticated computer-aided design (CAD), which has greatly shortened the design timescale, and reduced cost by 'getting it right first time'. In particular, the twin sciences of computational fluid dynamics (CFD) and multidisciplinary design and optimisation (MDO) have already combined to revolutionise the whole process. These twin design staples have likewise been seamlessly linked to the complementary technique of computer-aided manufacturing (i.e. CAD/CAM). Together with the fabrication of immensely strong and intricately mouldable non-metallic carbon composite, this combination has also removed many of the earlier restrictions imposed on the production of aerodynamic shapes.

Implementation of the corresponding physical aerodynamic solutions is increasingly reliant on hybrid mechanical and systemic elements, complemented by the classical ones of shape and form, most notably with 'fly-by-wire' (FBW) electronic control, stability augmentation and flight envelope protection systems superseding the traditional mechanical arrangements. However, due attention has to be paid to the risk of unreliability in complex equipment, especially where flight control is involved.

The new aerodynamic age now dawning
The complex industrial/technological relationship within which the aircraft design process now operates has brought the science of aerodynamics and its application to a new level of maturity. We stand on the threshold of a new age of advances, equivalent to the quantum step that came with the advent of the jet engine and swept wings half a century ago.

This new age is encapsulated in the new-generation aircraft expected to set the aerodynamic standards of the future. These are the Airbus A380 'Superjumbo' and Boeing Sonic Cruiser commercial airliners, a possible second-generation advanced supersonic transport (AST) and/or business jet (SSBJ), the Lockheed Martin F-35 Joint Strike Fighter (JSF), the Bell Boeing V-22 Osprey military convertible tilt-rotor transport; and the emerging new generation of unmanned combat air vehicles (UCAVs). These new products will be complemented by improvement of established lines through the application of new aerodynamic artifices, often resulting from the re-examination of old ideas. One example is the now almost universal use of winglets and/or special wingtip shaping on airliners.

Key areas of advanced research aiming to enhance the operational efficiency of these design leaders – and to lead into the next generation – include unconventional layouts such as the blended wing/body and joined-wing arrangements (claimed to have the potential for lift-to-drag (L/D) ratios in excess of 30); forward-swept wings; new lift-enhancing and drag-reducing measures, such as much greater boundary layer laminarisation through both passive and active flow control; variable camber; active control technology, such as to enable a civil transport to fly with a more rearward c.g. and hence reduced tailplane size, weight and drag, and hence fuel consumption; passive and active shock control; and more advanced automated flight management systems.

In the half-century since the advent of jet propulsion and supersonic flight, combat aircraft have passed though several distinct phases and fashions. Through the 1950s the initial trapezoidal swept shape progressed to the tailless delta wing, notably adopted by Dassault in France. A decade later there was widespread preference in the USA, Europe and Russia for the variable-geometry (VG) pivoted 'swing-wing' arrangement. With the wings spread it was possible to fly at slower speeds yet carry heavy loads from runways of modest length. At high-speed the fully-swept wing has low wave drag and is thus ideal for supersonic flight. Moreover, its small span promotes the advantage that a wing of correspondingly low aspect ratio, with the area distributed from front to rear rather than across a wide span, has a flat lift curve and thus exhibits only small changes in lift over large changes of angle of attack. Consequently a VG aircraft can fly at full power through turbulent air and at low level

FASTER, FURTHER, HIGHER

Variable geometry, exemplified by the Panavia Tornado GR.1, greatly expands an aircraft's performance envelope. Fully swept at 67 degrees the aircraft can attain supersonic flight and can fly through turbulent air and at low level with low gust response, while, with the wings spread, it can carry heavy loads from runways of modest length and fly at relatively low speeds.

Sukhoi's Su-47 Berkut, hitherto known by the manufacturer's designation S-37, has forward-swept wings incorporating 90% composite materials, and a tailplane as well as canard surfaces. A fifth-generation heavy tactical fighter, its outstanding manoeuvrability is enhanced by thrust vectoring.

NEW-AGE AERODYNAMICS

Supermanoeuvrability and short-field operation are imparted to the Eurofighter Typhoon by the all-moving foreplanes of its close-coupled, artificially stabilised canard configuration

with low gust response. This last factor is a principal consideration in the design of the wings of modern all-weather fighters, because it has long been thought that the safest way to penetrate hostile airspace is to fly as low as possible.

The next phase of fighter design was the reversion to fixed wings with more sophisticated shapes, given some VG capability by fitting full-span drooping leading edges and large trailing-edge flaps, as widely used in subsonic transport aircraft counterparts. These computer-controlled auxiliary high-lift devices continuously adjust the profile of the wing to suit the take-off, high-g manoeuvring, supersonic cruise and landing phases of the modern combat mission. They also incorporate relaxed stability and so result in a much smaller, lighter and more efficient overall design for a given role. The European industry has fostered three independently conceived close-coupled artificially stabilised canard fighters: the Swedish Saab Gripen, the French Dassault-Breguet Rafale and the Anglo-German-Italian Eurofighter Typhoon, with all-moving foreplanes affording reduced trim drag and the pitch control demanded for short-field operation and supermanoeuvrability.

However, this trend did not extend to the USA, because replacement fighters in Europe were required more quickly. The General Dynamics (now Lockheed Martin) F-16, the Northrop Grumman F-14 (swing-wing) Tomcat, Boeing F-15 Eagle and F/A-18 Hornet continue to be built in improved versions. (Though production of the revolutionary Lockheed F-117 Nighthawk stealth fighter, completed wholly within the company's legendary 'Skunk Works' experimental plant, has been completed.)

This range comprises a combination of fixed-wing and VG types designed to fulfil the specific fighter and fighter/tactical bomber roles that originated in the Cold War era, and have been demonstrated in the Gulf War and other subsequent conflicts. The longer view is now being taken for the much further advanced Lockheed Martin F-22 Raptor, the first true fifth-generation fighter of the twenty-first century, and the next-generation F-35 JSF. Meanwhile, the fractured Russian industry continues the slow development of its own Cold War equivalents – notably the MiG-29 Fulcrum and Sukhoi Su-27 Flanker – into the post-Cold War era. Aimed to maintain the industry's survival as a major player in military aerospace, the technologies being tested in the experimental forward-swept Su-47 Berkut and MiG 1.44/Mikoyan MFI multirole fighter demonstrators are expected to be incorporated in Russia's fifth-generation fighter. Much

FASTER, FURTHER, HIGHER

Low observable technology and a defensive avionics system give the lamda-winged Northrop Grumman B-2 extremely high survivability.

NEW-AGE AERODYNAMICS

smaller and lighter, and probably also incorporating thrust vectoring, to compete with the JSF.

In the strategic bomber field, the USA continues to rely on a combination of the 1950s Boeing B-52H for saturation bombing, together with the much later Rockwell B-1B Lancer supersonic swing-wing precision bomber and the Northrop Grumman B-2 Spirit supersonic lamda-winged stealth bomber (the military 'blended wing body'). The virtually identical Russian equivalent of the Lancer, the Tupolev Tu-160 Blackjack, also has a curious and complex all-moving cruciform tailplane and fin and rudder arrangement.

Transport aircraft developments

The late 1940s breakthrough into the pure jet age of civil aviation enabled an immediate doubling of cruising speed with corresponding savings in journey time, together with increases in size and operating range, as first demonstrated by Britain's aerodynamically first-rate de Havilland Comet. At the same time, the operational versatility and economy of the so-called 'interim' propeller-turbine engine, notably demonstrated by the Vickers Viscount, resulted in a new generation of faster and smoother regional transports.

The propeller-turbine engine continues in widespread use today. The propellers for the Airbus A400M four-engine turboprop military transport, for example, are being designed to absorb 10,000hp each; more than twice the power of the Rolls-Royce Tyne of the 1960s. This is achieved by the use of rapier-like, broad-chord multiple propeller blades, and hence high solidity rather than increased diameter or counter-rotation, because of the limitations of sonic tip speed and excessive noise and loading on both the propeller itself and the airframe structure.

Significantly, the smooth-running gas turbine engine in both forms also introduced a new operational concept of 'above-the-weather' passenger comfort and convenience, and with affordable operating costs and fares through increased speed and productivity.

Following the pioneering wartime German Messerschmitt Me 262 swept-wing, underwing podded-jet fighter (which some regard as the true forerunner of the modern airliner configuration), the Boeing B-47 and B-52 first-generation heavy jet bombers heralded the adoption of this formula in that company's archetypal 707 jetliner of the early 1950s. Thus the classical DC-4 type layout was brought into the jet age, with underwing

The Airbus A400M, depicted in this computer-generated image, has been designed to meet the needs of several European air forces for a next-generation strategic and tactical military airlifter.

FASTER, FURTHER, HIGHER

strut-mounted engines for efficient straight-though 'streamtube' intake and exhaust flow, kept away from the wing wake. Half a century later this arrangement is still the layout of Boeing and Airbus airliner families and the numerous military jet transport counterparts, albeit with improved turbofan engines, supercritical wing sections and many other aerodynamic enhancements.

The rear-engined alternative, inaugurated by the French Sud-Aviation Caravelle in the 1950s, has been adopted in both civil and military applications. But the much-debated T-tail of its successors resulted in the deep-stall phenomenon (disastrously so in the case of the BAC One-Eleven regional twinjet), requiring significant limiting control and stall-prevention devices. This phenomenon occurs when the aircraft gets uncontrollably locked within its own downwash at very high angles of attack, the tail being wholly blanketed and rendered completely ineffective. Nevertheless, the rear-engined T-tail layout has prevailed in most regional jetliners and business jets, principally because of its original dual benefits of an unobstructed high-lift wing and a much quieter passenger cabin.

This layout also enabled the tri-motor arrangement of earlier-generation piston-engine transports and bombers to be readopted using jet engines, to meet the trans-American 'engine-out over the Rockies' requirement. (Subsequently, the reliability of the large turbofan-powered twinjets became acceptable, not only for this type of operation, but also for transoceanic routes.) As exhibited in the medium-range short-field Boeing 727, this layout also allowed the use of the most powerful (non-blown) high-lift system ever used in a swept-wing airliner.

The ubiquitous and highly-versatile Lockheed C-130 Hercules turboprop military transport of the early 1950s incorporated the DC-4 layout in high-wing form, to permit truck-bed-height loading and paradropping. This set a pattern for this type of aircraft which has prevailed

The Ivchenko PROGRESS/Zaporozhye D-27 propfan engines of Antonov's An-70 each drive two coaxial contrarotating SV-27 propellers, the front fan having eight composite blades and the rear six.

through to the Boeing C-17 and the proposed Airbus A400M, differing only in the substitution of the longer-arm, smaller-area T-tail for the conventional low-fuselage-mounted unit of the Hercules). In Russia, Tupolev and Ilyushin also adopted the T-tail and the rear-engine layout, while Antonov retained a low-set tailplane with end-plate fins for its outsize high-wing military transports, presumably restricted owing to hangar height restrictions.

The quest for higher subsonic cruising speeds, and the unacceptable handling characteristics encountered with the onset of compressibility drag divergence, in turn required the application of a widening range of electro- and hydromechanical devices. Powered flying controls with artificial feel were the first major innovation. Most swept-wing jet airliners also exhibited the 'Dutch roll' phenomenon, which had to be controlled by yaw dampers. Mach trimmers were employed to ensure the necessary speed stability. Stick shakers and pushers were also devised to control stalling behaviour, plus sophisticated autopilots to cope with bad weather let-downs at higher approach speeds. Out of this progression has evolved the automated flight management and envelope protection systems of today, notably including provisions for full automatic landing.

High-speed airliner wing design

The dominant aerodynamic consideration in the evolution of the swept-wing jet airliner has been aerofoil design to enable operation at the higher subsonic speeds (around Mach 0.8 to 0.9) and near the drag divergence Mach number. Sonic speed over a wing is actually reached when the aircraft speed is only around Mach 0.8, and a shock wave is formed immediately aft of the crest of the aerofoil, causing a sudden rise in aerodynamic drag. The primary objective is to delay the critical drag rise due to the shock-wave-induced separation of the boundary layer. This has been achieved by using so-called 'supercritical' aerofoil sections to exploit the benefits of using relatively large regions of controlled supersonic flow over the upper surface at high subsonic free-stream Mach numbers. The main advantage of these aerofoil profiles is that, for a given thickness:chord ratio, the critical Mach number stays the same but the compressibility drag divergence Mach number can be delayed.

Compared with the hitherto conventional American NACA six-digit series of laminar-flow aerofoils, these supercritical sections have a much flatter upper surface shape, a large extent of supercritical flow, terminated by a weak shock wave. This reduces both the extent and the strength of the normal shock wave, as well as the adverse pressure behind the shock, thereby reducing aerofoil drag.

To compensate for the reduced lift on the upper surface resulting from the reduced curvature, increased camber has been introduced towards the trailing edge to increase the 'rear loading' in a region where the airflow on both surfaces becomes safely subsonic again. This produces extra lift by means of additional positive pressure on the lower surface to avoid the more sensitive boundary layer on the upper surface.

Higher cruising speeds with the same power are achieved either with thinner wings or greater sweepback, both being undesirable on several counts. A thinner wing requires a much thicker skin to achieve the required strength and stiffness, and is thus heavier, and also reduces the available fuel capacity. A sharp leading edge is also unsuited to lift at high angles of incidence and needs complex mechanical means to modify the aerofoil's nose shape. Increased sweepback means reduced low-speed performance with longer take-off and landing distances and possible control deficiencies.

The reshaped supercritical aerofoil sections have a bluffer leading edge, a flatter top, greater depth with a bulged underside and a downward cusped rear portion. While the free-stream air is accelerated more slowly over the leading edge, instead of reaching a massive peak, it then retains this speed back across the top of the wing, and then moves at a much higher speed as it traverses the aft section. Hence, instead of generating intense lift across a narrow strip near the leading edge, as in the earlier peaky wing concept, the 'roof-top' or 'rear-loaded' supercritical aerofoil generates powerful lift over virtually the entire wing area. The spanwise line of centre of pressure (i.e. the point at each aerofoil section through which the resultant lift force is considered to act) is translated from near the leading edge to around mid-chord or even further aft.

These key developments in high-speed airliner wing-section design, in progress since the late 1950s, are now fully implemented in the newest members of the Boeing and Airbus families. They will also feature, with further refinements, in the new Airbus A380 and Boeing Sonic Cruiser.

The British, American, and more recently European, civil aviation regulatory authorities — Britain's Air Registration Board (ARB) and Civil Aviation Authority (CAA); the USA's Federal Aviation Administration (FAA); and the European Joint Airworthiness Authority (JAA) — have also set many of the airworthiness and performance requirements and standards that have widely influenced and guided those of other nations as well. Their military authorities and air forces have done likewise in the military sphere.

The new-age product and technology portfolio

The technical and industrial implications of the new age of aerodynamics now dawning pose further challenges in scale, speed, mission envelopes, competitive performance margins and, most significant of all, sustainable cost.

The main design objective of the 555-seat Airbus

Typically seating 555 passengers in first, business and economy class in greater space and comfort than its competitors, the Airbus A380 is intended to achieve 15-20% lower operating costs. To enable it to use existing aprons and gates at major airports it embodies several distinctive features, including a double-deck fuselage.

A strong selling point for the Airbus family is commonality of flight deck layout, fly-by-wire control laws, systems and handling characteristics throughout the range. Seen here, fronted by the A300-600ST 'Beluga' special transport conversion, are the A340-300, A330-200, A300-600, A310, A321, A320 and Airbus Corporate Jetliner. The A380 will benefit from the same similarities.

A380 airliner is to achieve 15-20 per cent lower operating costs within the same operational framework as the latest Boeing 747-400. Three main aerodynamic challenges are being confronted: sheer size; the need to adopt a double-deck twin-aisle fuselage to keep the overall dimensions of the aircraft within the prescribed 80m (244ft) square to use existing major airport apron and gate spaces; and the need to have the same runway performance as the Boeing 747, i.e. around 11,000ft (3,350m). Retaining the conventional layout and aerothermodynamically efficient high-bypass-ratio turbofan engines (and with much copied from the latest four-engine A340), allowance has prudently been made within the dimensional constraints for a 20 per cent future growth in fuselage length and capacity. The double-deck fuselage has also resulted in a redistribution of side areas, and hence fineness ratio and tail arm, and the need for compensating stability augmentation and c.g. position control measures to minimise the size of the tail. The same considerations have affected the fin size. The wing design avoids the need for folding tip sections to fit the prescribed 'apron span' limit. The incorporation of a corresponding allowance in wing area for future performance developments has also resulted in sufficient rear-spar depth to enable simpler and lighter single-slotted flaps to be used, and the achievement of approach speeds of around 145kt (265km/h) at maximum landing weight. An effort is being made to achieve minimum climb-away noise levels and to keep the wake vortex and turbulence no stronger than that of the 747. There is also the difficulty of predicting aeroelastic behaviour and the consequent implications to wing aerodynamics and control effectiveness with such a large structure, and the great mass and moments of inertia are requiring further advances in control systems and the use of a 5,000lb/in^2 hydraulic power-actuation system. A major 'aerodynamic selling point' of the Airbus family is the commonality of the flight deck layout and FBW control laws, systems and handling characteristics, to enable cross-crew qualification, across all models, and now to embrace the A380.

Having created the mould of the modern jet airliner configuration with the 707, Boeing is now breaking it with its radically changed Sonic Cruiser concept, with a service availability date of 2008. Designed to cruise at speeds of up to Mach 0.98, and hence considerably faster than today's airliners operating in the Mach 0.8 and 0.9 bracket, very close to the still-uncertain transonic flow region, it would be operating much further up the compressiblity drag rise curve, though not, it is claimed, producing the troublesome sonic boom. This much higher speed will result in a time saving of around an hour on a 3,000-mile (4,830km) trip, and hence four to five hours on the current 23-hour London-Sydney journey. (Boeing also claims that a second speed 'sweet spot' has been identified at around Mach 1.02, and several potential operators are believed to have asked for speeds up to Mach 1.6 to be evaluated).

The 200-250-seat baseline design is intended for use between regional airports with medium-sized runways, thereby drastically reducing the journey time taken to fly to a major hub and change, while also having a range of up to 9,000nm (15,770km). The maximum cruising altitude will be the mid-40,000ft (12,200m) level or higher, 10,000ft (3,050m) above the rest of the traffic to achieve, economically, the higher cruising speed and to get there very quickly.

This performance envelope has demanded an innovative configuration that is a substantial step towards an SST but still only half the speed of Concorde and the AST studies. Only two 777-size engines would be required (but of wholly-new design to cope with need to rebalance the bypass ratio, overall pressure ratio and turbine inlet temperature relationship to embrace the extended operating regime, derivative engines no longer being an option). The use of four smaller engines would result in excessive mass, drag and fuel burn and hence operating cost.

Impressions show a broad-span cranked-arrow delta wing using a 777-type outer plane with raked tips in lieu of winglets, to provide adequate aspect ratio to meet the

NEW-AGE AERODYNAMICS

range requirement, and a sharply-swept gloved inner section with two underwing buried engines. There are large shoulder-mounted fixed canard foreplanes on the fuselage nose, and twin inward-canted fins at the rear. Boeing claims that this exotic configuration has now been validated in all leading respects by both low- and high-speed windtunnel testing with comparatively little change (apart from the shape, exact location and movement of the canards), and with linear pitch characteristics and no wing buffet up to Mach 1.08. Significantly, this process has already produced remarkable confirmation of the original CFD predictions and the accurate portrayal of the latest methodologies. Indeed, Boeing confidently claims that in CFD it now has a design tool which can be used to optimise and conduct configuration trade-offs on the aircraft shape and for derivative developments. Moreover, it has also enabled the abandonment of the earlier notion of the possible need to fly a sub-scale version of the aeroplane as the principle aerodynamic data validating tool.

Despite the higher unit cost of the more sophisticated airframe and the custom-designed engine, the higher specific fuel consumption (and possibly emissions), and hence operating cost, the correspondingly higher productivity could result in a smaller fleet size for a given route network, and therefore a lower investment. Boeing has also devised a means whereby successive stretching of the body, together with variable placement of the foreplanes, could be accommodated for a progressive family development.

The basic aerodynamic concept of the slender ogival-shaped delta-winged layout of the Anglo-French Concorde, so brilliantly conceived by Britain's Royal Aircraft Establishment more than forty years ago, set the broad pattern. The cranked slender delta has remained the preferred wing planform in virtually all of the multitude of AST and SSBJ feasibility studies that have emerged ever since.

The essential requirements for long-duration supersonic flight dictate a low-drag design with high installed thrust and a relatively high fuel fraction (the weight of fuel as a proportion of the aircraft's all-up weight), a slender fuselage and a large fuel tankage in the wings, for which the aerodynamically thin but physically thick slender delta is well suited. However, the VG 'swing-wing' alternative of the USA's B-1B and Russian Blackjack supersonic bombers is not seen as being any more credible now than when it was impractically incorporated in the American contemporary of Concorde, the moribund Boeing 2707.

Research into ASTs continues in the UK and France (now through Airbus) and in the USA, most recently with NASA and the quiet supersonic platform (QSP). Boeing also used a re-engined Russian Tupolev Tu-144 for a period of flight testing. The work is especially directed at determining feasible powerplant and aerodynamic means of reducing the airport noise footprint and

Boeing's twin-engine, 200-250-seat Sonic Cruiser is designed to cruise at speeds up to Mach 0.98 at a significantly greater altitude than other airliners without causing the sonic boom, but offering a considerable time saving on long-range flights.

FASTER, FURTHER, HIGHER

This projected 250-passenger advanced supersonic transport was the product of a European Supersonic Research Programme involving Aerospatiale, British Aerospace Airbus and Deutsche Aerospace Airbus in the late 1980s. The 138ft (42m)-span, 292ft (89m)-long aircraft, which had a foreplane and a double-delta wing planform, was designed to cruise between Mach 0.95 and Mach 2, and to 'meet noise and emissions standards current at entry into service', which was set at 'around the year 2007'.

sonic boom. It is also proceeding most determinedly in Japan, with interchange with Boeing on the Sonic Cruiser. The general view is that the basic specification would require a capacity of 250-300 passengers and a minimum range of 5,500 miles (8,850km) at cruising speeds of between Mach 2.05 and 2.4. Thus it would be a broadly similar vehicle to Concorde, flying at a similar speed, but 50 per cent larger and twice as heavy because of the increased passenger complement and range. It would still have a light-alloy airframe, rather than using the steel and titanium needed to cope with the kinetic heating at higher cruising speeds.

Relative to Concorde, the achievable design objectives indentified by the Working Party on Second Generation Supersonic Transport formed of manufacturing and airline specialists by the Royal Aeronautical Society (RAeS), and reporting in 1997, can be summarised as: 35 per cent higher supersonic lift/drag (L/D) ratio, i.e. around 10 cf 7.3 at Mach 2.0 (ten per cent higher than that achievable using today's technology); 20 per cent higher subsonic L/D; 40 per cent lower structural weight; 10 per cent lower supersonic and 20 per cent lower subsonic cruise specific fuel consumption; no higher approach and terminal manoeuvring speeds; and dimensions suited to airports used by today's subsonic aircraft. A 300-seat fuselage would be about 230ft (70m) long (well within the apron limit specified for the A380). Landing characteristics will require special attention, and some form of flight control surfaces at each end of this 'long cigar' will certainly be needed. Range growth must also be demonstrable. Various engine options require fur-

ther examination, and development of the very-aero-thermodynamically challenging variable-cycle engine may be necessary. Reduced jet velocity would also be required, to meet airport noise limitations.

Proposals to date for SSBJs have dispensed with the droop-nose arrangement on Concorde to save weight. Such aircraft will undoubtedly be fitted with fighter-style wing leading-edge devices, so the high nose-up attitude on approach will be less than that of Concorde. The view overall could be supplemented by military-style artificial vision, though there are doubts that this has a place on a civil aircraft.

Full stability augmentation will be included on any AST or SSBJ. This will enable tailored control laws to dominate handling qualities and flight envelope protection in the same way as on the heavily-augmented current subsonic transports.

Instead of being conventionally led by design and engineering adventurousness, Lockheed Martin's F-35 JSF, which probably represents the last generation of piloted combat aircraft, has been conceived to meet the mission needs of five air arms in two countries with a single design. The conventional take-off-and landing (CTOL) version is for the United States Air Force (USAF); the carrier version (CV) is for the US Navy (USN); and a short take-off and vertical landing (STOVL) version is for the US Marine Corps (USMC) and Britain's RAF and Royal Navy. The central aerodynamic challenge of this specification is the extremely demanding 'STOVL-to-supersonic (Mach 1.5) cruise' mission performance. The American competitive indus-

NEW-AGE AERODYNAMICS

trial bidding process culminated in the contrasting hybrid aero-mechanical solutions embodied in the prototypes built by Boeing, with the X-32, and Lockheed Martin (teamed with Northrop Grumman and BAE Systems) with the X-35, the latter emerging as the eventual winner.

The programme began in 1994 with a US congressional decision to combine a USAF/USN requirement for a strike aircraft with a USAF/USMC requirement for a light fighter/attack aircraft to be built in conventional and STOVL forms. The American Defense Advanced Research Projects Agency (DARPA) advanced the preferred solution of a radically new jet-lift powerplant concept. This used the main propulsion engine to shaft-drive a cold-air lift fan which would magnify the available lifting force and produce a relatively low-energy jet efflux (reducing ground erosion and hot-gas recirculation problems), and in its V/STOL form used the lift fan and a deflecting final nozzle. This was the hitherto untried solution adopted by Lockheed Martin.

The main disadvantage of this arrangement was that, when not used in the STOVL mode, the lift fan would become dead weight. In the winning Lockheed Martin submission, however, there are hidden benefits. The space occupied by the lift fan can be used for fuel in the CV and CTOL versions, both the USAF and the USN having greater range requirements than the USMC. However, the lift fan has two major advantages over the direct-lift concept employed by Boeing. Firstly, it greatly improves the thrust recovery from the engine, and, secondly, it avoids many of the problems caused by hot-exhaust re-ingestion. The lift fan blows a cushion of cold air under the aircraft while in the hover, thus preventing most hot gas from reaching the main engine intakes, which could seriously degrade engine performance. The aerodynamic performance of the F-35 will also be enhanced by the requirement for conformal carriage of the weapon load.

The X-32's configuration was related to the 'direct lift' concept long proven with the Harrier/Pegasus inherited from the former McDonnell Douglas AV-8B programme and USMC experience. Its two-dimensional nozzle is closed for STOVL, most of the gases exhausting through two rotatable nozzles just aft of the engine, with the fan efflux being ejected in the form of a transverse jet screen

In purely aerodynamic terms the Lockheed Martin F-35 Joint Strike Fighter, the world's first supersonic STOVL aircraft, incorporates two major advances to complement the primary aero-mechanical means of achieving the major part of its demanding mission. The considerable advances in stealth technology since the conception of the multi-faceted Lockheed F-117 Nighthawk have resulted in a sculptured yet practical design which gives an extremely low radar signature, as already demonstrated on the preceding Lockheed-Boeing F-22 Raptor air-superiority fighter. The design of the main engine intakes, traditionally accomplished with moving ramps and baffles within the intakes to reduce the incoming air to subsonic speed at the engine face, includes a new fixed feature attached to the fuselage sides and without moving parts (understandably classified in detail).

FASTER, FURTHER, HIGHER

Closely related to the 'direct lift' concept proven in the Harrier, the configuration of Boeing's X-32A submission for the JSF competition featured a two-dimensional nozzle that was closed for STOVL operations, most of the exhaust gasses being emitted through a pair of rotatable nozzles just behind the engine, with the fan efflux being emitted as a transverse jet screen.

to minimise hot-gas recirculation. But, overall, it was a straight 'lift versus weight' solution for all versions and without the extra load-carrying capability of the F-35 in the CV and CTOL versions, with extra fuel capacity in place of the lift fan.

The Bell Boeing V-22 Osprey convertible tilt-rotor transport represents the first major breakthrough in the design of rotary-wing aircraft since the original introduction of the helicopter. The idea of an air vehicle capable of switching from a vertical-rising and hovering helicopter mode to a conventional forward-flying transport aircraft mode has challenged aerodynamicists for around fifty years. Bridging the gap between the helicopter and the conventional turboprop utility aircraft, the tilt-rotor craft combines the vertical operations of the former with cruising speed and range approaching those of the latter.

Stemming from the original Bell XV-3 and XV-15 tilt-rotor concept demonstrators of the late 1950s and late 1970s respectively, the Osprey was first flown in March 1989. Strongly supported by the USMC for the military assault transport role, it was conceived to provide the technical breakthrough for a family of both military and civil tilt-rotor transports with two and four large rotatable engine/propeller units. The switch is accomplished by rotating the engine/propeller units from vertical to horizontal flight, and the efficiency of the propeller is com-

promised in accommodating both. The extra power required for the VTOL capability makes the Osprey an expensive aircraft. High complexity is also involved, not only in the rotation of the wingtip-mounted engines and rotors in pitch, but also because of the crucial need to ensure full mechanical interconnection and redundancy provisions so that both rotors continue to function in the event of an engine failure. The latter aspect has caused considerable difficulty, showing that the tilt-rotor concept is more susceptible to 'power settling' phenomenon or 'vortex ring state' (VRS) than the conventional helicopter. In essence, if a horizontal rotor moving at low forward speed accelerates to a descent rate similar to the velocity of its own downwash, its lift is reduced and it tends to sink even faster. While the threat of VRS can be a eliminated by appropriate flight rules, their implementation probably requires the development of an accurate airspeed sensor, taking a cue from the fixed-wing V/STOL Harrier, which required just such accurate sensing and display of sideslip angles at very low speeds.

With the Osprey's side-by-side configuration there is also the added possibility of encountering high asymmetric downwash (especially if one rotor is wholly over the side of a sea-going launch platform), with potentially catastrophic results. In this context, a considerable further aerodynamic difficulty was explicitly imposed on the

NEW-AGE AERODYNAMICS

V-22 by the restriction of rotor diameter for operation from the decks of amphibious vessels. The use of four or five rotor blades (rather than the current three) may also be preferable to reduce noise, and possibly to alleviate the VRS problem.

The Osprey requires further assessment of the flight-control software, the mechanical interconnections and the whole flying envelope – certainly before the planned civil version is technically and publicly acceptable for certification and service.

The tilt-rotor concept has also been envisaged in Europe, but in a different form that entails tilting the outer wing section and nacelle together. Whereas the full tilt-wing arrangement (as in the LTV Hiller Ryan XV-142 of the 1960s) results in poor hover control in strong winds, this newest concept would minimise downwash masking in the helicopter mode and allow for a more efficiently-sized rotor in the aircraft mode. This hybrid tilt-wing/rotor concept was first offered in 1999 by Agusta of Italy, as the basis for a 20-seat transport named Erica. The Eurocopter-led Eurotilt consortium has proposed a similar-sized vehicle in which only the forward part of the nacelle and the rotor would translate.

Following the successful deployment of the aerodynamically simplistic low-level General Atomics RQ-1A Predator, armed with precision-guided missiles, and the high-level-reconnaissance Northrop Grumman RQ-4A Global Hawk long-duration remotely-piloted and satellite-linked unmanned combat air vehicles (UCAVs) in the Afghan campaign, a renewed interest in this type of military aircraft has prefaced the appearance of the first X-series advanced technology full-capability combat aircraft demonstrator, the Boeing X-49A. This radical-looking aerodynamic proof-of-concept design is about half the size of the JSF. Its principal advantages would be the ability to carry a comparable weapons load through hostile enemy environment without risk to the human pilot, and higher g sustainability beyond the current 9g combat-aircraft design level. A range of UCAVs is currently being evaluated for land and carrier-based use. However, the development of UCAVs should not place heavy new aerodynamic demands beyond today's combat aircraft, since they would still have to be designed for high performance with precision and appropriate handling characteristics.

Reverting to the absolute roots of aerodynamics, bird and insect flight have now also been investigated in con-

The Bell Boeing V-22 Osprey's troubled and protracted flight-test programme, which began in 1989, has not deterred the US Navy, Marine Corps and Air Force, which still want this unconventional tilt-rotor transport and multi-purpose aircraft, able to carry 24 troops or 12 litters plus attendants.

FASTER, FURTHER, HIGHER

The General Atomics RQ-1A Predator unmanned combat air vehicle, able to deliver precision-guided missiles, proved its worth in the Afghan campaign. Simpler, lighter and more manoeuvrable than manned aircraft, UCAVs avoid the need to put pilots' lives at risk.

siderable detail. Insects have revealed at least one novel method of sustaining flight, the so-called 'clap and flap' techniques, whereby very high lifts can be developed. It is thought that these studies could have an aeronautical application in micro-flying vehicles equipped with correspondingly miniaturised sensing devices through the rising science of nanotechnology, for the remotely-controlled surveillance of the interior of hostile buildings and caves and in the aftermath of earthquakes.

Hypersonica

The next step for aerodynamics will be the hypersonic regime, at speeds of Mach 5 and above. The USAF already envisages hypersonic combat strike and reconnaissance aircraft capable of extending into space operations in the post-Stealth era. There is also a perceived need for hypersonic missiles for attacking possible simi-

A long-endurance, high-altitude aerial reconnaissance aircraft, the unmanned Northrop Grumman Global Hawk has an integrated sensor system comprising an all-weather synthetic-aperture radar/moving target indicator, a high-resolution electro-optical digital camera and a third-generation infrared sensor, all operating through a common signal processor. It can provide military commanders with high-resolution, near-real-time imagery of large geographic areas.

lar hostile missiles and transient land and sea targets. In civil operation, the ability to embrace these sorts of speeds will be essential in the development of vehicles that can traverse conventionally from Earth to space, or 'wave riders' travelling on the edge of space that could reduce the semi-global London-Sydney journey time to around only two or three hours.

The key enabling device for hypersonic flight is the long-mooted 'scramjet' (supersonic combustion ramjet), which avoids the high inlet temperatures and normal shock losses associated with the subsonic ramjet. Significantly, the scramjet will also change the shape of both aircraft and missiles through its ability to generate a substantial amount of lift by the pressure in the intake and exhaust systems.

Russian scientists are believed to have been the first to achieve supersonic combustion, in 1991, with a small ramjet-powered ballistic missile boosted to Mach 5.5, the threshold of the hypersonic speed regime. Two years later Rockwell, in the USA, reached Mach 8 with gun-launched ramjets, and Pratt & Whitney later conducted tunnel tests simulating flight conditions of up to Mach 18.

In the USA the NASA Langley Research Center for military aircraft launched a new hypersonic flight vehicle research programme in 1995. The first stage is the development of the Boeing-built hydrogen-fuelled X-43A Hyper-X scale model, intended eventually to achieve flight speeds of Mach 10 at 120,000ft (36,000m), following a boost to around Mach 7 at 95,000ft (29,000m) after being air-dropped from a B-52 at around 19,00ft (5,800m) over the Pacific. This slender rectangular-wedge-shaped vehicle with twin outward-canted fins is 12ft (3.66m) long, spans 5ft (1.52m) and weighs 2,800lb (1,270kg). The subsequent X-43B is intended to link an enlarged Hyper-X with the NASA Marshall Space Flight Center's Integrated System Test for an Air-breathing Rocket (ISTAR), which will function as a liquid-fuelled rocket at low speeds, as a ramjet at intermediate speeds and as a scramjet at hypersonic speeds. Further hydrocarbon and hydrogen fuelled versions are envisaged to extend the flight-test envelope to Mach 15.

Industrial benefits and reciprocity

Much has been made of the value of technological spin-off from aerospace to other high-technology industries. The process actually started as early as 1903 at the National Physical Laboratory at Twickenham in England, with the examination of the airflows around, and the forces exerted on, buildings and bridges, the spread of smoke pollution and the understanding of what was described as the 'atmospheric boundary layer'.

Much more recently, the automobile industry has benefited from cross-fertilisation with aerodynamic technology. Body shaping and styling are becoming increasingly important in drag and noise reduction, road holding, fuel

The NASA/Boeing unmanned combat air vehicle (UCAV) aerodynamic demonstrator prefaces the larger X-45B, which will, at the behest of the US Defense Advanced Research Projects Agency, be much nearer the operational system being urgently developed for the USAF and incorporating the same-size internal weapons bay as the JSF.

economy and competitive performance. The benefits are now widely visible in 'top range' saloon car styling, and extend well beyond the purely aesthetic.

Aerodynamic reciprocity is especially evident in Formula One (F1) motorsport. Major aerospace companies in the UK and France have deliberately forged formal alliances with the leading motorsport manufacturers. This has not only resulted in worthwhile reciprocal benefits in aerodynamic design, but also in the expertise that the F1 fraternity has developed in computer-aided rapid prototyping between successive events.

In the late 1960s racing cars began sprouting various aircraft-type devices designed to eliminate lift without creating excessive drag. In 1969 there were absurdly high wings, which increased downforce but tended to break and result in horrendous accidents. A decade later various alternative aviation-derived aerodynamic concepts were introduced, including side skirts and venturi bodywork sections within the wheelbase to reduce 'ground effects' in the relentless attempts to optimise downforce while also reducing driving effort. Despite ever-tightening competition and safety regulations, aerodynamic fine-tuning remains critical in the relentless battle against drag and lift. The motorsport companies are now employing their own high-grade aerodyamicists, and using aircraft-derived CFD technology, supercomputing and their own windtunnels for scale aerodynamic testing of car bodywork.

An F1 racing car has been likened to an upside-down

aeroplane flying close to the ground. Negotiating sharp bends and chicanes as well as the straights at speeds of up to 200mph (320km/h) has required sophisticated aircraft-style computer controlled power steering and the ability of the driver to withstand forces greater than 3g. But the downforce and track grip that these vehicles are able to command is the 'holy grail' of the whole design. Consequently, there has been increasing reliance on the adaptation of aerodynamic technology, notably with a venturi shaped body with jet-fighter-style engine air intakes and upside-down aerofoil sectioned wings at the front and rear. These wings impart a down force of four times the weight of the car, giving four times the grip on the straight and allowing much faster speeds through the corners. At the same time the vehicle is dangerously unstable and moves at a ride height of only around an inch (25mm) off the ground to enable the maximum volume of fast air at low pressure under the car, thus exploiting the classic Bernoulli effect.

Although aircraft-style computer controlled 'drive-by-wire' technology has also been widely transferred, it is now some time since F1's last major technological breakthrough in 'airflow management'. Hence the aerodynamic benefits of plasma technology, used experimentally on fighter aircraft to produce an ionised force field around the airframe that allegedly reduces drag substantially and renders the aircraft invisible to radar, is also being used in Fl cars to stiffen structures, thereby maintaining optimum aerodynamic body shapes.

Industrial aerodynamics now embrace such diverse subjects as the design of high-speed trains and ocean-going racing yachts, as well as civil engineering construction and wind-turbine energy-generation farms. Medical fluid dynamics has also developed, both respiratory and blood flows receiving particular analytical attention. In sporting circles, aerodynamic principles and windtunnel testing have been employed to explain the swing of the cricket ball, thereby enhancing the technique of swing and spin bowling and ball construction. Likewise, golf ball's faceted surface is being aerodynamically refined for more effective flight and longer driving distances.

Significant cross-fertilisation of aerodynamic technology is also accruing within the now much more broadly based aerospace and defence industry itself. BAE Systems in the UK and Northrop Grumman in the USA have also become their respective nations' leading national shipbuilders. Engaged in design and construction of new-generation warships and aircraft carriers, they can apply their extensive aerodynamic knowledge and resources to the streamlined shaping of the superstructures and the most efficient aero-hydrodynamic compromise in the shaping and performance of the hulls.

High-speed airliner-type rear-loaded aerofoil sections have most beneficially migrated to the design of helicopter rotor blades. Thanks to the imaginative British Experimental Rotor Programme (BERP), an otherwise standard Westland Lynx raised the world helicopter speed record to a remarkable 249mph (401km/h), compared to the generally prevailing helicopter top speed limit of around 150mph (242km/h). These rotor blades had thin sweptback tips, a rear-loaded outer section, and a reflexed inner section where downwash was reduced. Modern aerodynamics, enriched in its aviation domain by more than a century of research, design and service, is now repaying the debts incurred long ago from hydrodynamics.

Back to the future

The basic concept of the modern aeroplane, devised by Cayley, has prevailed throughout the era of powered flight with only minor variations, and is still discernible in the latest supersonic and hypersonic vehicle shapes. The canard variation, adopted by the Wrights, has reappeared in the modern-day fighter generation and the next-generation transonic and supersonic transport proposals. Yet in little more than a century advances have been made in aerodynamic science and its application in the manifold extensions of the capabilities and performance envelopes of all kinds of aerial vehicles.

Further significant advances will be constrained by the confluence of the limiting asymptotes that are extending the likely timescales of incorporating them into aircraft designs well into the new century. These limitations include the proximate approach to the uncertain transonic regime; the inability of human physiology to withstand excessive gravitational forces and to comprehend interception speeds in supersonic military combat manoeuvres; and the escalating scale of the costs of research and implementation.

Nevertheless, much is possible, particularly with the aid of the sophisticated computational fluid dynamics and multidisciplinary design optimisation techniques and computing power now available. The discipline of aerodynamics will thus continue to be richly challenging and rewarding.

Finally, it is an interesting thought that the realisation of a transglobal hypersonic transport (HST) aircraft, and the replacement of the US Space Shuttle with a horizontal take-off and landing (HOTOL) single-orbit reusable aerospaceplane, could overcome the ultimate limits of the 'Faster, Further, Higher' theme of this volume and bridge both domains simultaneously within the first half century of the new millennium.

Bibliography

Abzug, M J, *Computational Flight Dynamics*, (AIAA Education Series, USA, 1998). A detailed instructional kit including software illustrating digital solutions to problems in aircraft dynamic stability, control and flight performance.

Baker, Dr D, 'Aerodynamics and Structures', in Putnam's History of Aircraft, *The Modern War Machine: Military Aviation Since 1945* (Putnam, London, 2000).

Barfield, Dr N, 'Airliner Aerodynamics', Putnam's History of Aircraft, *Modern Air Transport:Worldwide Air Transport from 1945 to the Present* (Putnam, London, 2000).

Beamont, R P, *The Years Flew Past: 40 Years at the Leading Edge of Aviation* (Airlife, Shrewsbury, Lincolnshire, 2001). Includes a most graphic and informative first-hand account of the onset of compressibility and the 'sound barrier' in the transonic 1940s and the supersonic 1950s.

Braybrook, R, *Supersonic Fighter Development* (Haynes, Sparkford, Somerset, 1987). A broad survey of the conceptual and operational aspects of the supersonic fighter genre.

—— 'Today's Future Classics', *Air International*, July 2001, pp36-41.

Chambers, J R, *Partners in Freedom: Contributions of the Langley Research Center to US Military Aircraft of the 1990s*. NASA History Series No 19; NASA SP-2000-4519 (History Division of the National Aeronautics and Space Administration office of Policy and Plans, Washington, D.C., USA, 2000). A very informative and mainly aerodynamically-orientated survey of twenty-three combat, transport and research aircraft, and of missiles and windtunnels, plus an extensive reference bibliography for each type.

Donald, D, 'JSF Goes Vertical', *Combat Aircraft* Vol 3 No 6 September 2001, pp518-524. (AIRtime Publishing, Norwalk, Connecticut, USA). A detailed description of the alternative solutions offered by the competing Boeing X-32 and Lockheed Martin X-35 technology demonstrators to the multinational JSF requirement.

Edwards, Sir George, 'The technical Aspects of Supersonic Civil Transport Aircraft'. Review lecture delivered to the Royal Society on 29 March 1973. Probably the most perceptive and cogent coverage of the subject, and notably the derivation of the aerodynamic characteristics, at its point of highest practical activity.

Green, J E, 'Air Travel – Greener by Design': *The Technology Challenge, Report of the Technology Sub-Group, July 2001*. (Society of British Aerospace Companies. Produced under the aegis of the UK Government Department for Trade and Industry's Foresight Programme. Includes significant aerodynamic content. (Published in full in the 'Greener by Design' issue of *The Aeronautical Journal*, RAeS, February 2002).

Gunston, B, *Airbus: The European Truimph* (Osprey, London, 1988).

—— *Combat Arms: Modern Fighters and Modern Helicopters* (Salamander, London, 1988 and 1990 respectively).

—— *Faster Than Sound: The Story of Supersonic Flight* (Patrick Stephens/Haynes, Sparkford, Somerset, 1992).

—— with Gilchrist, P, *Jet Bombers from Messerschmitt Me 262 to the Stealth B-2* (Osprey, London, 1993). Collectively, the aforementioned five volumes constitute a valuable contribution to the description of the aerodynamic context of all types of contemporary civil and military aircraft.

Hall, M G, 'On Innovation in Aerodynamics' (*The Aeronautical Journal*, 1,000th Issue, December 1996, pp463-470. RAeS, London).

Hassell, P, 'Advances in Aerodynamics', Putnam's History of Aircraft, *Biplane to Monoplane: Aircraft Development 1919-39* (Putnam, London, 1997).

Howe, D, *Aircraft Conceptual Design Synthesis* (Professional Engineering Publishing, Institution of Mechanical Engineers, London 2000). A landmark analysis of the modern aircraft design process, notably including the contextual significance of aerodyamics. Derived from the author's long association with the aeronautical faculty of the Cranfield College of Aeronautics and Cranfield University. Complete with computer disc software for spreadsheet presentations.

Huenecke, K, *Modern Combat Aircraft Design* (Airlife, Shrewsbury, UK, 1987). An excellent descriptive and illustrated survey, especially in aerodynamics, including most of the types in active service today.

Jupp, J, 'Wing Aerodynamics and the Science of Compromise', RAeS Lanchester Lecture, 2001 (*The Aeronautical Journal*, November 2001, pp633-641, RAeS, London). A concise discussion of the newest refinements in the wing design of the Airbus airliner family.

Kuchemann, D, *The Aerodynamic Design of Aircraft* (Pergamon, Oxford, 1978). A classic text by this world-leading aerodynamicist and former head of the aerodynamics department of the Royal Aircraft Establishment, Farnborough.

Kumar, A, and Hefner, J N, 'Future Challenges and opportunities in Aerodynamics' (*The Aeronautical Journal*, August 2000, pp365-373, RAeS, London). The latest view from the US NASA Langley Research Center.

Lake, J, 'A Supersonic Future?' (*Air International*, March 2001, pp154-156). A perceptive summary of the prospects for a future supersonic airliner.

McMichael, T, et al, 'Aerodynamic Technology – The role of aerodynamic technology in the design of modern combat aircraft' (*The Aeronautical Journal*, 1,000th Issue, December 1996 pp411-424, RAeS, London).

—— *F-117 Stealth Fighter* (Aerofax, Arlington, Texas, USA, 1990). Contains definitive details of this uniquely-distinctive American stealth fighter concept.

Mullins, J, 'Plasma Magic' (*New Scientist*, 28 October 2000, pp26-29). Examines the possible potential of an original Russian invention to reduce drag, repel shock waves and increase stealth.

Norris, G, and Wagner, M, *Modern Boeing Jetliners* (Motorbooks International, Osceola, Wisconsin, USA, 1999). Covers the technical aspects of all current members of this primary American airliner family, and includes contextual references to key aerodynamic features.

Owen, K, *Concorde: Story of a Supersonic Pioneer* (Science Museum, London, 2001). A fully updated edition of the author's classic work, *Concorde: New Shape in the Sky*, published in 1982 and now expanded to cover more than twenty years of airline service.

Poll, D I A, 'Aerospace and the Environment – What contribution can aerodynamics make?', (*The Aeronautical Journal*, July 2000, pp321-323. RAeS London).

Richardson D, *Stealth Warplanes: Deception, Evasion and Concealment in the Air* (Salamander, London, 2001). A concise treatise on this now centrally-significant aerodynamic factor.

Rogers, E W E, 'Aerodynamics – Retrospect and Prospect?' (*The. Aeronautical Journal*, February 1982, pp43-67, RAeS London). A perceptive overview by the then deputy director/A of the Royal Aircraft Establishment, Farnborough.

Shane, B, 'Modern Winglets: A Lift for Commercial Airliners', (*Airliners: The World's Airline Magazine*, No 72, Nov/Dec 2001, pp34-37).

Sobieszczanski-Sobieski, J, 'Multidisciplinary Design Optimisation [MDO] methods: their synergy with computer technology in the design process' (*The Aeronautical Journal*, RAeS, August 1999, pp373-382, Special MDO issue). From NASA Langley Research Center, USA, and with further extensive reference sources.

Spick, M, *Designed for the Kill: The Jet Fighter – Development and Experience* (Airlife, Shrewsbury, UK, 1995). The definitive work in its field.

Sweetman B, *Advanced Fighter Technology – The Future of Cockpit Combat* (Airlife, Shrewsbury, 1988). An informative aerodynamically-orientated volume by this well-known author in the field of combat aircraft design.

Trubshaw, B, *Concorde: the Inside Story* (Sutton, Stroud, Gloucestershire, 2000). The definitive account of the flight development and route-proving of the Anglo-French supersonic airliner.

'Unmanned Futures: Unmanned Air Vehicles' (*Flight International* supplement, 30 January-5 February 2001). A comprehensive survey of the current status of this increasingly-important type of military air vehicle.

Whitford, R, *Fundamentals of Fighter Aircraft Design* (Airlife, Shrewsbury, 2000). A highly instructive survey from historical origins to the newest concepts, notably including reference to the implicit aerodynamic factors and features.

—— 'Fundamentals of Airliner Design', *Air International*, Part 4 Aerodynamics 1, August 2001, pp108-114; Part 5 Aerodynamics II, October 2001, pp234-239; Part 6 Aerodynamics III, December 2001 pp360-365; and Part 7 Stability and Control I, February 2002.

Wilby, P G, 'Shockwaves in the Rotor World – A Personal Perpective of 30 Years of Rotor Aerodynamic Developments in the UK', RAeS 1997 Cierva Lecture, *The Aeronautical Journal*, march 1998, pp113-128, RAeS London).

2
Technology for the Supersonic Era
Richard P. Hallion

Supersonic flight constituted one of the two great challenges to aeronautics at the mid-twentieth century; the other was the refinement of reaction propulsion, particularly the gas turbine (jet) engine. Both revolutions were inextricably interlinked, and both were accompanied by some of the most remarkable engineering ever evinced in the entire aerospace revolution. Moreover, both dramatically transformed the design and performance of aircraft. After the jet and high-speed aerodynamic revolution, the high-performance aeroplane itself looked very different from its predecessors of the preceding forty years. Gone were propellers, blunt cowlings and thick, straight, broad-span high-aspect-ratio wings. Now the future belonged to the sleek and pointed jet, typified by the graceful swept or delta wing. Less obvious were equally significant changes to the aeroplane's interior and exterior. The adjustable tailplane and (later) the all-moving tail; notched or sawtooth leading edges; wing fences; vortex generators; 'blown' flaps; yaw dampers and (later) stability augmentation systems foreshadowing the era of computer-controlled fly-by-wire flight; pressurised cabins; ejection seats or ejection capsules; full-pressure pilot protection suits; and, on the ground, the genuine transonic and supersonic windtunnel, the necessary adjunct to producing successful supersonic aeroplanes.

Supersonic flight is now commonplace, but even over the last forty years only three aircraft have 'thrived' in the supersonic arena, capable of sustained flight beyond Mach 1: the Lockheed A-12/YF-12A/SR-71 Blackbird; the Concorde supersonic transport (SST); and the Lockheed-Martin F-22A Raptor. This is as much a tribute to their propulsion systems as to their aerodynamics. Some others, notably the English Electric Lightning, demonstrated remarkable performance for their time, limited only by available fuel. The ease with which modern military aircraft exceed Mach 1, with a bare flicker of the Machmeter betraying the passage from the subsonic to the supersonic realm, belies the very real problems that accompanied the achievement of supersonic flight in the late 1940s.

Supersonic flight was born of a crisis, namely the problem of aerodynamic compressibility. At subsonic speeds air can be considered an incompressible medium. In reality, of course, the compressibility of air as it bunches up ahead of an aircraft approaching the speed of sound leads to significant changes in aerodynamic and aircraft performance. Over time, careful design has minimised these changes. But, in the late 1930s, so drastic and serious were these departures from the norm of aircraft performance that aerodynamicists debated whether, in fact, the speed of sound could ever be penetrated by a piloted, winged aeroplane. British aerodynamicist W F Hilton well captured the prevailing climate of opinion when he stated bluntly that the speed of sound loomed 'like a barrier against future progress'.

The classic tool of the aerodynamicist, the windtunnel, proved of surprisingly little value during this great debate. In the pre-slotted throat tunnel era, windtunnel testing gave at best mixed and highly debatable results. The windtunnel furnished accurate and reliable data to Mach 0.75, and beyond Mach 1.25, but in the all-important transonic region between these two firmly subsonic and supersonic boundaries the tunnels 'choked'. The airflow became violently distorted and turbulent, shock waves formed on models and then reflected back-and-forth across test chambers, and reliable performance prediction became impossible. Eventually, in the late 1940s, the 'slotted' or open-throat windtunnel resolved such difficulties, but when the compressibility crisis first appeared, that tunnel was approximately fifteen years in the future.

The transonic crisis was real, possibly serving as a barrier to future supersonic flight; of that there could be no doubt. In the mid-1930s, as all-metal, high-performance fighters emerged that were capable of exceeding Mach 0.7 in dives, producing accelerated airflow over their wings in excess of Mach 1 and thus generating standing shockwaves on the wing with associated flow distur-

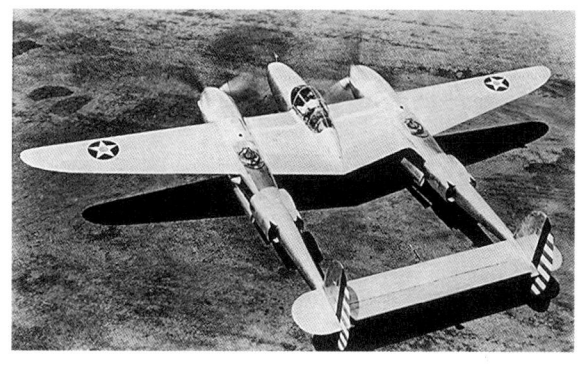

One of the Second World War aeroplanes that succumbed to the violent disturbances at the speed of sound while diving was the Lockheed YP-38 Lightning, which broke up over California in November 1941, killing company test pilot Ralph Virden.

FASTER, FURTHER, HIGHER

Germany's Messerschmitt Me 163 rocket-propelled interceptor exhibited viciously dangerous characteristics in high-speed dives, as well as being prone to explode on the ground or during landing, owing to the hypergolic nature of its fuel mixture.

bances, buffeting, and trim and control changes, the stage was set for the onset of the crisis. The first victim of such problems was Messerschmitt test pilot Dr Kurt Jodlbauer, killed when a diving Messerschmitt Bf 109 plunged into a lake near the Rechlin test centre on 17 July 1937. Perhaps the best-known victim was Lockheed test pilot Ralph Virden, killed when his Lockheed YP-38 Lightning broke up over Glendale, California, on 4 November 1941, during a dive recovery test. The year before, the National Advisory Committee for Aeronautics (NACA) had undertaken dive tests of a thoroughly instrumented Brewster XF2A-2 Buffalo at Langley Memorial Aeronautical Laboratory. Although the tests concluded successfully, one engineer recalled: 'We were left with the strong feeling that a diving airplane operating close to its structural limits was not an acceptable way to acquire high-speed research information'.

Eventually, most of the significant Second World War high-performance propeller-driven and jet aeroplanes exhibited handling qualities limitations and performance degradation ranging from severe (e.g. risk of injury or death, and compromising the ability of the aeroplane to fulfil its military mission) to acceptable (e.g. careful handling required under certain circumstances, but otherwise minimal impact upon flight safety or military utility). These include virtually all the 'big names'; the Supermarine Spitfire, Messerschmitt Bf 109, Focke-Wulf Fw 190, Hawker Typhoon, Lockheed P-38, Republic P-47 Thunderbolt, North American P-51 Mustang and, at the end of the war, the Gloster Meteor, Messerschmitt Me 163 and Me 262 and Lockheed P-80.

Some of these aeroplanes, particularly the Me 163, had viciously dangerous characteristics, and all experienced abrupt drag rise, buffeting, trim changes, and changes in flight loads, control forces, and control effectiveness, sometimes with disastrous results. Even the most highly

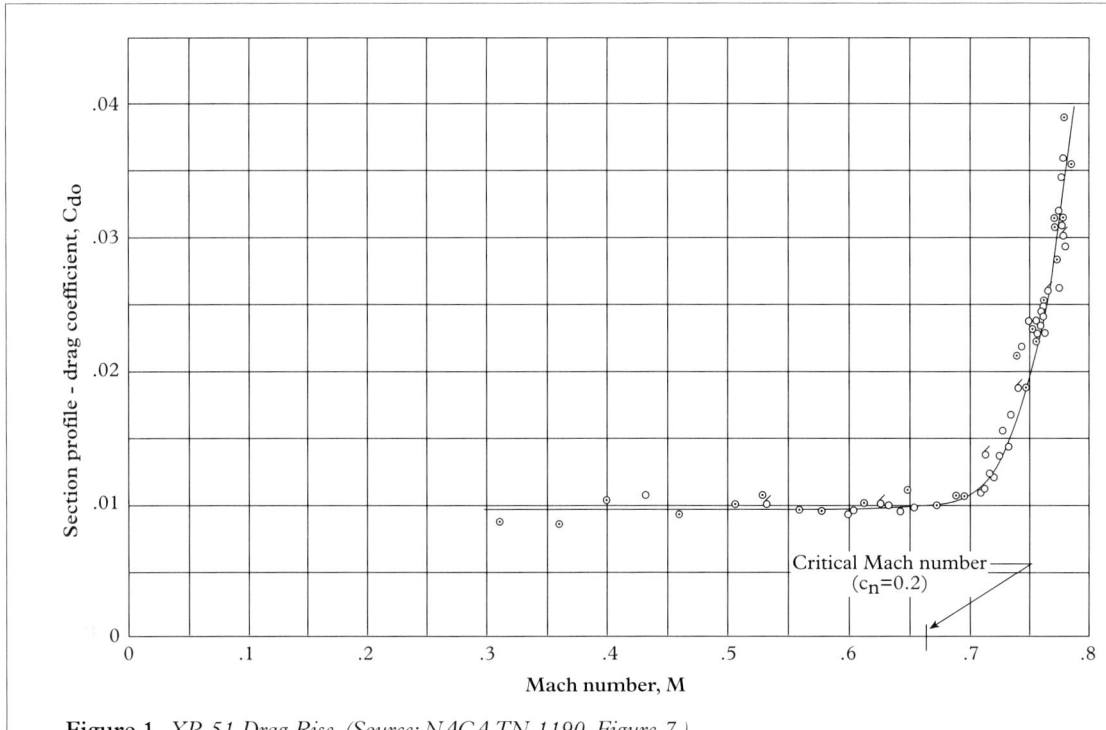

Figure 1. *XP-51 Drag Rise. (Source: NACA TN 1190, Figure 7.)*

refined conventional aircraft designs were afflicted by seemingly uncontrollable drag rise, typified by the drag plot shown in Figure 1, taken during actual flight test measurements of an instrumented North American XP-51 fighter test-flown by the NACA in 1943.

The North American P-51 Mustang was aerodynamically the 'cleanest' propeller-driven fighter developed by the USA during the Second World War, and the nemesis of the Luftwaffe and other Axis air arms. Nevertheless, note how the drag of its wing section rises abruptly at approximately Mach 0.66 (the so-called 'drag divergence' Mach number) from a drag coefficient of .01 to a drag coefficient of .04 at Mach 0.79, close to the aeroplane's limiting Mach number. At this point, to see it from the pilot's perspective, the aeroplane is buffeting, with pronounced trim changes and greatly reduced control effectiveness. In short, the pilot is concentrating on flying the aeroplane — indeed surviving — rather than fulfilling any sort of military mission.

It was those very limitations — the risk of a dive into the dense lower atmosphere in an aeroplane of doubtful flying qualities, the inadequacies of the existing windtunnel technology — that led to the development of four stop-gap flight-testing methodologies to confront the problems of high-speed flight. These were the dropping weighted, instrumented bodies (and radar-tracking them as well); firing instrumented rocket-propelled models, diving high-speed research aeroplanes with small models mounted on their wings, taking advantage of the accelerated supersonic local flow over the wing; and, finally, undertaking comprehensive in-flight testing of piloted research aeroplanes.

But the roots of the postwar supersonic revolution predate the Second World War, and, indeed, the 'compressibility crisis' of the late 1930s. In pre-First World War

The extraordinary series of ramjet-powered aircraft developed by René Leduc in France is represented here by the Leduc 021 No 02, mounted on the Languedoc mother aircraft which carried it aloft.

France the French aviation pioneers René Lorin and René Leduc had already conceptualised using reaction-propelled aircraft for high-speed flight, and Leduc notably extended this to include ramjet-powered designs in the interwar years, one of which, the 0.10, would undoubtedly have flown much earlier than it did but for the tragedy of the Second World War. In 1918-1924

The NACA wing flow test system is exemplified in this picture of a tiny scale model of a Lockheed P-80 mounted on the wing of a North American P-51 Mustang for high-speed-dive tests.

This prominently marked Lockheed YP-80 was used for transonic dive testing during the type's development.

teams of researchers at the US Army Air Service's Engineering Division, the US Bureau of Standards, the Royal Aircraft Establishment (RAE), and the National Physical Laboratory first detected the characteristic drag rise, loss of lift and formation of shock waves and turbulence associated with a wing approaching the speed of sound, from tests of propellers. The turbojet, ramjet and rocket propulsion work of a variety of international researchers in the 1930s, including Frank Whittle, Hans von Ohain, the previously mentioned Leduc and Lorin, Eugen Sänger, Lovell Lawrence, James Wyld, Roy Marquardt, Boris Stechkin, Yuri Pobedonotsev and Igor Merkulov, set the stage for exploiting the aerodynamic revolution that occurred in that same time period, thanks to an equally distinguished group of aerodynamicists, including Bennett Jones, Jacob Ackeret, Arturo Crocco, Eastman Jacobs, John Stack, R T Jones, and Adolf Busemann, the latter all disciples preaching the gospel of the streamline aeroplane.

The generalised 'streamline revolution' of the late 1920s and 1930s, typified by high-performance racing aircraft (which, in many ways, functioned as the 'X-series' of their day) inspired various designers to conceptualise idealised high-speed aircraft that could reach and exceed 500mph (805km/h). Figure 2 shows the first notable 'compressibility' research aircraft proposal, advanced by John Stack of the NACA in 1934.

Stack's proposal was, of course, an idealised concept. However, it inspired great interest in the NACA at a time when the agency was already firmly establishing its reputation as the world's leader in refined aerofoil and shape development, as typified by the end of the decade by the evolution of its symmetrical low-drag aerofoil families. By mid-1941 NACA aerodynamicists believed that a thin NACA wing section would have little difficulty passing through the speed of sound.

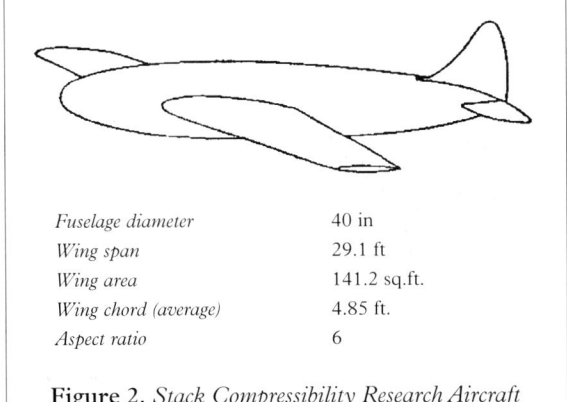

Fuselage diameter	40 in
Wing span	29.1 ft
Wing area	141.2 sq.ft.
Wing chord (average)	4.85 ft.
Aspect ratio	6

Figure 2. *Stack Compressibility Research Aircraft Proposal. (Source: 'Effects of Compressibility on High-Speed Flight,'* Journal of the Aeronautical Sciences, *January 1934.)*

The NACA was not, however, the only leader at this time in ground-based transonic and supersonic research. Aggressive facilities development led to the first supersonic tunnels entering service in Europe (in Switzerland, Italy, and Germany) well in advance of their appearance in the USA. Not surprisingly, then, the first serious examination of advanced wing and design configurations for transonic and supersonic flight took place in Europe as well. By the end of the 1930s Alexander Lippisch was already refining his thoughts on tailless swept-wing aircraft. His lower-aspect-ratio configurations would approach the delta configuration now so closely associated with his name (though his wings were considerably thicker than those subsequently used by Convair, Dassault, Gloster and Avro. In addition, Lippisch had already taken the old notion of the variable-sweep wing (dating to Clément Ader, who advanced it as a means of effecting roll control in his Éole and Avion III aircraft) and applying it to transonic aircraft as a means of giving them good high-speed and low-speed characteristics. Adolf Busemann, likewise, had taken the older swept wing of John Dunne and G T R Hill, used by them as a means of producing inherent longitudinal stability, and had postulated from it the transonic 'arrow wing' (swept wing) as a means of effectively raising the critical Mach number of a wing (the point where it began to experience drag divergence), with the added benefit of moving the wing's control surfaces within the Mach 'cone' formed around an aeroplane. Eugen Sänger and his research associate (and future wife) Irene Bredt had conceptualised the Silbervogel, a rocket-boosted space station supply vehicle(!) with sharp-leading-edge wedge supersonic aerofoils (later, in the 1940s, they would 'upgrade' it as a proposed *Raketenfernbomber*).

Overall, Nazi Germany led the world in high-speed aerodynamic facilities development, research and application in the early 1940s. The facilities issue was a critical one. In 1945 only a handful of supersonic tunnels existed in America and England, but Germany had no fewer than eight in service in four research complexes. Six of these tunnels could exceed Mach 3, and one could exceed Mach 4. Sir Roy Fedden, the leader of one postwar Allied technical intelligence team that went into Germany, noted subsequently that: 'Data obtained from research in these high-speed windtunnels and elaborate laboratory equipment, combined with jet engines, had revolutionised German ideas on design for high-speed military aircraft, and provided a confident approach to the problems of flight at sonic speed'. As a result, the German aircraft industry witnessed an explosion of interest in high-speed flight, with a plethora of proposed designs having swept wings, delta wings, trapezoidal layouts, and all powered by jets, rockets, pulsejets and ramjets. Many were extraordinarily bizarre, but others were far more practicable. None had the potential to reverse

TECHNOLOGY FOR THE SUPERSONIC ERA

the war for Nazi Germany, for the true stakes in the war were really development of the atomic bomb, and here, thanks to the doctrinaire belittlement of so-called 'Jewish physics', Nazi Germany was most fortunately behind the Allies. Had greater research focus and direction characterised the German acquisition process, rather than the muddle imposed by Göring, Udet and Milch (and, of course, Hitler himself), the Allied air war could have been far more difficult to prosecute, and Allied air supremacy (recognised by planners at the time as a necessary precondition for the invasion of Europe) far more problematical to achieve than was actually the case.

German efforts did result in one significant vehicle that did routinely exceed the speed of sound beginning in October 1942. This was the A-4 ballistic missile, better known as the 'Vergeltungswaffe Zwei' (Revenge Weapon Two), the infamous V2. In the late 1930s the Germany army had made an extensive investment in rocket technology as a means of sidestepping restrictions on the size of artillery imposed by the Treaty of Versailles, and the result was this Wagnerian terror weapon. Germany's high-speed tunnels played a key role in the development of the V2 and its generalised shape and tail-fin design, and of a winged variant of it, the A-4b, which, in 1944-1945, made three abortive boost-glide flight attempts. Two flight attempts failed at launch, but in the course of the third, on 24 January 1945, the A-4b became the world's first winged vehicle to exceed the speed of sound, as it roared aloft from its Peenemünde launch pad. Following engine shutdown, it coasted into the upper atmosphere, then plunged earthwards. One wing failed as it re-entered, and the missile broke into pieces. If researchers could take heart from the penetration of the sonic region during launch, the return had been a dismal failure.

By the middle of the Second World War, Nazi Germany and Great Britain had undertaken the first two piloted aircraft projects developed explicitly for supersonic flight: the German DFS 346 and the British Miles M.52. In 1942 the Deutsche Forschungsanstalt für Segelflug (DFS) initiated a series of design studies for high-altitude and high-speed military aircraft. Out of this operationally-oriented thinking sprang the DFS 346, a rocket-propelled swept-wing aircraft with a high T-tail which, under way too late to fly before Germany's collapse, formed the basis for the Soviet Union's first serious foray in supersonic flight testing after the war. In 1943 the RAE had begun a comprehensive programme in high-speed research, using instrumented Mustang and Spitfire fighters dived to above Mach 0.80. One of these, Spitfire PR.XI EN409, flown by Squadron Leader A F 'Tony' Martindale, eventually reached Mach 0.89 ± 0.01. But the RAE recognised, as did the NACA, that such dive trials were both fraught with risk (Martindale lost the propeller and reduction gear from his fighter, and had to make an emergency dead-stick landing, which he carried off with aplomb) and very brief in duration and research productivity (as the pilot uneasily confronted the terrain growing inexorably larger and clearer in view). Accordingly, the British government released Specification E.24/43 for an aeroplane to fly at 1,000

Nazi Germany's A4B rocket, a winged variant of the A4 (V2) ballistic missile, was the world's first winged vehicle to exceed the speed of sound, though it suffered a structural failure and broke up during its descent.

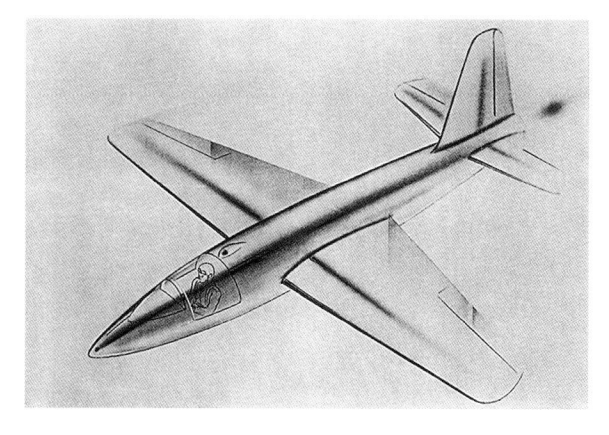

A 1944 design study by Major Ezra Kotcher of USAAF Wright Field for a 'Mach 0.999' transonic research aeroplane.

33

FASTER, FURTHER, HIGHER

Britain's Miles M.52 might have become the world's first supersonic piloted aircraft. This is a metal model for high-speed windtunnel tests. Other tests were conducted with a larger air-launched model, and a Miles Falcon fitted with an M.52 wing for low-speed trials. A mock-up had been built and construction of the actual aircraft had begun when the project was abruptly cancelled.

mph (1,609km/h) in level flight at 36,000ft (10,970m) (approximately Mach 1.5), awarding a development contract to the Miles Aircraft Company in October 1943. This bold requirement led to the abortive Miles M.52, a remarkably advanced and interesting aircraft regrettably cancelled in 1946.

The USA did not first start down its own path towards a supersonic research aeroplane until December 1943, when the NACA hosted a Washington conference following a presentation by William S Farren, then director of the RAE. The chairman, Dr William Durand, asked what best could America do with the jet engine brought to the USA from Great Britain. A young Bell engineer, Robert Wolf, recommended developing a transonic research aeroplane. Together with independent work in the two military services (particularly by Ezra Kotcher and Theodore von Kármán within the United States Army Air Force (USAAF)), this evolved into a two-pronged USAAF (later United States Air Force; USAF) and US Navy effort to build two flying supersonic research aero-

The Douglas D-558-1 Skystreak supersonic research aircraft, which first appeared in 1947, had unswept flying surfaces and a simple circular-section fuselage housing an axial-flow turbojet.

planes, each carrying a comprehensive package of flight-test instrumentation.

The first American early supersonic research aircraft were the Air Force-sponsored Bell XS-1 (later X-1), first flown in January 1946, and the two-phase Navy-sponsored Douglas D-558, the first of which, the D-558-1 Skystreak, was completed in early 1947. Both drew heavily on the NACA for technical support, and both represented very different philosophies about achieving supersonic flight. The XS-1 was an air-launched, all-rocket, straight-wing aircraft, while the D-558-1 was a straight-wing, ground-take-off jet. Its stablemate, the D-558-2, was a swept-wing, ground-take-off jet-and-rocket design, later modified for all-rocket air-launch operation. Both reflected extreme simplicity and the most basic of assumptions about the transonic and supersonic regimes. The XS-1 had a fuselage based on the shape of a 0.50-calibre bullet (known from ballistics tests to have stable deceleration characteristics as it slowed from supersonic to subsonic speeds); the D-558-1 was a simple body of revolution wrapped around an axial-flow turbojet engine, and the D-558-2 was a ogival body of revolution having, in its original incarnation, a flush canopy (later abandoned owing to poor visibility) similar to that of the XS-1. But both shared some important new characteristics, particularly thin, low-aspect-ratio wings; a high fineness ratio; and powered, adjustable tailplanes giving them, in effect, 'slab' tails for enhanced transonic and supersonic longitudinal control above Mach 0.90.

Thus, with windtunnels incapable of meeting the challenge for reliable and useful transonic design information (though they remained useful for clearly subsonic and clearly supersonic measurements), various nations—particularly the USA, Great Britain and Nazi Germany, and later the Soviet Union, turned to developing instrumented high-speed research aeroplanes that could use the sky as a laboratory and, hopefully, exceed the speed of sound. Of all these efforts, only the USA succeeded

A windtunnel model of the Bell XS-1 is prepared for tests in NACA's 8ft transonic tunnel. Although the XS-1 was straight-winged like the D-558, it was rocket powered and air-launched. Its fuselage was based on the shape of a 0.50-calibre bullet.

FASTER, FURTHER, HIGHER

The Bell X-1 is carried aloft by its parent Boeing B-29 for a high-altitude air launch. The B-29's pressurised crew compartments made it ideal for such tasks.

within a reasonable time, in some ways more by fortune than careful planning, when USAF test pilot Captain Charles E 'Chuck' Yeager first exceeded Mach 1 on 14 October 1947, flying the first Bell XS-1 (46-062) to Mach 1.06 at 43,000ft (13,100m). Table I shows the relative times when the USA, Great Britain and the Soviet Union first exceeded Mach 1.

Aside from Nazi Germany, whose research efforts were most fortunately brought to an abrupt end by the Allied victory in May 1945, both Great Britain and the Soviet Union came close (and the Soviets apparently extremely close) to being the first to build a piloted aircraft capable of flying through the 'sound barrier'. In the summer of 1946 de Havilland began tests of its second D.H.108 Swallow, TG306, the series' first high-speed semi-tailless technology demonstrator. Sadly, it crashed from transonic pitch divergence and subsequent structural failure on 27 September 1946, having exceeded approximately Mach 0.87 at an altitude of only 7,500ft (2,290m), killing its pilot, Geoffrey de Havilland Jr. The more refined third D.H.108, VW120, did not fly until after the XS-1 had first broken the sound barrier, though, admittedly, it was a very poor-flying transonic aircraft in any case, and fated to soon meet its own tragic end. In the spring of 1947 the Soviet Union began flight testing its own variant of the DFS 346, air-launched from a former USAAF Boeing B-29.* On one flight in May the Samolyot 346-1 approached Mach 0.93, but test pilot Wolfgang Ziese, who had been a test pilot in the

Table I: Supersonic Pioneering

Country	Date & Mach Number	Aircraft Type	Pilot
USA	10/14/47, Mach 1.06	Bell XS-1	C E 'Chuck' Yeager
Great Britain	9/6/48, Mach 1.04	de Havilland D.H.108	John Derry*
Soviet Union	12/26/48, Mach 1.02	Lavochkin La-176	I V Federov

* Derry was the first British pilot to exceed the speed of sound in a British aeroplane. Earlier, flying an experimental North American XP-86 in mid-1948, British test pilot Roland Beamont had exceeded Mach=1.0 in a dive at Muroc Dry Lake, California (now Edwards Air Force Base).

TECHNOLOGY FOR THE SUPERSONIC ERA

Apart from its maker, the Douglas D-558-2 Skyrocket had little in common with the D-558-1. As its name suggests, it was rocket powered, and it had swept flying surfaces. Like the Bell XS-1 it had a powered, adjustable 'slab' tailplane. The use of a thin wing necessitated an undercarriage that retracted into the fuselage.

Luftwaffe during the Second World War, encountered 'violent vibrations' that forced him to decelerate. Five months later Chuck Yeager whiplashed the Mojave with a sonic boom, inaugurating the era of piloted supersonic flight.

His matter-of-fact report of that flight belies the very real drama that attended it:

> With the stabiliser setting at 2 degrees the speed was allowed to increase to approximately .98 to .99 Mach number where elevator and rudder effectiveness were retained and the airplane seemed to smooth out to normal flying characteristics. This development lent added confidence and the airplane was allowed to continue to accelerate until an indication of 1.02 on

---

* (This Superfortress, 42-6256, *Ramp Tramp*, had made an emergency landing at Vladivostok on 29 July 1944, the first of an eventual three to arrive in Siberia after raid mishaps over Japan. As the Soviet Union and Japan were officially neutral at that time, the aircraft were interned as international law required.)

In England, de Havilland built three D.H.108 Swallow tailless research aircraft, one for slow-speed trials and the others for high-speed research. The design was a poor transonic flyer, and one suffered transonic pitch divergence after exceeding Mach 0.87 and killed its pilot.

the cockpit Machmeter was obtained. At this indication the meter momentarily stopped and then jumped to 1.06 and this hesitation was assumed to be caused by the effect of shock waves on the static [pressure] source. At this time the power units [e.g. the Reaction Motors Inc XLR-11 6,000lb (2,725kg)-thrust rocket engine with its four individually ignited rocket chambers] were cut and the airplane allowed to decelerate back to the subsonic flight condition. When decelerating through approximately .98 Mach number a single sharp impulse was experienced which can best be described by comparing it to a sharp turbulence bump.

The XS-1 and D-558-1/D-558-2 constituted but the first of over forty 'X-series' piloted and unmanned aircraft and rocket systems which have been funded to date, the most notable of which have been research vehicles for high-speed flight. Others, not bearing the X-series designation but nevertheless functioning as such, were also developed on occasion by manufacturers, either on their own or in partnership with military services and research agencies. These had their own equivalents in many foreign designs, but only the USA had a clearly definable evolutionary 'family' of such craft, due to both the strength of the American economy and research and development funding after the Second World War, and the strong unified direction imparted to American aeronautical research and development by the military Services and the NACA (subsequently reorganised and expanded as the National Aeronautics and Space Administration - NASA). The military services and the NACA (and later NASA) ran these programmes through the NACA Research Aircraft Projects Panel (RAPP) and, later, through special joint steering committees formed via memorandums of understanding between the services and the NASA.

The American 'X-series' high-speed flight research aircraft, particularly the 'Round One' X-series (the X-1, X-2, X-3, X-4, X-5, D-558-1 & -2 and XF-92A) made many diverse and profound contributions to transonic and supersonic technical development and theoretical knowledge alike. Their legacy consisted of four broad categories of research:

Aerodynamics, including:
- Validating and interpreting tunnel test data,
- Aerodynamic heating,
- Lift and drag studies, and
- Inlet and duct studies

Flight loads, including:
- Load distribution
- Effect of wing sweep on gust loadings
- Gustiness at high altitudes
- Buffeting
- Aeroelastic effects
- Effect of stability reduction upon flight loads

Stability and control, including:
- Longitudinal (pitch) control
- Effectiveness of blunt-trailing-edge control surfaces
- Alleviation of aerodynamic pitch-up by wing devices
- Effect of the inertial axis upon lateral stability
- Exhaust jet impingement and directional stability
- Inertial (roll-yaw-pitch) coupling
- Directional instability at increasing Mach numbers
- Reaction controls in low dynamic pressure("q") flight

Operations, including:
- High-speed flight test exploration
- Speed loss in manoeuvring flight
- High-altitude problems
- Pressure suit research and use
- Airspeed measurement
- Variable wing-sweep operation

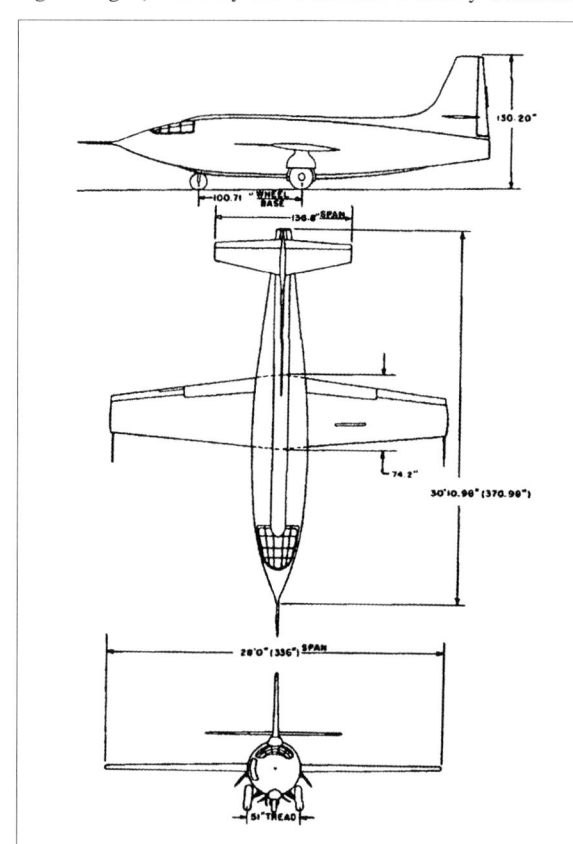

Figure 3: *Bell XS-1 research aeroplane. (Source: USAF, Air Force Supersonic Research Airplane XS-1, Report No.1, 9 January 1948, p.2).*

Additionally, they evaluated the merits (and demerits) of various design configurations including thick versus thin

TECHNOLOGY FOR THE SUPERSONIC ERA

An assortment of 'X-series' experimental research aircraft at NASA's Flight Research Center at Edwards Air Force Base, California. Clockwise from top right they are: the variable-sweep Bell X-5; the Douglas D-558-II, the first aircraft to exceed Mach 2; the semi-tailless Northrop X-4 Bantam; The Bell X-1A; the Douglas D-558-1 Skystreak and the Convair XF-92A. In the centre is the Douglas X-3 Stiletto.

wings, sweepback versus straight, semitailless versus tailed configurations, low versus high-placement of tailplanes, and the transonic control effectiveness of fixed versus moveable tailplanes. Figure 4 shows the effects of wing sweepback (L) and reduced wing thickness (as measured by thickness/chord ratio, t/c) on the transonic performance of straight, moderately swept, and sharply sweptwing planforms. Note that the 'unbroken' drag rise depicted in Figure 1 is clearly a thing of the past; these wings (typifying the design technology of aircraft such as the North American F-100, Convair F-102 and Lockheed F-104) are clearly quite comfortable in the transonic-supersonic regime, a measure of the advance of aerodynamics over the decade after 1944.

The Round One X-series research aircraft acted as technology demonstrators and, as a rule, their performance capabilities were generally at least half a generation beyond contemporary design practice. Traditionally, technology demonstrators evaluate or showcase innovations that are typically first adopted piecemeal in fighter, then bomber, and finally transport aircraft. For example, in the pre-First World War era Louis Bechereau's Deperdussin Monocoque Racer functioned in such a role. In the post-Second World War era the swept wing progressed from the Bell L-39 to the F-86, and on to the B-47 and then the Boeing Model 367-80, the progenitor of all subsequent Boeing jetliners. Likewise, the adjustable 'flying tail' (which played such a major role in defeating the MiG-15, which lacked such a tail, in Korea) was first tested and validated on X-series aircraft. More recently, in the computer era, electronic flight control technology was first applied to experimental aircraft, then on the F-16 and B-2, and finally to commercial aircraft.

Over time, this performance disparity grew larger, particularly after the mid-1950s, by which time the 'altitude-airspeed' envelope still occupied by most aircraft today had been mapped out. Figure 5 indicates the relationship between American supersonic research aircraft, operational fighters and fighter prototypes from the perspec-

FASTER, FURTHER, HIGHER

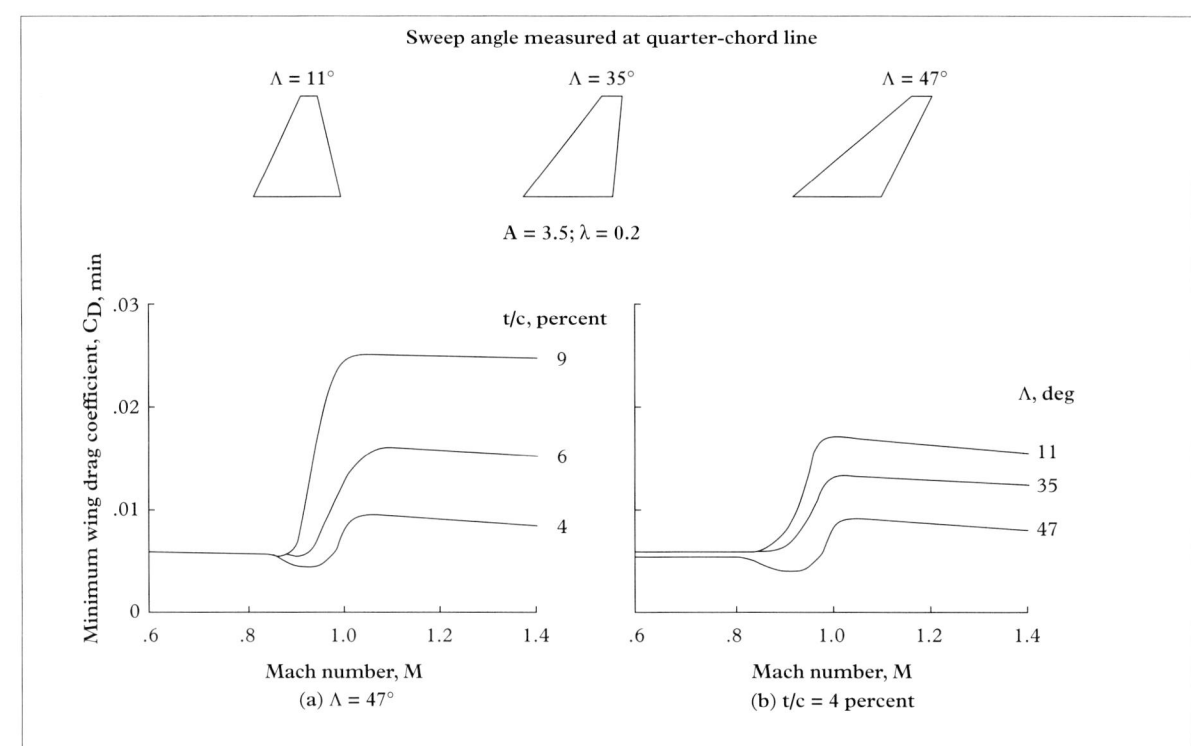

Figure 4: *Relative drag reduction attainable by reducing wing thickness-chord ratio and increasing wing sweepback angle. (Source: Laurence K Loftin, Jr,* Quest for Performance: The Evolution of Modern Aircraft *(Washington: NASA, 1985), Figure 10.12, p.253.)*

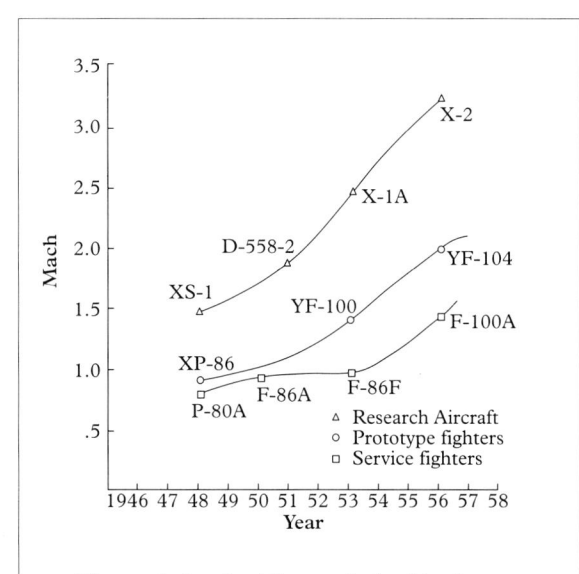

Figure 5: *Leader-follower relationships between research aircraft, prototypes and operational fighters. (Source: R P Hallion,* On the Frontier: Flight Research at Dryden *(Washington: NASA, 1984), p.81.)*

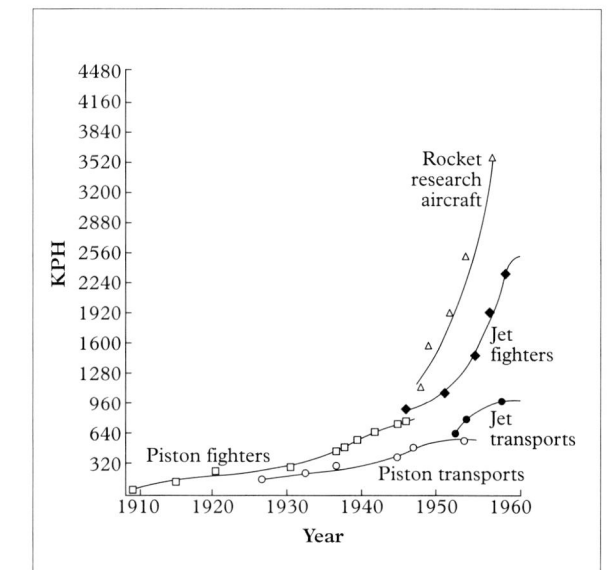

Figure 6: *Leader-follower relationships of selected operational piston and turbojet fighters and transports, and rocket-propelled research aircraft. (Source: Hallion,* On the Frontier, *p.82.)*

TECHNOLOGY FOR THE SUPERSONIC ERA

Table II: Selected American high-speed flight test vehicles flown since 1945

Designation	Research purpose	Performance	Propulsion system
Bell XS-1 (X-1)	Exceed Mach 1	Mach 1.45	air-launch; rocket
Bell X-1A/B/D	Mach 2+ aerodynamic research	Mach 2.44	air-launch; rocket
Bell X-2	Sweptwing & aerofoil research	Mach 3.2	air-launch; rocket
Douglas X-3	Mach 2+ turbojet & config. study	Mach 1.21	ground take-off; jet
Northrop X-4	Semi-tailless behaviour	Mach 0.88	ground take-off; jet
Bell X-5	Variable-sweep wing behaviour	Mach 0.95	ground take-off; jet
Convair XF-92A	Delta wing behaviour	Mach 1.0	ground take-off; jet
Douglas D-558-1	Transonic configuration studies	Mach 1.0	ground take-off; jet
Douglas D-558-2	Supersonic sweptwing studies	Mach 2.005	ground/air; jet/rocket
Lockheed X-7	M=4+ aerodyn. & ramjet perf.	Mach 4.31	air-launch; ramjet
NAA X-15	Hypersonic & high alt. research	Mach 6.72	air-launch; rocket
Lockheed X-17	Hypersonic re-entry testing	Mach 14.4	four-stage rocket
NAA XB-70A	Sustained Mach 2.5+ cruise	Mach 3.1	ground take-off; jet
ASSET	Hypersonic aerothermodynamics	Mach 18.4	Thor-Delta booster
PRIME	Manoeuvring; ablative studies	Mach 25.4	Atlas booster
BGRV	Manoeuvring w. flaps, thrusters	Mach 18.0	Atlas booster
Shuttle *Columbia*	Piloted Lifting Entry, 4/12-14/81	Orbital	Solid/liquid rocket

tive of the most interesting time period of the supersonic breakthrough, that of the late 1940s to mid-1950s. At the beginning of this period one early supersonic research aircraft (the X-1) was almost twice as fast as the fastest operational USAF fighter (the P-80), and over a third faster than the fastest experimental fighter (the North American XP-86). At the end, the Bell X-2 was over twice as fast as the fastest operational USAF fighter (the F-100), and over a full Mach number faster than the fastest experimental fighter (the YF-104).

Figure 6 places this in the larger context of aerospace vehicle development from the era of the Wright brothers to the era of manned spaceflight. Here the dramatic technology accelerant made possible by the advent of the jet engine, rocket engine and supersonic flight is clearly apparent. Note also that by the mid-1950s the speed envelopes within which most aircraft continue to operate had been clearly mapped out: Mach 0.8-0.9 for jet transports, Mach 2.2-2.5 for jet fighters. Two 'point specific' aircraft developed since that time, the Lockheed A-12/YF-12A/SR-71A Blackbird and the Anglo-French Concorde, offer evocative exceptions. Both represent extraordinary technical achievements, but while the former constituted a notable national security success, the latter proved unsuccessful in winning widespread market support among the world's airlines, despite aggressive marketing and romantic appeal. Subsequent changes to this graph have only occurred in the hypersonic regime, with the North American X-15 extending piloted flight to Mach 6.72 (October 1967), and the Space Shuttle Columbia completing the first piloted hypersonic winged flight into and from space (April 1981).

The first European aircraft to exceed the speed of sound in level flight was the French-designed SFECMAS Gerfaut, which accomplished the feat on 3 August 1954. Produced to furnish data for a single-seat fighter, it was powered by a SNECMA Atar 101C turbojet.

FASTER, FURTHER, HIGHER

Table III: Selected transonic, supersonic and hypersonic flight test vehicles flown since 1945

Country	Programme	Purpose/notable characteristics
Great Britain	Vickers M.52 model	Rocket model of abandoned Miles M.52 jet; Mach 1.38
	De Havilland D.H.108	Transonic tailless research; 1st UK Mach 1 piloted aircraft
	English Electric P.1A/B	Swept-wing Mach 1.0-2.0; predecessor to Lightning ftr.
	Fairey F.D.2	Mach 1.5 delta testbed; modified as BAC 221
	BAC 221	'Ogee' wing aerodynamics to support Concorde SST
	Bristol Type 188	Mach 2.5 aerodynamic and propulsion research
France	Leduc 010/016/021/022	Air-launched; transonic/supersonic ramjet testbeds
	SFECMAS 1402 Gerfaut	Tailed delta; first French supersonic 'on the level' aircraft
	S.O.9000 Trident	Jet-and-rocket propulsion; rolling tail control behaviour
	S.E.212 Durandal	Supersonic delta aerodynamics
	Nord 1500 Griffon	Turbojet-ramjet; canard delta; first Euro. M=2+ a/c
Soviet Union	Florov 4302	Rocket; straight-wing; high-speed testbed; abandoned
	MiG I-270	Rocket; straight-wing; based on Ju 248; abandoned
	Samolyot 346	Rocket; swept-wing; air-launched; based on DFS 346
	Bisnovat B-5	Rocket; swept-wing; air-launched; abandoned
	Lavochkin La-176	Jet; sweptwing; experimental fighter; 1st Sov. Mach 1+ a/c
	Yakovlev Yak-1000	Jet; tailed delta; very low aspect ratio; abandoned
	Sukhoi S-1	Jet; swept-wing; development a/c for Su-7; Mach 2.05
	MiG Ye-50	Jet & rocket; tailed delta; exp. ftr.; abandoned; Mach 2.3
	MiG Ye-152-1 'Ye-166'	Jet; tailed delta; exp. ftr.; record-setter; Mach 2.8
	MiG-21I Analog	Jet; ogee wing testbed for Tu-144 SST development
	BOR-1, 2, 3 models	Hypersonic; suborbital models lofted on SS-4 boosters
	VKS orbital models	Hypersonic lifting-body models orbited on SL-13 boosters
	VKS full-size orbiters	Hypersonic lifting bodies orbited by SL-16 boosters
	BOR-4 orbital testbed	Modified VKS model for testing VKK thermal protection
	BOR-5 orbital testbed	Orbital re-entry testbed having VKK Shuttle configuration
	VKK *Buran* Shuttle	Unpiloted Soviet Shuttle; one orbital flight test 11/15/88

Table II (page 41) enumerates selected American research craft flown since 1945.

But any accounting of high-speed flight research must as well accord due credit and respect to those foreign research efforts, a surprising number of which flew and which shaped, in many cases, the future course of European and Soviet aeronautical development. Table III lists selected foreign test vehicles flown in the same period.

As both listings clearly indicate, high-speed research aeroplanes and unmanned technology demonstrators were particularly active in furtherance of the post-Second World War revolution in high-speed flight, from transonic velocities right up to hypersonic flight, orbital insertion and atmospheric entry from space. Again, it should be noted as well that this revolution, which took flight from the propeller-driven subsonic era to the jet-and-rocket-propelled supersonic and hypersonic era, did not constitute a 'singularity'. Rather, it resulted from the blending of the two great aeronautical revolutions that existed at mid-century, namely the gas turbine and rocket propulsion revolution, and the supersonic breakthrough,

The canard-delta Nord 1500 Griffon of 1955, another French product, was Europe's first Mach 2-plus aircraft. Developed from the earlier Gerfaut, it was initially powered by a 6,500lb-thrust SNECMA Atar 101F-2, which with a SNECMA 'eyelid' afterburner gave a thrust of 8,370lb. Subsequently this was replaced by a 7,710lb-thrust Atar 101F, augmented by a SNECMA ramjet, the aircraft, seen here, then being named Griffon II.

TECHNOLOGY FOR THE SUPERSONIC ERA

Another French creation was the Sud-Ouest SO.9050 Trident light target-defence interceptor of 1955, powered, as its name suggests, by a single SEPR 631 rocket motor and, in the version depicted, a pair of Turboméca Gabizo turbojets in the wingtip nacelles. It had been preceded by the similar SO.9000, also named Trident.

The twin-engined MiG Ye-152A was designed as a fast-climbing interceptor. A development, the Ye-152-1, with a bigger wing and a single engine, set a 100km-circuit speed record of 2,401km/h (1,491.9mph) on 7 October 1961, and a 15/25km record of 2,681km/h (1,665.89mph) on 7 June 1962.

Russia's straight-winged, T-tailed MiG I-270 of 1947 was based on the wartime German Junkers Ju 248. Powered by an RD-2M-3V rocket motor, it proved to have poor manoeuvrability, and with both chambers of the motor firing its endurance was a mere 255 seconds. It was soon abandoned.

FASTER, FURTHER, HIGHER

SAAB's diminutive 210 was a scale testbed aircraft used to investigate the aerodynamics of the distinctive double-delta wing adopted for Sweden's Draken interceptor. In this view it can be seen that the starboard wing is tufted for flow visualisation.

and drew upon others as well, particularly structures, materials, and the rise of digital electronics and computer-based avionics.

It must be noted that these two listings do not include a number of other aircraft that supported high-speed research studies after mid-1939, such as:

- early turbojet engine propulsion testbeds such as the Heinkel He 178, the Gloster-Whittle E.28/39 or the Bell XP-59A Airacomet;
- instrumented propeller-driven fighters of the Second World War (for example, the Supermarine Spitfire Mk.XI, Focke-Wulf Fw 190 or the North American P-51D Mustang) used for transonic dive testing above Mach 0.75;
- early jet fighters used as 'gap-filling' research aircraft (such as the Lockheed YP-80A Shooting Star and North American F-86A Sabre);
- prototypes, pre-production, or early production examples of the 'Century series' (F-100 to F-107) fighters

During research concerned with the development of the Concorde SST, the slow-speed characteristics of slender-delta wings were investigated using this aircraft, the Handley Page H.P.115. Wooden leading edges of different forms could be fitted, and coloured dye was emitted to leave a visible flow pattern on the wings.

TECHNOLOGY FOR THE SUPERSONIC ERA

and their foreign equivalents, such as the English Electric P.1B Lightning and the Dassault Mirage III);
- later high-speed aircraft applied to supersonic aerodynamic and propulsion studies (such as the American Lockheed YF-12A/C and SR-71A/B Blackbirds, the North American A-5A Vigilante and XB-70A Valkyrie, the General Dynamics F-111A TACT testbed, the NASA-Vought TF-8A supercritical wing testbed, the McDonnell-Douglas F-15A Eagle, the General Dynamics F-16XL, and the McD-D F-18 Hornet, together with foreign programmes such as the mid-1990s NASA-Russian Tu-144 SST co-operative supersonic cruise test programme;
- specialised handling-qualities testbeds of proposed designs, such as the Bell L-39, Saab 210, Short S.B.5, Handley Page H.P.115, NASA Ames-Dryden AD-1, Rockwell HiMAT, Grumman X-29 and Rockwell-MBB X-31;
- piloted approach-and-landing test beds of proposed lifting-body or winged spacecraft (such as the American Northrop M2-F1/2/3, Northrop HL-10, Martin X-24A/B; and the Soviet Mikoyan Lapot ['Wooden Shoe'], 105-11, BTS-01, and BTS-02) as these were not, per se, high-speed test beds themselves, even though the American craft could approach Mach 2 and reach altitudes in excess of 90,000ft (27,430m);
- and purely 'paper' projects that did not progress significantly down the airframe path, such as the Miles M.52, Boeing X-20, HOTOL, Hermes, the X-30 NASP, the X-33 Starclipper, and the X-34.

Although the discovery of the advanced state of German aerodynamics research at the end of the Second World War revolutionised European and Russian aviation, it

Consolidated Vultee's XF-92A delta-wing demonstrator was an offshoot of German Second World War research by Alexander Lippisch. Although it was not a success, it was the first powered delta-wing aircraft flown, and led to the F-102 and F-106 fighters that equipped the USAF.

Sired by the XF-92A, of which it was basically a scaled-up version, Convair's YF-102 Delta Dagger first flew in October 1953. Initial performance deficiencies were resolved by introducing a 'Coke-bottle' waisted fuselage; the first application of the area rule principle developed by Richard Whitcomb of NACA.

had far less influence on American aviation than is generally supposed. The most notable example of German design superiority, namely the swept-and-delta wing planform, is a case in point. In popular myth the USA 'discovered' the swept and delta wing amidst the rubble of Nazi Germany, quickly applying them to new aircraft (the F-86, B-47 and F-102). In fact, NACA scientist Robert T Jones had postulated the transonic and supersonic benefits of both, assessed and validated his conclusions by quick tunnel testing, and briefed both NACA and Air Force officials (including Theodore von Kármán) well before either the NACA or the USAF were exposed to the fruits of Nazi Germany's aeronautical research establishment, particularly the work of Busemann and Lippisch. What shocked American technical intelligence teams looking through the records and laboratories of the Third Reich was not the concept of the swept and delta wing (or any other of the accoutrements of high-speed flight), but the broad range of applications to which the Third Reich's design teams were already putting these discoveries, and the broad range of facilities available to German researchers. (This latter situation directly spawned the extensive investment in postwar transonic, supersonic, and hypersonic tunnel, shock tube and aeroballistic range facilities development, particularly the creation of the USAF's Arnold Engineering Development Center in Tullahoma, Tennessee, and, as well, the specialised Air Force Air Research and Development Command. Additionally, some of the German tunnel facilities themselves were dismantled and then shipped to the USA and elsewhere).

The revelations also confirmed the work of Jones and others in the USA, and accelerated the application of the swept wing on operational aircraft, notably the F-86 and the B-47, but they did not create American interest in, or alert American engineers to, the concept of the swept wing. As for Lippisch's delta work, Convair's proposed XP-92 rocket-ramjet interceptor, which spawned the XF-92A demonstrator, the YF-102, and the subsequent F-102 Delta Dagger, F-106 Delta Dart and B-58 Hustler

TECHNOLOGY FOR THE SUPERSONIC ERA

families (as well as the abortive XF2Y-1 Sea Dart), owed absolutely nothing to Lippisch and his work. In fact, when Convair's delta design team was made aware of it, they wisely rejected his thick-section aerofoils (as exemplified by the DM-1 glider brought to the USA from Germany after the war and tunnel tested by the NACA) as entirely unsuitable for the transonic and supersonic regimes.

In one significant case did German aerodynamic work alert the USA to a previously unrecognised opportunity; the Lippisch concept for variable wing sweep as a means of alleviating transonic drag rise and the poor low-speed behaviour characteristics of a sharply-swept planform. Here, the concept of Lippisch, coupled with the fortuitous recovery of an abortive German fighter project-turned-swept-wing-tested, the Messerschmitt P.1101, triggered American interest in developing a transonic variable-wing-sweep testbed. Out of this sprang the Bell X-5 programme, and another abortive fighter project, the US Navy's Grumman XF10F-1 Jaguar, both of which proved the concept, though their means of wing sweeping — translating the wing roots fore-and-aft rather than, as today, pivoting the outer wing panels independently from a wing-fuselage fixed-wing 'glove' — was entirely unsuitable for an operational aircraft. The notable outboard-pivot work in the 1950s of Barnes Wallis on his Wild Goose and Swallow concepts carried interest in variable wing sweeping a great deal further. Wallis had envisioned using variable sweep as a means of improving high- and low-speed behaviour and (Ader-like) of furnishing lateral control by differential applications. He followed with a proposed futuristic long-range supersonic (and eventually hypersonic) transport. Though never built, his concepts were bold and exciting, and as his biographer, J E Morpurgo, has noted, they constituted 'essays in aerodynamic efficiency of a kind that no other designer and no other nation had attempted'. Subsequently, Wallis's work, coupled with that of William Alford and Edward Polhamus of the NACA Langley Memorial Aeronautical Laboratory (later the NASA Langley Research Center), brought the variable-sweep concept to its logical conclusion. The blending of all three strains — the German, British and American work — found its fruition in aircraft such as the General Dynamics F-111, the F-111-inspired Sukhoi Su-24 Fencer, the MiG-23/27 Flogger, the Panavia Tornado, the

Based on the wartime Messerschmitt P.1011 project, the Bell X-5 was built to investigate the aerodynamic effects of changing the angle of wing sweep in flight. The first of two X-5s to be built made its maiden flight in June 1951, but the aircraft proved difficult to fly.

Rockwell B-1, and the B-1-inspired Tupolev Tu-22 Backfire and Tu-160 Blackjack.

In two serendipitous cases, X-series aircraft furnished more useful information on the basis of what they could not do than what they actually accomplished. Douglas's rakish and evocative X-3, aptly nicknamed the Stiletto, failed to attain its design objective of sustained Mach 2+ air-breathing flight, thanks to profound difficulties with its planned Westinghouse J46 engine system that forced reliance on two much lower-rated J34 engines instead. But it furnished critical information for the design of the Lockheed F-104 and, even more importantly, on the phenomenon of inertial coupling, a classic problem in the early supersonic era afflicting new generations of long, narrow (high fineness ratio and low aspect ratio) aircraft loaded primarily along their length than along their span. Likewise, the tiny Northrop X-4, which reflected a curiosity whether the semi-tailless swept-wing planform (as advanced by Lippisch and employed on his Me 163 Komet and subsequently employed by de Havilland on the D.H.108) could effectively eliminate the need for horizontal tail surfaces and furnish the requisite longitudinal stability for transonic flight, proved a great disappointment. The answer, in the pre-computer-controlled flight era was a firm 'No'. The X-4, like the Me 163 and the D.H.108, developed dangerous longitudinal pitching motions at Mach 0.88, pilots comparing its ride to the experience of driving on a convoluted 'washboard' road. Only testing the X-4 at high altitudes (and hence at low dynamic pressure) prevented it from experiencing the same kind of catastrophic failure that destroyed the ill-fated Swallow. Despite this, in X-3 fashion, the X-4 furnished a great deal of information on the stability and control of tailless configurations, and also on the flying characteristics of aircraft having low lift-to-drag ratios, for its speed brakes could be opened to various settings, giving the X-4 a lift-to-drag ratio akin to the modern Space Shuttle Orbiter.

Additionally, one important basic concept of high-speed flight, the area rule principle enunciated in the early 1950s by NACA scientist Richard T Whitcomb, sprang from Whitcomb's own conceptualisations, being then validated by tunnel testing (in the newly developed slotted throat tunnel era), and then being applied to two experimental jet fighters then plagued by disturbingly high transonic drag rise: the Convair YF-102 interceptor and the Navy's XF11F-1 Tiger jet fighter. Redesigned to incorporate Whitcomb 'wasp waisting' or 'Coke bottle' shaping, these two aircraft became transonic and supersonic success stories, and the F-102, in particular, went on to see extensive service as an air defence interceptor at the height of the Cold War.

As Table III clearly shows, the European nations and the Soviet Union pursued transonic and supersonic flight technology vigorously, even if they lacked the overall economic resources available to the USA. Surprisingly, Great Britain, which might have been expected to lead the transonic and supersonic revolution following on its early pursuit of the gas turbine and the collapse of German work in the rubble of the Third Reich, fell behind, despite a plethora of companies and many extraordinarily gifted designers. The Soviet Union, reaping the benefits of access to German technology, Western engine sales late in the 1940s, and perhaps the fruits of its espionage programme within the USA, quickly caught up with the USA by the early 1950s. Two countries, France and Sweden, made notable progress in high-speed aircraft design. A review of this history indicates at once the drive and insight that accompanied European and Soviet work in supersonics.

Great Britain's cancellation of the M.52 programme, coupled with the abandonment of the idea of a nationally supported coherent programme of experimental demonstrators and research aircraft, had a disastrous impact upon British aviation. It was exceeded, perhaps, only by the infamous Sandys' 1957 White Paper and the cancellation of TSR.2 and other advanced projects (such as the P.1154) in the 1960s. 'No single act', the distinguished British engine designer Sir Roy Fedden wrote a decade after Yeager's flight,

> … set back Britain's aircraft development quite so drastically, however, as the Government's decision in 1947 not to allow manned supersonic investigations …. The Miles supersonic aircraft project was abandoned, and we turned to an abortive programme of supersonic investigation through models powered by a rocket motor and launched in the air form a Mosquito. This was followed by another equally fruitless plan to launch models of possible supersonic shapes from the ground. Whilst the Minister responsible [Sir Ben Lockspeiser, the Ministry of Supply's Director-General of Scientific Air Research] was making humanitarian play out of a refusal to allow pilots to suffer the dangers of supersonic speed, the pilots themselves were actually being exposed to much greater risks as their machines entered the turbulent transonic region between subsonic and supersonic flight risks multiplied by the absence of suitable windtunnels in which to investigate the effects of the compressibility of air on an aircraft's behaviour. And so it was left to the pilots, such as Geoffrey de Havilland, John Derry, Neville Duke and Michael Lithgow, to find out for themselves. The lack of proper facilities for integration meant, too, that the industry had to wait for the prototype for full-scale testing. Our high-speed subsonic types succeeded in diving past the speed of sound, but the ban on manned supersonic investigation continued for four years. Its effect, however, was a greater loss than that. This unfortunate decision cost us at least ten years in aeronautical progress. It put us at least a generation behind the United States in fighters, and more than that in the basic knowledge necessary for the production of Mach 2 machines.

Despite this situation, British companies evidenced a remarkable interest and vigorous productivity of high-speed test beds and designs from the late 1940s onwards, characterised by the steady stream of prototypes and 'one-offs' from the likes of Hawker, De Havilland, Avro, Gloster, Bristol, Handley Page, Shorts, Saunders-Roe, Supermarine, Fairey, English Electric, Folland and the like. Several of these (the Hunter, Gnat, Lightning, Vulcan and Victor spring readily to mind) demonstrated excellent mastery of the transonic and supersonic regime, and went on to notable Service success and, at least in the case of the Hunter, to significant foreign sales as well. Pure research vehicles such as the Fairey Delta 2, the H.P.115 and the P.1127 worked their own influence upon subsequent aircraft development, for example the Concorde and the Harrier.

Throughout the Second World War, researchers at the Soviet Union's Central Aerodynamics and Hydrodynamics Institute (TsAGI), had undertaken their own efforts at pursuing an understanding of transonic flight. Like other nations, the USSR benefited from the technological ransacking of German research facilities at the end of the Second World War, and drew upon the German experience for its first postwar forays in jet and rocket aircraft, and in missile development as well. Overall, the Soviet Union entered the postwar world with a fragmented and disorganised approach to high-speed flight. As with the atomic bomb, the USSR started considerably behind the West in the field of high-speed flight. But, again as with the atomic bomb, Stalinist Russia quickly caught up, largely through some of the same means, including a vigorous programme of espionage directed against science and technology institutions in the West. In the case of supersonic flight, the Soviets benefited from having a brilliant young NACA engineer and supersonic tunnel designer, William Perl, serving as both a key agent and as a graduate student at the California Institute of Technology, working with Theodore von Kármán. Perl had access to the latest NACA reports and studies on supersonic flight, straight from the experience of the X-series programme, as well as the 'shop talk' that accompanied such work (and the work of the military services). It is likely that Perl was the key reason that Soviet high-speed research, at first fragmented like Nazi Germany's had been and, likewise, dependent to a great degree upon German work and inspiration, became much more focused and directed by the early 1950s.

Certainly there was a dramatic change from the era of the MiG-15 and MiG-17 (both excellent aircraft for their time, but with roots clearly in the German experience and, as well, with a great assist from the sale of British Nene engines to Russia in the late 1940s), to the era of the MiG-19. This aircraft, appearing simultaneously with the West's F-100, gave full evidence that Russian transonic and supersonic design had caught up with the West, a reality confirmed subsequently by the MiG-21. The subsequent history of Soviet aircraft design indicated that the Soviet Union mastered the basics of transonic and supersonic design technology very quickly. Although there were numerous false starts and disappointments, these were, by and large, no greater nor more numerous than those of the West. For example, the evolutionary progression of products from the MiG, Sukhoi, Ilyushin and Tupolev design bureaus indicates a number of aircraft types that went on to long and distinguished service both at home and abroad.

The experience of Sweden and France clearly indicate that the supersonic revolution was not exclusively the province of the major countries. Sweden, always possessing a strong national aircraft design capability, quickly moved into the turbojet and high-speed aircraft arena. A review of 'signature' Swedish aircraft success stories clearly indicates this: SAAB's imaginative modestly swept-wing Tunnen; the more complex Lansen; the bolder-still double-delta Draken; the exotic Mach 2+ Viggen, and now the Gripen of the fly-by-wire era. France, for its part, achieved notable success by the refined pursuit of the swept-wing, delta, and tailed-and-canard-delta concept, typified by the Dassault company line. First with aircraft such as the Mystere and Super Mystere, then the Mirage III, IV and V, the Mirage F-1, and on to the Mirage 2000, 4000 and the Rafale, Dassault's design teams have shown an extraordinary ability to take a basically simple configuration and adjust it to meet changing requirements and circumstances.

The wide diversity of supersonic designs springing from international companies and bureaus should not imply that all were great successes transforming the future of flight. Many (the F-104, Lightning and Mirage III come to mind) achieved supersonic flight only at the

The Fairey Delta 2 became the first aircraft in the world to exceed 1,000mph. Its high angle of attack at slow speed and during take-off and landing led its designer to give it a 'droop snoot' nose, as seen here, which could be lowered to enhance the pilot's forward view in this critical part of the flight envelope.

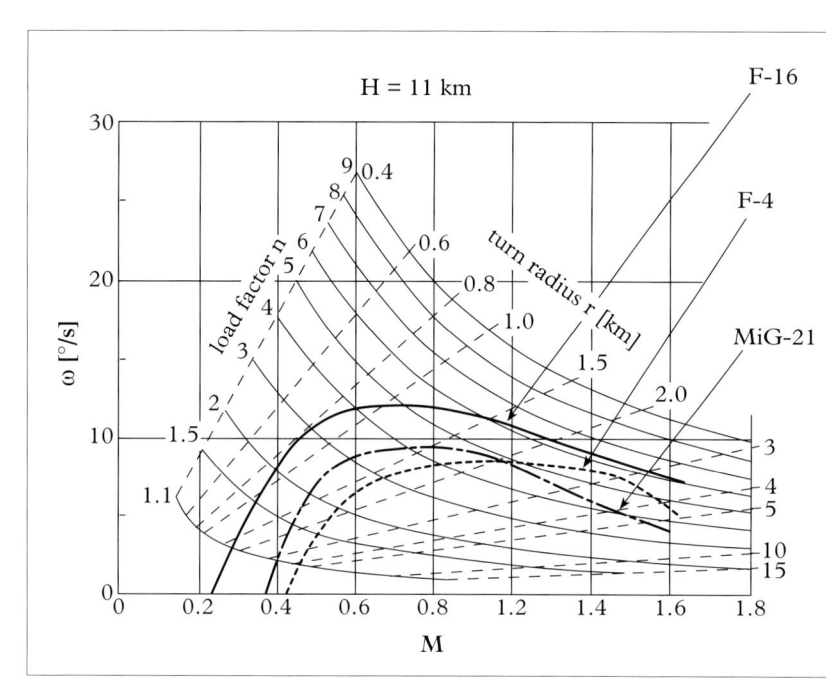

Figure 7: *Turning performance of the MiG-21, F-4 and F-16. (Source: Klaus Huenecke,* Modern Combat Aircraft Design *(Annapolis, MD: Naval Institute Press, 1987), Chart 9-14, p.207.)*

expense of military utility and the ability to undertake a wide range of missions with good success. A genuine few (the McDonnell F-4 and Northrop F-5/T-38, for example) were truly extraordinary and influential designs. Many were surprising disappointments, required extensive development times, or had other limitations that significantly reduced their ability to meet broader operational needs. The Convair B-58 Hustler and most of the so-called 'Century series' fall into this category. So, too, did the first efforts to develop economical supersonic transports, typified by the Concorde and the Tupolev Tu-144. (Though, in fairness, Concorde's problems stemmed not from the design itself but, rather, from the intensely controversial nature of sustained over-land 'boom' generating and possibly environmentally harmful supersonic flight.)

In any case, by the end of the classic era of transonic and supersonic flight, which can be said to have ended about the time of Sputnik in late 1957, a technology base and technology infrastructure existed throughout the world's leading industrial nations that could routinely produce safe and efficient supersonic aircraft. From this base would spring the two most evocative (at least to the public) supersonic designs yet flown: the Lockheed Blackbird and the graceful Concorde. By late 1957 the basic operating speed-and-altitude environment for supersonic aircraft had been mapped out. The next phase in the aeronautics revolution would be the joining of that revolution to the on-going revolution in computer-controlled flight. What would change would not be speed (nor altitude, for that matter) but, rather, the relative efficiencies of aircraft operations within this existing opera-

tional envelope. For example, the agility of modern fighters compared with their 1950s predecessors. Figure 7 offers dramatic proof of this, comparing the turning performance (w, the turn rate in degrees per second) of the late 1950s-early 1960s Mach 2 Soviet MiG-21 Fishbed and American-built F-4 Phantom II with that of the 1970s-era fly-by-wire F-16. A less dramatic (but no less significant) graph could likewise compare the economic efficiency of modern jet transports such as the Airbus family or the Boeing 757/767/777 to their Comet, Caravelle and Boeing 707/720/727 predecessors.

Today we enjoy unprecedented mobility for military and civil purposes, thanks to the high-speed aeronautics revolution. Key to that achievement has been the artful partnership of ground and flight test experimental methods, typified by the contributions of high-speed research aircraft to the aeronautics field. The proliferation of contemporary study efforts looking towards the transatmospheric frontier suggests strongly that this partnership will continue to create as fruitful a climate for aerospace science in the second century of flight as it has in the first.

Bibliography

American Institute of Aeronautics and Astronautics, *Evolution of Aircraft Wing Design* (AIAA, Dayton, Ohio, 1980).

Anderson, J D, Jr, *A History of Aerodynamics and its Impact on Flying Machines*, vol. 8 in the Cambridge Aerospace Series (Cambridge University Press, Cambridge, 1997). The classic history of the science.

Beamont, R, *Testing Early Jets: Compressibility and the Supersonic Era* (Airlife, Shrewsbury, 1990). A test pilot's personal experiences.

Becker, J V, *The High-Speed Frontier: Case Studies of Four NACA Programs, 1920-1950* (National Aeronautics and Space Administration, Washington, DC, 1980).

Crossfield, A Scott, *Always Another Dawn: The Story of a Rocket Test Pilot* (World Publishing, Cleveland, Ohio, 1960). The autobiography of an X-15 pilot.

Dryden, H L, 'A Half Century of Aeronautical Research', *Proceedings of the American Philosophical Society*, April 1954.

Duke, N, and Lanchbery, E, *Sound Barrier* (Cassell, London, 1953). An account of the challenges posed by high-speed flight.

Emme, E M, (ed), *The History of Rocket Technology: Essays on Research, Development, and Utility* (Wayne State University Press, Detroit, Michigan, 1964).

Eppley, C V, *The Rocket Research Aircraft Program, 1946-1962* (Air Force Flight Test Center, Edwards, California, 1963).

Fedden, R, *Britain's Air Survival: An Appraisement and Strategy for Success* (Cassell London, 1957).

Foxworth, T G, 'North American XB-70 Valkyrie', *Historical Aviation Album*, Nos. 7-8, 1969-70. A full account of the evolution and testing of this high-speed bomber.

Hallion, R P, 'Lippisch, Gluhareff, and Jones: The Emergence of the Delta Planform and the Origins of the Sweptwing in the United States', *Aerospace Historian*, vol. 26, No 1 (March 1979).

———, *On the Frontier: Flight Research at Dryden, 1946-1981* (National Aeronautics and Space Administration, Washington, D.C., 1983).

———, *Supersonic Flight: Breaking the Sound Barrier and Beyond – The Story of the Bell X-1 and Douglas D-558* (Brassey's, London, 1997). A very complete account of the two aircraft that spearheaded the USA's early research into supersonic flight.

Hansen, J R, *Engineer in Charge: A History of the Langley Aeronautical Laboratory, 1917-1958* (National Aeronautics and Space Administration, Washington, DC, 1987).

Hartman, E P, *Adventures in Research: A History of Ames Research Center, 1940-1965* (National Aeronautics and Space Administration, Washington, DC, 1970). The official history of this research centre.

Hunley, J D, ed., *Toward Mach 2: The Douglas D-558 Program* (National Aeronautics and Space Administration, Washington, DC, 1999). An official history of this programme.

Johnson, C L, 'Some Development Aspects of the YF-12A Interceptor Aircraft', *AIAA Journal of Aircraft*, July-August 1970.

Kármán, T von, *Aerodynamics: Selected Topics in Light of Their Historical Development* (McGraw-Hill, New York, 1963).

King, B, and Kutta, T, *Impact: The History of Germany's V-Weapons in World War II* (Sarpedon, Rockville Centre, NY, 1998).

Lacroze, J, and Ricco, P, *René Leduc: Pionnier de la propulsion à réaction* (Editions Larivière, Clichy, 2000). A biography of this French pioneer of jet propulsion, who concentrated on ramjets.

LeVier, T, with Guenther, J, *Pilot* (Harper & Row, New York, 1954). The autobiography of a Lockheed test pilot.

Lippisch, A, *The Delta Wing: History and Development* (Iowa State University Press, Ames, Iowa, 1981). A first-hand account of the development of delta-wing aircraft.

Lithgow, M, *Mach One* (Allan Wingate, London, 1954). The memoirs of a Supermarine test pilot.

Matthews, H, *D.H.108: The Saga of the First British Supersonic Aircraft* (HSM Publications, Beirut, 1996). A carefully researched account of these early tailless jets.

——— *Samolyot 346: The Untold Story of the Most Secret Postwar Soviet X-Plane* (HSM Publications, Beirut, 1996).

Miller, J, *The X-Planes: X-1 to X-31* (Aerofax, Arlington, TX, 1988). The most complete account of all of the USA's 'X' series.

Morpurgo, J E, *Barnes Wallis* (Longman, London, 1972). An authoritative biography of the Vickers company's famous inventor/designer.

Neufeld, J, et al, (ed.), *Technology and the Air Force: A Retrospective Assessment* (Air Force History and Museums Program, Washington, DC, 1997).

D, with Darlene Lister, D, *Wingless Flight: The Lifting Body Story* (National Aeronautics and Space Administration, Washington, DC, 1997). An authoritative account of the lifting-body research programme.

Rotundo, L, *Into the Unknown: The X-1 Story* (Smithsonian Institution Press, Washington, DC, 1994).

Saltzman, E J, and Ayres, T G, *Selected Examples of NACA/NASA Supersonic Flight Research* (National Aeronautics and Space Administration/Dryden Flight Research Center, Edwards, CA, 1995).

Stanley, R M, and Sandstrom, R J, 'Development of the XS-1 Supersonic Research Airplane', *Aeronautical Engineering Review*, vol 6 no 8, August 1947.

Shahrov, V B, *History of Aircraft Construction in the USSR, vol 2: 1938-1950* (Mechanical Engineering Publishers, Moscow, 1978). A Soviet aircraft designer's history of aviation's development in the Soviet Union.

Stillwell, W H, *X-15 Research Results* (National Aeronautics and Space Administration, Washington, DC, 1965).

Thompson, M O, *At the Edge of Space: The X-15 Flight Program* (Smithsonian Institution Press, Washington, DC, 1992).

—— and Peebles, C, *Flying Without Wings: NASA Lifting Bodies and the Birth of the Space Shuttle* (Smithsonian Institution Press, Washington, DC, 1999).

Trubshaw, B, and Edmondson, S, *Brian Trubshaw: Test Pilot* (Sutton Publishing, Phoenix Mill, 1998). The autobiography of a Concorde test pilot.

Tomayko, J E, *Computers Take Flight: A History of NASA's Pioneering Digital Fly-by-Wire Project* (National Aeronautics and Space Administration, Washington, DC, 2000).

Twiss, P, *Faster than the Sun* (Macdonald, London, 1963). An autobiographical account of the first flight to exceed 1,000mph, made by the Fairey F.D.2.

Williams, W C, and Drake, H M, 'The Research Airplane: Past, Present, and Future', *Aeronautical Engineering Review*, vol 17 no 1, January 1958.

Yeager, C, and Janos, L, *Yeager: An Autobiography* (Bantam, New York, 1985).

Young, J O, *Meeting the Challenge of Supersonic Flight* (Air Force Flight Test Center History Office, Edwards, CA, 1997).

White, A, 'The XB-70 From the Pilot's Point of View', in *Society of Experimental Test Pilots, 1998 Report to the Aerospace Profession: Forty-Second Symposium Proceedings* (Society of Experimental Test Pilots, Lancaster, CA, 1998).

Wood, D, *Project Cancelled: British Aircraft that Never Flew* (Bobbs-Merrill, Lancaster, CA, 1975). A classic survey of the numerous cancelled postwar British projects.

3
Extending the Reach
Colin Cruddas

Historical background

Ever since the inception of heavier-than-air powered flight, designers have grappled with the perpetual problem of how to maximise both range and payload. While feasible solutions continued to tax the technically minded, the possibility of achieving these aims by refuelling an aircraft in flight was prophetically made public in a satirical cartoon published in a 1909 edition of *Punch* magazine. In 1918, in an attempt to avoid transporting American-built bombers across the perilous Atlantic by ship, US Navy Reserve officer Godfrey Cabot proposed a novel system of aircraft picking up fuel containers from vessels. Using an elastic rope and catching device he eventually achieved limited success, but the end of the war had precluded the need for further development and, as with some other crude and unrecorded pioneering attempts both before and during the First World War, Cabot's method aroused little official interest. In 1921 however, Alexander Seversky, having earlier defected from a Russian Naval Mission to remain in the USA, and now working for the controversial General 'Billy' Mitchell, registered the first patent application relating to air-to-air refuelling (AAR). That same year, barnstorming stuntman Wesley May demonstrated the world's first aerial fuel transfer when, at 3,500ft (1,070m) over Long Beach, California, standing on the top wing of a Lincoln Standard J-1 biplane with a five-gallon can of fuel strapped to his back, he caught the wingtip skid of a Curtiss JN-4 and hauled himself aboard. After threading his way through the maze of struts and bracing wires he then transferred at least some of this load into the Jenny's fuel tank — but most of it seems to have disappeared in the slipstream.

In 1923, military awareness of AAR's potential was awakened when, in the USA, Lieutenants L H Smith and J P Richter, flying a DH-4B, established a refuelled endurance record of 37hr 15min, and in Europe, Captain P Weiss and Adjutant Van Caudenberg of the French l'Armée de l'Air also undertook a successful series of AAR demonstration flights. These events, coupled with a proposal for refuelling commercial aircraft in flight put forward at the International Air Congress by Boulton Paul's chief designer, J D North, prompted a response from Britain's Air Ministry. The Royal Aircraft Establishment (RAE) at Farnborough duly received a request to carry out a number of refuelling experiments, which were done in 1924 with a pair of Bristol F2B Fighters using water instead of fuel, but despite a prom-

This cartoon, recognising the need to extend the aeroplane's capabilities, appeared in the October 1909 issue of Punch.

Wesley May's change-of-plane stunt is popularly regarded as the first recorded case of fuel transfer in flight!

FASTER, FURTHER, HIGHER

Flying a DH-4B from Rockwell Field, San Diego, in August 1923, Lieutenants Smith and Richter achieved a record of 37hr 15min. The flight, which involved fifteen separate contacts with another DH-4B for fuel and other consumables, represented the US Army Air Service's first official interest in endurance flying.

ising outcome, no subsequent action was taken. On mainland Europe, too, some interest in AAR was being shown, for some four years later, in June 1928, a D.H.9 flown by Adjutant Aviator Louis Crooij and Sergeant Aviator Victor Groener of the Belgian Aeronautique Militaire raised the refuelled endurance record to 60hr 7min, but this was soon to be overtaken by a significant event that took place in America.

On 1 January 1929 an Atlantic C-2A (an American built Fokker F-VIIA/3m) named *Question Mark* (because of the uncertain nature of the venture) took off under the

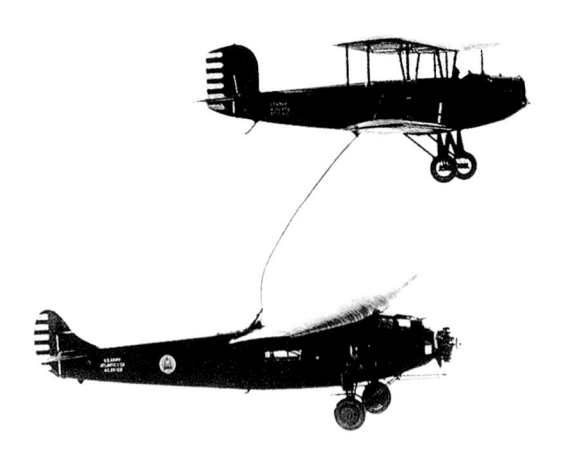

On 1 January 1929 the US Army Air Corps Fokker C-2A Question Mark *took off to fly up and down the Californian coast on what was to become a new record endurance flight of over 150hr. Engine lubrication problems caused the venture to be terminated after seven days.*

command of Major Carl Spatz* from Los Angeles Metropolitan Airport. In a flight lasting nearly seven days (150hr 40min) it received fuel on thirty-four separate occasions and, during additional contacts, oil, food and other consumables from two Douglas C-1 tanker aircraft. This milestone achievement attracted not only the attention of the world's aviation authorities, but also presented a challenge for many publicity-driven individuals determined to make their mark (and hopefully fortunes) in the official record books.

The following year saw more than forty attempts on the refuelled endurance record, with the longest time spent aloft rising to 420hr. The figure continued to climb until it reached the 653hr achieved by the Key brothers in 1935. After the achieving of this record, the Fédération Aéronautique International withdrew the endurance with refuelling category but, apart from the wartime period, national record seeking still continued. At Las Vegas in 1959 Robert Timm and John Cook remained airborne for 1,500hr, having their Cessna 172 refuelled from a tanker ploughing along the highway and from fuel cans passed by hand from an automobile. Although the method was unorthodox and of little practical value, it was certainly effective in promoting publicity, fundraising and the achieving of a record which, not taking account of refuelled spacecraft endurance, stands to this day.

Although interest had to some extent been stimulated in other countries such as the Soviet Union, Japan and especially Germany, where the Aeronautical Research Institute collaborated with Lufthansa in developing 'AAR apparatus', it was only in Britain that a serious and continuing programme of research took place in the 1930s. With the Air Ministry now wanting to determine the feasibility of refuelling large flying boats in the air, testing again got under way at the RAE, using D.H.9A and Vickers Virginia aircraft to establish methods of contact. These usually entailed an observer in the receiver aircraft using a crooked stick to retrieve a line dangling below the tanker, which, having been caught, then allowed a fuel hose to be drawn down. A measure of success having been achieved using this method, the first public demonstration of AAR in the UK was then given at the Royal Air Force Display at Hendon on June 25-27, 1931. In 1932 the Hon Mrs Victor Bruce, already famous for record-breaking exploits on land, sea and in the air, made an attempt on the women's endurance record. For this she devised the novel technique of standing up in the front cockpit of a Saunders-Roe Windhover flying boat and catching a cord stretched between two Bristol Fighters. Having secured the line, she then hauled a fuel hose aboard to complete the refuelling operation. After eight refuelling contacts and 63hr in the air, the redoubtable lady's efforts were brought to a premature

---

*Spatz later changed the spelling of his name to Spaatz.

end when a burst oil-feed pipe caused a forced landing in Portsmouth Harbour.

It was now that Sir Alan Cobham entered upon the scene. His long-distance route-proving flights in the 1920s had convinced him of the need for aircraft requiring shorter and less risky take-offs. If such aircraft could then be refuelled in flight, greater range and/or increased payloads would be a natural consequence. In addition, the economic advantages of not having to provide fuel or personnel at difficult locations en route would, he reasoned, be especially appealing to airline operators. Accordingly, having already conducted early experiments with his flying-display aircraft during the winter stand-down periods, he set about organising a non-stop AAR flight to Australia. It was the Air Member for Supply and Organisation, Air Vice-Marshal Sir Hugh Dowding, who, when approached to provide Service assistance en route, suggested India as a less-expensive alternative that would still serve to prove Cobham's point. But this ambitious endeavour ended in disappointment when, after leaving Portsmouth on 22 September 1934, and having successfully undertaken two refuellings, the throttle linkage on Cobham's Airspeed Courier monoplane became disconnected, resulting in a dead-stick landing at Malta. Further tragedy occurred when the Handley Page W.10 used to refuel the Courier immediately after take-off later crashed, killing all on board. Notwithstanding this misfortune, which he later claimed set back the case for AAR by ten years, and indeed perhaps enhanced the case for achieving range/payload benefits from other means such as aircraft carried 'piggy-back' fashion or catapulted from ships, Cobham turned his furnace-like energy towards the advancement of the AAR art. In October 1934 he founded Flight Refuelling Limited to investigate techniques and develop appropriate equipment. His work, conducted at Ford Aerodrome, soon coincided and competed with experiments already being undertaken by the RAE. This delicate situation was further compounded when Flight Lieutenant Richard Atcherley, a serving Royal Air Force officer who, after witnessing successful refuelled endurance flights in America, wanted to develop improved contact procedures, joined the RAE team in 1935. His new approach, called the 'crossover'

Although longer flights subsequently took place, the Hunter brothers, flying a Stinson SM-1B Detroiter, established the last 'world refuelled endurance record' before the category was withdrawn in 1932.

Sir Alan Cobham undertook his first aerial refuelling experiments in early 1933. He is shown here flying de Havilland D.H.60G Moth G-ABAF, exploring contacting techniques with Airco D.H.9 G-AACR, piloted by Squadron Leader Helmore.

method, required the receiver, trailing a weighted cord and grapnel, to fly above and diagonally across a similar cord streamed behind the tanker. Once both grapnels had engaged it was then a simple matter to haul the refuelling hose across to the receiver. But it was inevitable that, with Cobham's team pursuing development along broadly similar lines, technical confrontation would arise. The Air Ministry, finding itself caught between two camps, called for a comparative demonstration of the respective techniques. A decision was made soon afterwards that all of the RAE's test findings and trials aircraft should be transferred to Flight Refuelling Limited, as the company was judged to be more suited than Interair Ltd, a holding company of which Atcherley was a main subscriber, for subsequent commercial exploitation. Accordingly, in 1938, and now temporarily equipped with a variety of test aircraft that included the Vickers B.19/27, a Boulton Paul Overstrand, a Fairey Hendon, the Handley Page H.P.51 and Armstrong Whitworth A.W.23, Flight Refuelling began refuelling trials with Imperial Airways S.23 flying boat *Cambria* in preparation for a proposed non-stop transatlantic service.

The following year a wingtip contact method was investigated, but this was later abandoned when a refined version of the crossover system was introduced. This was the so-called 'looped hose' technique, in which an ejector gun fired a steel projectile carrying the tanker's connecting cable across the receiver's line. This seemingly simple improvement immediately overcame the major problem of aircraft losing sight of each other in turbulent or cloudy conditions during the crossover period. This new procedure enabled Imperial Airways' S.30 boats *Cabot* and *Caribou*, refuelled by Flight Refuelling's Handley Page Harrow tankers over Ireland and Newfoundland, to carry out a highly successful series of non-stop mail-carrying flights to and from America in August–September 1939. Such was the promise offered by Flight Refuelling's new system that Air France Transatlantique also planned to operate a transatlantic service using Farman 2234 airliners. Unfortunately this came to nought when, in 1940, during the Second World War, the German invasion of France caused a terminal interruption in contract negotiations.

The success achieved with the C-class flying boats did not however, carry over into the production of Short Sunderlands with an AAR capability. Had this been the case, it would have allowed this aircraft to have remained for longer periods of convoy protection over the notorious mid-Atlantic U-Boat hunting grounds. The Air Ministry had recognised the potential benefits afforded to Service aircraft when, in the mid-1930s, it included an AAR requirement for the first time, in Specification B.1/35 which eventually matured as the Vickers Warwick. Furthermore, some consideration was also given to installing tanker equipment in the de Havilland Albatross airliner and receiver refuelling equipment in the Short Stirling and Avro Manchester bombers, but, apart from two flights carried out at Boscombe Down by the H.P.51 and A.W.23 to evaluate a 'taking off in contact' technique, little RAF interest was shown in AAR until much later in the war.

Just before the Second World War an attempt was made to convince naval officials that advantages in enabling aircraft such as the Fairey Swordfish to take off with increased war loads could be gained from AAR. But what had seemed to be promising discussions led

Squadron Leader Helmore, standing in the cabin of an Airspeed Courier, receives fuel from a Handley Page W.10 tanker. The Courier, piloted by Sir Alan Cobham, later attempted a non-stop flight from England (Portsmouth) to India (Karachi), but a disconnected throttle linkage made necessary a forced landing at Malta on 22 September 1934.

EXTENDING THE REACH

nowhere when it was officially stated that: 'Fleet Air Arm aircraft have to be prepared to leave the deck under full load, and one thing we want to avoid is the addition of any weight or fittings which are not 100% essential'. The possible effect on the crew members of open-cockpit machines if subjected to extended hours on patrol in unfavourable conditions is not recorded.

Following the Japanese attack on Pearl Harbor, the US Army Air Force (USAAF), using a Consolidated B-24 Liberator tanker and a Boeing B-17 Fortress receiver, evaluated Flight Refuelling's trailing-hose equipment as a possible aid to mounting a long-range retaliatory raid on Tokyo. However, the industrial effort and training involved was eventually considered too great, and the morale boosting response was eventually carried out by carrier-based non-refuelled North American B-25s led by Lieutenant Colonel Jimmy Doolittle. At this stage of the war the USAAF investigated several other methods of transferring fuel in flight. One series of trials that held brief promise involved a B-24H Liberator bomber towing a fuel tank that was intended to be caught and locked in position by a reception device located between the cockpit and starboard engine of the twin-boom Lockheed P-38 Lightning. However, urgent though the need was to increase the range of its escort fighters, the Air Technical Service Command considered the system essentially unsafe and testing was discontinued.

Another means of range extension that came under USAAF scrutiny had first been tried out in Russia in the mid-1930s. This involved the towing of a glider that used the connecting (pipe) line to pass fuel forward to the parent aircraft. Several flights took place during which a Cornelius XFG-1 dubbed the 'flying fuel tank' replenished a Boeing B-29, but the proposal lost favour after the

This Armstrong Whitworth A.W.23 prototype bomber-transport was one of several types loaned by the Air Ministry to Flight Refuelling Ltd for refuelling development work in the late 1930s. Note the hose-drum unit protruding from the underside of the fuselage. Sir Alan Cobham (with peaked cap) is in the foreground.

57

FASTER, FURTHER, HIGHER

inherently unstable glider crashed, killing the pilot. This method of refuelling also aroused passing interest in Germany. During late 1944 a test programme was undertaken in which a V1 flying bomb, with engine, tailplane and warhead removed and modified to become a fuel container, was towed behind an Arado Ar 234 jet bomber. But, as with the Cornelius experiment, instability problems soon caused the project to be abandoned. Other Luftwaffe trials included the use of a Heinkel He 111 as a tanker with Dornier Do 217, Gotha Go 244 and Focke-Wulf Fw 58 receivers, and the Junkers Ju 390 was proposed as a tanker for use with Ju 290A reconnaissance aircraft, but little effort was made by the Axis powers to introduce AAR into full-scale wartime operations.

Undoubtedly the most ambitious AAR project undertaken during the Second World War saw large-scale preparations put in hand to convert 600 Avro Lancasters and 600 Avro Lincolns to tankers and receivers for use in the Far East after the war in Europe had ended. However, the rapid advance of US forces through the Pacific released airfields that brought air attack of the Japanese mainland within the range of unrefuelled B-29s, and the training of RAF aircrew and production of refuelling equipment for 'Tiger Force', already well under way, was brought to an end in April 1945. But post-war events were soon to provide an impetus to development that even that astute AAR protagonist, Alan Cobham, could hardly have foreseen.

Post-war development

Swallowing his disappointment that AAR had not played a major part in wartime operations, and having already submitted a number of schemes for air-refuelled airliners to the Ministry of Civil Aviation, Cobham turned his attention to securing government support and funding for his range-extension proposals. Accordingly, during the winter of 1946-47, the first of three separate trials took place aimed at proving the viability of AAR under representative commercial operating conditions. Using the newly developed Rebecca/Eureka radar system for

Flight Refuelling Limited's Handley Page Harrow tankers, based in Ireland and Newfoundland, provided a tanker service for Imperial Airways 'C' Class flying boats during transatlantic trials in August-September 1939.

interception, Lancaster tankers and receivers originally modified by Flight Refuelling for 'Tiger Force' made contact over the Channel on forty-three occasions, including seventeen at night. With BOAC unable at this time to release aircraft or personnel, British South American Airways provided the crews for the receiver aircraft for this, and also for a follow-up series of eleven weekly return non-stop flights from London to Bermuda in which AAR took place from Flight Refuelling's tankers positioned over the Azores. Encouraged by the South Atlantic Trials, as they were officially known, a third and more exacting test phase took place in 1948, when more powerful Lancastrian tankers, this time operating from Shannon in Ireland and either Gander or Goose Bay in Newfoundland, linked up with a BOAC Liberator over the North Atlantic. Although technically successful, the North Atlantic Trials were reported on with only limited enthusiasm by BOAC's pilots, and with civil aircraft of greater range and performance due for delivery, such as the Boeing Stratocruiser and Canadair C4 Argonaut, AAR was regarded as unlikely to meet the airline's safety and economic criteria.

In the event, Cobham's vision of his company's tankers servicing airliners plying the world's major trade routes would soon fade, to be replaced by what at last became an urgent military need. At this time in the USA, bitter long-standing high-command rivalries were heightened with the birth, on 18 September 1947, of the United States Air Force (USAF). The question of funding to pay for the Strategic Air Command's (SAC) new Convair B-36 intercontinental bombers, which in the Air Force's view was the only immediate and practical way to deliver an atomic bomb, became a burning inter-Service issue. The US Navy (USN) presented a strong case for a projected 70,000-ton (71,123-tonne) aircraft carrier that could launch an atomic attack from anywhere on the high seas, and the long-delayed arrival of the B-36 and its early unpromising performance posed a severe threat to the USAF generals. Faced with securing a useful future role, the decision was taken to extend the range of the existing B-29 bomber force to enable it to reach any target within the Soviet Union. The only readily available means of achieving this involved the use of AAR, and orders were quickly placed with Flight Refuelling for equipment for ninety-two KB-29M looped-hose tankers and 131 B-29 and B-50 receivers. Throughout 1948 the USAF-USN battle continued, with a Congressional inquiry being undertaken into the adversaries' conflicting claims. Then, with the political argument reaching its climax, the USAF dramatically exploited the anniversary of the Japanese attack on Pearl Harbor to show off its new global capabilities. On 7 December 1948 a Boeing B-50 bomber, having flown from Carswell Air Force Base (AFB), Fort Worth, Texas, carried out an undetected mock attack on the US Navy's Pacific Fleet. Having dropped a dummy 10,000lb (4,540kg) atomic bomb on Hawaii it returned to base, completing a 9,400-mile (15,128km) round trip that impressively underlined the virtues of AAR.

Only two months later another B-50, *Lucky Lady II* of the 43rd Bomb Group, took off, again from Carswell, and, with the aid of sixteen KB-29M tankers (including reserve aircraft) deployed along the route to provide four refuellings, carried out a non-stop globe-encircling flight in 94hr 1min. These demonstrations, although conducted with crews inexperienced in AAR, provided timely if somewhat desperate proof that the USAF could meet its obligations, and in so doing sealed the fate of the USN's supercarrier. Although construction of the USS *United States* had already begun, the cancellation axe fell resoundingly in April 1949. After the requirement for refuelling the RAF's Far East Tiger Force fell away in 1945, Cobham had elected to buy large stocks of the equipment already produced, to meet anticipated future civilian orders. Although these did not materialise, the placing of orders for the USAF B-29s came to his rescue. Unaware of the existence of the vast store of refuelling hoses, cables and other equipment, the Americans were continually surprised by the trouble-free supply of parts, which frequently arrived well ahead of their scheduled delivery date.

Despite this, the Flight Refuelling method of refuelling, which had barely progressed beyond that employed prewar, was soon judged by the USAF to be cumbersome, and the Boeing Airplane Company was charged with finding other ways of transferring fuel in flight. The outcome of this investigation was the 'Flying Boom' concept, which required a dedicated refuelling operator positioned at the rear of the tanker, directing, by means of moveable control surfaces, a rigid telescopic fuel delivery tube into a reception coupling on the receiver aircraft. A parallel approach was also made by the USAF to Sir Alan Cobham to devise a simpler, automatic system of delivering fuel to single-seat fighter aircraft. But whereas time had befriended Cobham in his first dealings with the USAF, it now worked against him, for, with only a few months to design, test and demonstrate a method he had optimistically stated to be already well in hand, his company's reputation was clearly on the line. This challenge led Flight Refuelling Limited's design team, then unaware that American inventors were also thinking along similar lines, to invent the 'probe-and-drogue' method which, first produced in 1949, is still in widespread use today. On 7 August 1949, and unknown to Cobham, who was then visiting the USA, Flight Refuelling's two senior pilots decided to make an attempt on the jet endurance record. Tom Marks, flying one of the company's Lancaster tankers, and Pat Hornidge in its Gloster Meteor III receiver development aircraft, then took off from Tarrant Rushton, the company's airfield.

FASTER, FURTHER, HIGHER

Ten refuellings and twelve hours later Hornidge landed, having successfully established a new jet endurance record. When handed a newspaper recounting the event, Cobham was unable to convince his USAF hosts that he knew nothing about it!

Shortly afterwards, four B-29s and two Republic F-84 fighters were sent over from the USA to undergo conversion. One B-29 was converted to become a YKB-29T three-point tanker with pumping units installed in pods under each wing and a similar unit in the fuselage. Another was equipped as a single-point tanker, and the remaining pair, along with the two jet fighters, became receivers. The F-84s, piloted by USAF Colonels Dave Schilling and Bill Ritchie, achieved fame in September 1950 when they became the first jet aircraft to undertake a successful AAR-assisted crossing of the Atlantic. The AAR conversion work did not, however, go according to plan. Working to a highly ambitious and unrealistic fixed price, Flight Refuelling's workforce encountered great difficulty in coping with the more advanced systems and structures inherent in the American machines.

Eventually, with the conversion work completed, albeit well behind the promised delivery dates and well over budget, Cobham had little option but to sell the design rights of the new invention to the USAF for an amount equal to the overspend. His company thus lived to fight another day, but it had been a close-run affair, and a golden opportunity to stay well ahead of overseas competition appeared to have been lost. It was perhaps small consolation to Cobham to know that, after initially deploying KB-29M tankers fitted with looped-hose equipment and subsequently eight KB-29T tankers hurriedly converted at Tarrant Rushton to incorporate the so called 'Quickie' probe-and-drogue conversion kits, the USAF claimed that his AAR system played a significant and successful role in the Korean War. This procedure allowed Lieutenant Colonel Harry Dorris to carry out a remarkable exploit over North Korea on 28 September 1951. In a flight lasting 14hr 15min his Lockheed F-80A, carrying a full warload of bombs, rockets, guns and camera, undertook sixteen separate refuelling contacts, eight for each wingtip tank fitted with a probe, while radio

Flight Refuelling's test pilot, Pat Hornidge, flying a Gloster Meteor Mk III, achieved a 'probe and drogue' refuelled jet endurance record of 12hr 1min on 7 August 1949. Ten refuelling contacts were made with the company's Lancaster tanker, flown by Tom Marks.

silence was observed throughout.

Following the demise of Tiger Force, the Royal Air Force's interest in AAR continued to ebb and flow. Bomber Command now regarded it as of little operational use, and Transport Command, faced with a massive post-war task, was unable to devote its aircraft to a large-scale modification programme, but Coastal Command expressed its desire to carry out Service trials with meterological reconnaissance aircraft. Contracts were placed in 1946 for the conversion of ten Handley Page Halifax VIs and three Lancaster IIIs as looped-hose receivers and tankers but, as with the Avro Shackleton, for which similar provision for receiver equipment was made in the first prototype, changing operational requirements brought about a late cancellation of orders. In 1948 the Air Staff briefly considered the possibility of providing an AAR capability for its first-generation jet fighters, the Meteor and Vampire, but, with RAF interest merely lukewarm, Flight Refuelling's development work progressed only as a private venture.

It has been suggested that the differences in opinion held by British and American Services chiefs regarding AAR stemmed from vastly different operating scenarios in the Second World War. The RAF had rarely to exceed a radius of action in Europe exceeding 800 miles (1,290km), whereas the campaign fought by the USAAF's 20th Air Force in the Pacific frequently involved round-trip distances of well over 3,000 miles (4,830km). Although it is debatable, these facts, coupled with the ever-present fiscal restraints, may well have contributed to the initial reluctance of the British to embrace AAR.

Consideration of the probe-and-drogue method of contact was not confined to Britain and America as the 1950s began. In the quest for escort fighter range extension, Russia's Yakovlev design bureau, noting the USAF's experiments with F-84s, both suspended below the fuselage of a B-36 fuselage and carried on wingtip coupler panels, devised a novel towing system. This required a 'harpoon' probe mounted on the fighter receiver, to engage in a cone- or drogue-shaped reception coupling trailed behind a towing aircraft. Initially tested in 1949 using a Tupolev Tu-2 (B-25) to tow a Yakovlev Yak-25, the system, designated 'Burlaks' (Hobblers) was further developed to allow Mikoyan MiG-15 fighters to be towed behind Tupolev Tu-4 (B-29) long-range bombers, both types being then in wide service use. Shutting down the engine during the tow did, however, cause depressurisation of the fighter's cockpit, and in being asked to overcome the problem, the design bureau was also requested to investigate other improvements, including provision for AAR using a similar method. It would transpire that all future development of AAR systems in the Soviet Union (and latterly Russia) would centre on hose-based systems contained in the tanker's fuselage or in wingtip-mounted pods.

Throughout the 1950s Cold War period the Soviet VVS-DA (Long Range Aviation Group) operated the Myasishchyev Mya-4 (Bison) as a single-point refueller. Later, the Tupolev Tu-16 (Badger) played a prominent tanker role, employing a wingtip-to-wingtip contact method. This highly unusual technique, seemingly based on a wingtip coupling method briefly experimented with by Flight Refuelling in the late 1930s, was modified to allow lateral pick-up and fuel transfer. In recent years the Ilyushin Il-78 (Midas), equipped with the Russian-designed UPAZ refuelling pod, has provided tanker services by more conventional means to a wide range of receivers. It is of interest to note that, during the bargaining negotiations of the Strategic Arms Limitation Talks (SALT II), the Soviets agreed not to install AAR receiving equipment in the Tupolev Tu-22M (Backfire) bomber, in an attempt to assuage the fears of American analysts that the Tu-22M could make round trip flights to the continental USA.

It was not until 1951, with the probe-and-drogue system clearly making a big impression on the Americans (and the Russians), that the British Air Ministry sponsored a proper Service evaluation. To this end, sixteen Meteor 8s of 245 Squadron, based at Horsham St Faith, were equipped with nose-mounted probes and refuelled in their UK coastal patrol role by Lincoln and Lancaster tankers supplied by Flight Refuelling. Operation *Pinnacle* was carried out over a period lasting several months, and two of the squadron's aircraft were also used in conjunction with Flight Refuelling's own Meteor III to demonstrate the newly converted YKB-29T's multi-point tanking capability. The breakthrough hoped for by Cobham did not, however, occur, for the Air Staff decided that, with limited Service funds available, the acquisition of specially equipped tankers would mean a balancing cutback in the number of fighters that could be purchased. It was not until the arrival of the fuel-thirsty Gloster

Soviet Tupolev Tu-16 'Badgers' employed the novel wingtip-to-wingtip method of refuelling introduced in the late 1950s.

FASTER, FURTHER, HIGHER

The Gloster Javelin FAW.9Rs of 23 Squadron were the first Royal Air Force fighters capable of being refuelled in flight, albeit for ferry purposes only. The Valiant BK.1 shown here belonged to 214 Squadron.

Javelins and English Electric Lightnings in the early 1960s that probe receiving equipment became a standard fitment on Fighter Command aircraft. But Cobham's determination to 'bend the ear' of the Service decision makers finally bore fruit in 1952, when it was announced that AAR provision would be made for the new V-bomber force. Accordingly, an English Electric Canberra was loaned to Flight Refuelling for testing AAR equipment at representative heights and speeds. In 1954, with Air Ministry confidence in the system now established, the requirement was laid down that all Vickers Valiants, Handley Page Victors and Avro Vulcans were not only to be capable of receiving fuel in flight, but also readily convertible to tankers should the necessity arise. Paradoxically, after the Valiant satisfactorily completed trials as a tanker, the Victor and Vulcan were then no longer considered for this role (although, due to circumstances entirely unforeseeable at this time, both would successfully serve as tankers later in their careers).

In 1958 214 Squadron, equipped with single-point Valiant B(K) Mk 1s, became the RAF's first squadron to be tasked with tanker service evaluation, and, after having revealed its new-found skills at the SBAC flying display at Farnborough, it further extended its wings with non-stop AAR-assisted demonstration flights to South Africa, Canada and the Far East.

Following the acquisition of rights to manufacture Cobham's probe-and-drogue equipment, US industry was soon invited to provide its own versions, and Beech, Fairchild, Douglas, Sargent Fletcher and the Hayes Corporation became leading suppliers of AAR devices to the USN and US Marine Corps. Meanwhile, General Curtis LeMay, head of SAC, had decreed that his force would best be served by Boeing's flying-boom method. So it has remained to this day, with the respective proponents of each system claiming significant advantages in performance and operational technique over the other.

In 1950, however, the need for total flexibility, or what is now termed 'interoperability' lay well into the future, and as with so many of its technical innovations, Britain's attempts to stay in the AAR game faced a mighty challenge. Cobham still persevered in trying to convince British Overseas Airways Corporation that AAR had a part to play, and, despite much airline reluctance following the North Atlantic Trials the previous year, flight tests took place in late 1950 and early 1951 in which the first prototype de Havilland Comet, fitted with a dummy nose probe, attempted to make dry contacts with a Lincoln tanker. After several unsuccessful prods at a reception basket swinging wildly on the end of a small-diameter hose, it was quickly deduced that such contacting manoeuvres would have little appeal to the fare-paying public or to airline pilots traditionally accustomed to keeping well clear of other aircraft. In a final attempt to work round the problem, Flight Refuelling then proposed a 'reverse refuelling' solution in which, to avoid the airliner (i.e. receiver) having to manoeuvre, it trailed the hose and reception coupling while the tanker made the contact and pumped fuel 'uphill'. This was to no avail, for with longitudinal stability at low-speed proving marginal on the early Comets, the addition of a hose reel unit in the rear fuselage would clearly have exacerbated the condition. Apart from design studies that briefly considered the provision of AAR for Concorde, this proved, finally, to be the swan-song of AAR for civil applications.

The military use of AAR was fast gaining ground. At the outset of the 1950s SAC operated 126 KB-29M looped-hose tankers, one squadron of which, the 43rd Air Refuelling Squadron based at RAF Marham, became the first USAF tanker unit to be deployed in the UK. However, with war clouds increasing over Europe and the Far East, the operational limitations imposed by this outmoded technique now focused USAF attention on the

The Boeing-designed 'flying boom' refuelling system was adopted by the US Air Force's Strategic Air Command in the late 1940s and first installed on the KC-97 tanker. Here, a Republic F-84F Thunderstreak is shown in contact with a KB-29P tanker.

EXTENDING THE REACH

Although flight trials were conducted with the prototype de Havilland Comet sporting a dummy fuel receiver probe, air-to-air refuelling was considered unsuitable for commercial operations.

more flexible probe-and-drogue method, and three squadrons of the 136th Fighter-Bomber Wing had their F-84Es and F-84Gs fitted with probe equipped wingtip tanks, the latter becoming the first type to become a dual-system receiver. Codenamed Operation *Hightide*, the evaluation not only underlined the almost self-evident advantages of AAR, but also stressed the importance of providing a properly engineered single-point refuelling system in the receiver. The exercise also led to the eventual introduction into Tactical Air Command (TAC) of 136 KB-50J and KB-50K probe-and-drogue tankers that, with SAC now irrevocably wedded to the flying–boom system, led to a lack of standardisation that would inevitably cause operational friction in the years to come.

Boeing's KC-97, introduced in 1950 and based on the C-97 cargo aeroplane, proved to be an excellent piston-engined tanker but, inevitably, its performance fell short when called upon to supply SAC's fast moving Boeing B-47 and B-52 front-line jet-bomber force. The KC-97G's altitude limit of some 20,000ft (6,100m) meant that a B-47 normally cruising at 35,000ft (10,700m) had to descend for replenishment and then climb back to its operating height after completing the refuelling. This was clearly an inefficient and expensive way to conduct refuelled missions with, it was estimated, the need to consume up to 50% of the fuel just taken on-board to complete the return to its original height. A turbojet powered tanker therefore became a top priority, but although Boeing proposed a jet powered version of the KC-97 it failed to interest the military. In the mid-1950s a substantial test programme was conducted in which a KB-47F tanker supplied a KB-47G receiver, however, at $2.7 million per copy the cost of conversion was considered prohibitive and the project was shelved in 1957. The B-47 had in many respects shown itself to be a satisfactory tanker, but it fell short of USAF requirements in one critical area. Unlike the KC-97, it could not offer a workhorse capability to carry support personnel and equipment.

The answer to the problem was provided almost inevitably by Boeing, which, in a bid competition with Lockheed and Douglas, submitted a tanker design based on its Model 367-80 (the prototype 707), which had first flown on 15 July 1954. Boeing's success was not, however, a foregone conclusion, as both of its competitors' proposals were initially judged to be technically superior. But the USAF, unable to wait for new prototypes to be developed, placed an interim order for twenty-nine of the almost readily-available Boeing tanker variant now designated KC-135. More orders followed for a further 363 interim models and, with the eventual elimination of the supposedly winning Lockheed design, Boeing went on to deliver a total of 732 KC-135 tankers. For some twenty years after its entry into service the KC-135 and its company stablemate, the B-52, were inseparable components within America's nuclear deterrent system, until that awesome delivery role was taken over by the intercontinental ballistic missile.

The SAC's change-over from piston- to jet-powered tankers gathered pace as the 1960s began. In 1961 it employed a total of 1,095 machines with the piston-powered KC-97 well outnumbering its jet successor (651 to 444). However, by the time the Cuban missile crisis arose the following year, the peak figure had reduced to 1,018; fewer aircraft, but more jets (503 to 515).

Wingtip mounted fuel tanks modified to accept fuel receiver probes enabled US Air Force F-84s and F-80s (shown here) to be replenished by KB-29M tankers during the Korean War.

63

FASTER, FURTHER, HIGHER

Earlier, the air war over Korea had soon demonstrated to the USN and Marine Corps jet fighter squadrons the need for increased range and endurance, and they had readily adopted the probe-and-drogue method as the most flexible way of meeting requirements. The first carrier-borne aircraft to be operated as a tanker was the North American AJ-1 Savage, which had a hose drum unit produced by Flight Refueling Inc, a company formed in the USA to overcome the government's reluctance to become dependent on an overseas supplier. The success of this aircraft in supplying the Grumman Panthers and Cougars and McDonnell Banshees led to the arrival in 1956 of the Convair R3Y-2 Tradewind flying-boat tanker. Although it was capable of refuelling four aircraft at once from wing-mounted fuel pods, problems with the Allison T-40 turboprop engines unfortunately caused this mighty machine's premature retirement from service.

In Britain, the use of what had become called the 'buddy-buddy' method of AAR by the USN and Marines had not escaped the notice of Royal Navy tacticians. This system allowed the rapid on-deck conversion of a fighter or fighter-bomber into a pod-equipped temporary tanker which could then refuel aircraft already low on fuel and facing an obstructed landing area. Armed with this precedent, in which the Douglas AD-6 Skyraider and subsequently the Douglas A4D-2 Skyhawk equipped with that company's D-704 refuelling pod were the most widely used examples, Flight Refuelling now submitted its Mk 20 refuelling pod for ministerial and Service approval. No opportunity was lost to impress upon naval planners that the means now existed to multiply the range and striking capability of a relatively small force significantly, and this effort ultimately resulted in the decision to equip as tankers a third of the Supermarine Scimitars and de Havilland Sea Vixens then about to enter service. It was then a natural step forward to provide the production Blackburn Buccaneer S.2 with a similar performance-enhancing capability. Interestingly, a design study was also embarked upon using the bomb-bay unit originally intended for the Buccaneer to convert the Sud-Aviation Caravelle airliner into a tanker, but the only practical aerial refuelling work carried out in conjunction with the French at this time involved Flight Refuelling's Canberra test aircraft supplying French Air Force Etendard and Vautour receivers at Istres in the late 1950s.

As the 1960s got under way the Air Staff took the decision to retire the Victor Mk 1 from its bomber role and use it to replace the Valiant as a tanker. It was 55 Squadron that first received the Victor B(K).1 interim two-point tanker conversions, full entry into service starting in June 1965. By the end of the year a second Victor squadron, 57, had also been formed at Marham, along with a Tanker Training Flight. This acceleration of the Victor tanker force followed the finding of fatigue failures which grounded the Valiant and caused its withdrawal from service in early 1965. By 1967 both squadrons, along with 214, were re-equipped with ten Victor K.1

In contrast to the USAF, the US Navy and Marine Corps regarded the probe-and-drogue method of refuelling as more suited to carrier-borne operations. Here, a North American AJ-1 Savage refuels a Grumman F9F-5 Panther from the Naval Air Test Center at Patuxent River.

EXTENDING THE REACH

two-point and fourteen K.1A three-point tankers, and the build-up of the Marham Tanker Wing was then complete. Major deployments of RAF fighters to the Middle and Far East were now regularly undertaken with the aid of the Victor tanker force, and the rapid reinforcement of local air defence in this manner was underlined by the term 'Force Extender' which had begun to enter the Service vocabulary. A study report showed that a tanker fleet could, for example, increase the operational effectiveness of a fleet of eighteen airborne warning and control systems (AWACS) aircraft to that of twenty-six — a 44% improvement — by simply extending the time on station and increasing the coverage of the surveillance areas. Hence, AAR also became synonymous with the term 'Force Multiplier'!

At that time the Vietnam war also provided an opportunity to demonstrate the flexibility of AAR. Although by this time rotary-wing aircraft provided a means of assault, they were still regarded in many military circles as essentially an instrument for rescue. But when Sikorsky HH-3E 'Jolly Green Giant' helicopters were equipped with probes to receive fuel from Lockheed HC-130P Hercules tankers, the increased range and

Undoubtedly the most impressive machine used as a tanker by the US Navy was the Convair R3Y-2 Tradewind flying boat. Fitted with Sargent Fletcher Model FR250 wing-mounted refuelling units, it was capable of supplying four aircraft simultaneously. In what appears to be a formation approach exercise, only three of the McDonnell F2H-3 Banshees have refuelling probes extending from the gunport position.

The Handley Page Victor served as the Royal Air Force's mainstay tanker for a thirty-year period. First introduced into the tanker role in 1965, it was operated by 55, 57 and 214 Squadrons before its retirement from service in 1994.

FASTER, FURTHER, HIGHER

Several companies supplied refuelling systems to the US forces. Using a 'boom' refuelling pod unit developed by Republic Aviation in the early 1960s, an F-105A Thunderchief passes fuel to an F-105B Thunderchief receiver.

In 1982 the Falklands Conflict required the deployment of the Royal Air Force's Victor tanker force to the South Atlantic. To enable the United Kingdom's home-based NATO commitments to be met, six Avro Vulcans of 50 Squadron were equipped with Flight Refuelling's Mk.17B Hose Drum Units to become 'stop-gap' tankers.

EXTENDING THE REACH

endurance that became available greatly enhanced their value. The Air Rescue Service retrieved 3,383 downed aircrew from enemy-held areas during the course of the war, and in the wake of this success the USN, USMC and Coast Guard quickly adopted AAR for their helicopter rescue fleets.

In 1973 the Israeli Air Force used AAR to help its strike aircraft bypass the heavily defended Canal Zone for missions deep in Egyptian territory. But it was the Argentinian invasion of the Falkland Islands in 1982 that truly impressed upon a watching world the crucial importance of AAR in a major trans-global operation. Britain's military response was obviously highly dependent on maintaining its extended lines of communication, and all air operations from the nearest forward base, Ascension Island, had to rely totally on AAR. Vulcan B.2s, hurriedly prepared for long-range action, had probe receiving systems reactivated that had not been used for some considerable time, and a top-priority exercise was undertaken to fit Flight Refuelling's Mk 17B hose drum units into the fuselages of six Vulcans and six C-130s. These conversions were carried out to ensure that, once the Victor tanker force had left the UK for Ascension Island, the RAF, working alongside the USAF's KC-135s, could still meet its home-based NATO commitments. The work programme also included fitting refuelling probes for the first time to the Hercules and the Nimrod MR.2 maritime reconnaissance aircraft.

The Falklands conflict brought the Tanker Force effort very sharply into focus for the first time since its formation nearly a quarter of a century earlier, and the public at large became aware of its strategic importance because of the extensive media coverage of its role. A brief glance at the operational statistics reveals its enormous contribution to the overall effort.

During the ten-week period of hostilities, 55 and 57 Squadrons' Victors flew some 3,000 hours in carrying out 600 refuelling sorties and transferring more than twelve million pounds of fuel. The vital part played by Victor refuellers in the single Vulcan's 8,000-mile (12,875km) round-trip attack on Stanley airfield (*Black Buck 1*), drew special Service praise. Eleven Victors were launched to support the primary and reserve Vulcans, with the actual machine that completed the raid taking on fuel on ten occasions. In addition, many contacts were made between the tankers themselves as, in diminishing numbers, they continued to support the attacking aircraft. Although the military value of what proved to be the longest bombing mission ever undertaken has frequently been questioned, the operation did provide final proof of AAR's capability to stretch the range of the V-bombers as had been envisaged some thirty years previously.

Perhaps even more impressive in terms of tanking force logistics was the effort to support the USAF strike against Libya in April 1986. As a punitive measure for sustained terrorist attacks against its civilians and establishments, Operation *Eldorado Canyon* was mounted as a joint USN-USAF operation involving carrier-based aircraft attacking Benghazi and British-based USAF General Dynamics F-111s concentrating on Tripoli. Routeing complications soon indicated the need for AAR, and SAC had to make available twenty-three McDonnell Douglas KC-10s, nineteen of which served as refuellers and one as an airborne command post. But even this formidable force could not, on its own, provide the fuel required by the eighteen F-111s and three EF-111A Raven electronics countermeasures aircraft taking part in the 4,500-mile (7,240km) mission. The solution was to have ten KC-135Rs refuel the KC-10 force, so that they could provide a full 'top-up' service to the attackers as they entered the Mediterranean. This they did with entirely satisfactory results.

The USAF has always been, and will continue to be, the largest user of AAR-equipped aircraft, with upgraded versions of the KC-135 Stratotanker and KC-10 Extender expected to remain in front-line service for many years to come. This capability is backed by large numbers of KC-135s serving with the Air Force Reserve units and Air National Guard. The USAF's impressive tanker force spearheaded the western world's response when Iraqi forces invaded Kuwait in 1991. The Coalition countries conducted two separate major operations during the Persian Gulf war. The first, *Desert Shield*, saw the build-up of forces to protect Saudi Arabia. The second, *Desert Storm*, provided the combination of air and ground forces necessary to liberate Kuwait. These activities placed an enormous responsibility on the joint alliance tanker resources and, with the need for 'interop-

The mid-1960s saw the French Air Force take delivery of Boeing C-135F tankers. Although they were fitted with a boom system it was possible, using the Boom Drogue Adaptor, to refuel aircraft such as the Mirage IV, shown here, equipped with fuel probes.

FASTER, FURTHER, HIGHER

Following the retirement of the Victor, the BAC VC10s of 101 and 10 Squadrons, along with the Lockheed TriStars of 50 Squadron, have provided the Royal Air Force's tanking support. Here, a VC10 K.Mk3 slakes the thirst of two Harrier GR.3s during an Arctic winter exercise.

erability' never greater, the fundamental differences between the probe-and-drogue and flying-boom refuelling systems could have proved an unsurmountable problem. Assistance came to the fore in the form of the boom drogue adaptor (BDA). First introduced in the 1950s, this was an inelegant but effective device which, via a short length of hose-and-drogue reception coupling, provided the means to supply the swarms of probe-equipped tactical receivers from 'boom' tankers.

The RAF's ageing Victor force was supported by the newer Vickers VC10 K.2s and K.3s of 101 Squadron that, following the Falklands conflict, had been converted from civil standard and Super VC10 variants. In addition, a small number of ex-British Airways Lockheed L.1011 TriStar 500s, now equipped as single-point TriStar K.1s, completed the UK's large tanker complement. Together with probe-and-drogue tankers from the Saudi, Canadian and USN forces, these aircraft managed to keep in the air the RAF's Tornadoes, Buccaneers, Jaguars, Hercules and Nimrods, as well as American A-7 Corsairs, AV-8Bs, EA-6B Prowlers, F-14 Tomcats, F-18 Hornets and S-3A Vikings. To highlight the level of international co-operation even more, French Mirage 2000s, Saudi and Italian Tornadoes and Canadian CF-18 Hornets were also on the receiving end of the tanker supply lines. When considering the wealth of statistics that emerged from *Desert Storm*, perhaps one of the most significant is that the transferring of some 716 million pounds of fuel involved almost 20% of the operational missions!

By the time the Gulf conflict erupted in 1991, Curtis Emerson LeMay, the forceful architect of SAC's boom-tanker force, had died; six weeks short of his eighty-fourth birthday on 3 October 1990. Just over a year after the Gulf War ceased, SAC itself was dissolved and its assets divided between three new commands, the Air Mobility Command gaining most of the tankers.

The old order was indeed changing, and in the UK, the second largest operator of tanker aircraft, 1994 saw the RAF's Victors finally withdrawn from service after thirty years, leaving its tanking legacy to the VC-10 and TriStar squadrons based at Brize Norton. The Buccaneers, which now transferred from naval duties to the RAF, still offered a buddy-buddy tanking capability.

In the late spring/early summer of 1999, NATO forces were brought together to halt a deteriorating situation in Kosovo. Codenamed Operation *Allied Force*, the air campaign against the Serbs proved to be the first time that concentrated air power became the sole instrument of victory. As in previous recent conflicts, AAR was a vital element in this success, some 180 large tanker aircraft, deployed in Italy, the Netherlands, Hungary, France, Spain, Germany and the UK, undertaking 7,500 missions; the ratio of 20% of the total number of sorties carried out being identical to that obtained in Gulf War operations some eight years previously.

The identification of significant problems and the dispersion of information relevant to AAR operators is conducted at regular forums co-ordinated by the Aerial Refuelling Systems Advisory Group (ARSAG). This organisation was first brought into being in 1978, when it became clear that standardisation of coupling equipment was necessary if refuelling mishaps were to be avoided. Since then, much-needed expert attention has been focused on the need for a common approach to the design, development and in-service use of equipment; work likely to continue for many years to come.

Today, the old distinction between tactical and strategic roles is less well defined, as aircraft designed for short-range operations can be projected rapidly across the globe. With this capability, allies can be reassured and potential enemies reminded that massive air power can be brought to bear in crisis situations within a matter of hours. AAR has been the key element in making this possible.

Bibliography

Butowski, P, 'Russian Flight Refuelling', *Air International*, August 1998.

Chant, C, *B-29 Superfortress Superprofile* (Haynes, Sparkford, England, 1983). The evolution of the B-29 is covered in fine detail.

Cruddas, C, *In Cobham's Company* (Cobham plc, Wimborne, England, 1994). The history of Flight Refuelling Limited.

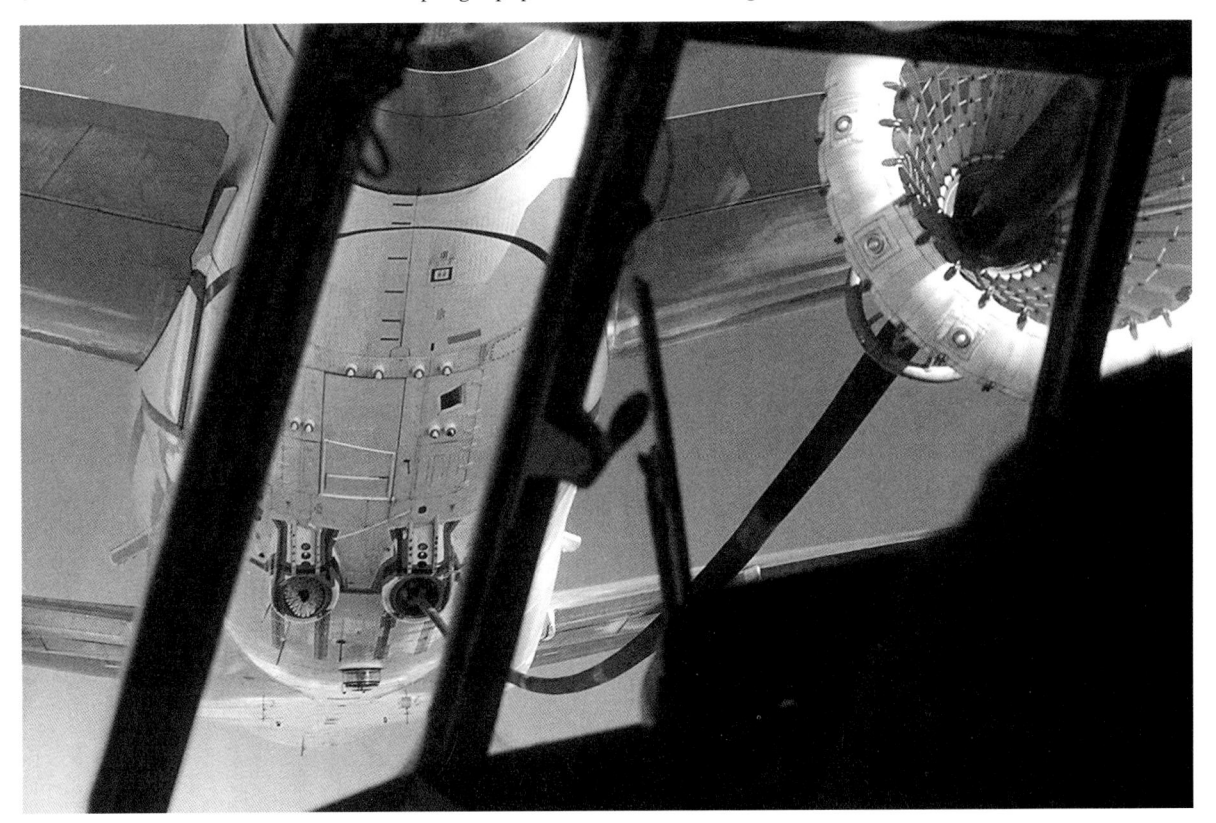

The Flight Refuelling Limited Mk.17T Hose Drum Unit installed in the Lockheed TriStar K.1 tanker incorporates duplicated hose assemblies to provide an extra safeguard when 100% mission reliability is essential.

An Italian Air Force Panavia Tornado refuels from a Boeing KC-135E Stratotanker via a Boom Drogue Adaptor during an Operation Allied Force mission over Kosovo.

Flight Refuelling Limited Air-to-Air Refuelling Methods (Flight Refuelling Limited, Wimborne, England, 1997). The boom and probe-and-drogue methods compared.

Gardner, B, *Aerial Refuelling at Farnborough* (Air-Britain, Tonbridge Wells, England, 1999). A definitive coverage of the work carried out at the RAE during the interwar period to develop AAR.

—— 'Skytanker – The Story of Air-to-Air Refuelling', *Air Extra* No 49, July 1985

—— 'Origins of Refuelling in Flight', *Flypast*, September 1995

—— 'Flight Refuelling, the Wartime Story', *Air Enthusiast* No 25, 1984

—— 'Air Refuelling in the RAF', *RAF Yearbook*, 1981

—— 'When You Need a Buddy: The Development of Air Refuelling in the United States Navy', *The Hook*, fall 1983

—— 'Refuelling in Flight — the History and Development' (unpublished).

—— 'Review of World Air Forces using Ait-to-Air Refuelling' (unpublished).

Holder, B, and Wallace, M, *Range Unlimited* (Schiffer, Atglen, USA, 2000). Provides an historical overview of AAR, with major emphasis on work done in the USA.

Smith, R K, *75 Years of Inflight Refuelling Highlights, 1923-1998* (US Government Printing Office, Washington DC, 1998). This publication was produced ahead of a full-length book (still in preparation) commissioned by the USAF.

EXTENDING THE REACH

System Descriptions

The looped-hose system

This was the method developed in Great Britain in the late 1930s by Flight Refuelling Limited. It was used operationally in the postwar period by the USAF until superseded by the probe-and-drogue and boom methods. The refuelling procedure was as follows:

i) With a small grapnel attached to its end, a 250ft (76m) steel hauling line was trailed behind the receiver.
ii) A projectile fired from the tanker flying alongside carried a light contact line in front of the arced hauling line. A physical link was established when the projectile and grapnel became engaged.
iii) After contact the tanker climbed to a position above the receiver and the trailed lines were wound in. The tanker's refuelling operator then separated the grapnel and projectile and secured the receiver's hauling line to a fuel hose nozzle.
iv) The fuel hose was then pulled across to the receiver and locked by hydraulically operated catches into a receptacle. The system was then purged with nitrogen before fuel was transferred under gravity at approximately 100gal/min (455lit/min).
v) After refuelling, the hydraulic locks were released and the tanker banked to one side, pulling the fuel hose, still with hauling line attached, clear of the receiver. At a predetermined point in the breakaway manoeuvre, a weak link in the hauling line snapped to allow complete separation of the tanker and receiver. Operators in both aircraft, who had communicated with each other throughout the operation by flags or Aldis lamps, then rewound the trailing elements back into their respective aircraft.

Principal users of looped-hose equipment:

Imperial Airways/Flight Refuelling Limited
Transatlantic trials 1939 with non-passenger, mail-payload-only aircraft

British South American Airways/Flight Refuelling Limited
Channel trials 1946-47
South Atlantic trials 1947

British Overseas Airways Corporation/Flight Refuelling Limited
North Atlantic trials 1948

USAAF
System Evaluation 1943

USAF – SAC 1948-1956
Far East Air Force 1951-1957
Used to equip first dedicated AAR tanker force

1. The probe-and-drogue system

First invented in 1949 by Flight Refuelling Limited, the system – successor to the company's looped-hose method – was conceived in response to a USAF requirement for a simple automatic means of refuelling single-crew fighters.

The principle of a receiver's probe engaging a drogue-encased reception coupling at the end of a hose trailed behind a tanker has applied to wing- and fuselage-mounted units supplied by manufacturers in several countries. However, the following description of Flight Refuelling's Mk 32B series AAR pod provides a typical example of probe-and-drogue equipment in world-wide service today.

i) Suspended below the wing on a pylon, the pod can contain up to 96ft (29m) of refuelling hose. A ram-air turbine at the front of the pod drives and controls the output pressure of a pump that receives fuel from the aircraft's main system. The fuel then flows through a control valve before entering the delivery hose. The control valve, depending on its selected position, also directs fuel to a vane pump/motor that in pump mode forms part of the hose braking system during hose trail. In motor mode it drives the drum to rewind the hose after a refuelling operation.

A key feature of Flight Refuelling's hose reel fuel supply units is the Tensator unit, which consists of a number of coils of spring steel connected directly to the hose drum. These store energy as the hose is trailed and provide a rewind torque to the hose drum. This provides a mechanism for eliminating hose slack when a receiver engages the drogue. It also assists the hose rewind operation.

A hose jettison facility is provided, allowing the hose to be disengaged when at full trail should an emergency arise.

ii) Pod functions are controlled by a digital refuelling control unit (DRCU) which monitors inputs from various pressure, temperature, flow, position and speed sensors within the pod. Operator demands from the aircraft control panel and system information from the DRCU are relayed via an ARINC serial data link. Built-in test equipment provides a continuous self-monitoring capability and allows the operator to check control system integrity rapidly, either pre/post or during flight. ➤

FASTER, FURTHER, HIGHER

Air-to-air refuelling enables both fixed- and rotary-wing aircraft to carry out a diverse range of operations. Shown here is a Sikorsky MH-53 Pave Low IV helicopter as it approaches a Lockheed MC-130P Combat Shadow tanker over flooded Central Mozambique. This mission formed part of Operation Atlas Response, *the US military relief effort following the torrential rains and flooding experienced in southern Africa in 2000.*

iii) The pod can deliver 540gal/min (2,455lit) at altitudes up to 35,000ft (10,670m) within a refuelling speed range of 140-280mph (225-450km/h).

iv) The receiver aircraft features a probe mast, either fixed or retractable, terminating in a 'male' fuel connection. A weak link is also provided which, if abnormally heavy loads are encountered, allows the end connection to shear and be retained in the drogue receptacle. The probe mast, usually located on the fuselage or wing, has, on occasions, been mounted on external drop tanks.

2. The flying-boom method

This system was originally produced by Boeing in response to a request from the USAF for a high-flow-rate alternative to the looped-hose method.

i) It consists of a telescopic fuel tube (boom) attached to the underside of a tanker's rear fuselage, which, when extended, engages in a receptacle in the receiver aircraft.

ii) Extension/retraction of the boom and contact with the receiver is governed by the tanker's refuelling operator, located in a special station allowing full view of the boom and the formating receiver.

iii) Final guidance of the 'male' nozzle into the receiver's receptacle is achieved by 'ruddervator' directional control surfaces located at the end of the boom. Once engaged in the receiver, the refuelling nozzle is locked in position by hydraulically operated toggles.

iv) Although descriptive of the Boeing boom systems fitted to the KC-135/707 range of tankers, this principle of operation also applies to the installation designed by McDonnell Douglas for the KC-10.

v) The boom system was conceived before the need for inter-Service interoperability became an overriding requirement, and its inflexibility was largely overcome by the introduction of the boom drogue adaptor. Although it needed to be fitted before a refuelling mission, the BDA provided an interim means for probe-equipped aircraft to be supplied from a boom tanker. Today, both the probe-and-drogue and boom systems are likely to be incorporated in the same machine, thus providing an overall refuelling capability.

4
The Upper Reaches
Jerry Scutts

Fixed-wing aviation was still very much in its infancy when designers realised that one of the most significant considerations to be borne in mind when creating viable aircraft was the potentially dangerous stresses that flight placed on the human body. Little had been done in the first decade of the twentieth century, as not only were aeroplanes little more than dangerous 'boxkites' with almost more drag than engine power, but few of the pioneers possessed any real knowledge of flying. The tremendous natural forces an aeroplane encountered in merely staying aloft for any length of time and landing safely had yet to be understood. But if aeroplanes were ever to carry passengers on extended flights, then the comfort of the people they were to bear aloft would become paramount.

After the Wright brothers' tremendous achievement at the start of the twentieth century, aircraft construction slowly became more robust and engines more reliable, and by the early months of the First World War the limiting factor of flying, particularly at altitude, remained the vulnerability of the human frame. This was particularly true in the universal 'open cockpit' era of aviation, because the pilot, other crewmen or any passengers carried would not only freeze without adequate clothing, but pass out and eventually die if starved of oxygen at altitudes significantly greater than 10,000ft (3,000m).

This did not deter the quest by pioneers to attain ever higher altitudes, irrespective of risk, but if commercial air travel were to expand, the medical condition of progres-

The German pilot Otto Linnekogel, complete with oxygen mask, in the Rumpler Taube 4C monoplane in which he set a world altitude record of 6,560m (21,552ft) in July 1914. The use of oxygen at this early date is unusual. His 'protective clothing' amounts to little more than a warm coat, a scarf, goggles and the reversed cloth cap typical of the time (though many aviators were by now wearing proper flying helmets).

sive hypoxia, manifest in popular terms in the three stages of severity as 'grey-out', 'red-out' and 'black-out', had to be prevented. Other detrimental symptoms resulting from unprotected flight at high altitude were revealed as early as 1783, when the occupant of a balloon experienced head pains violent enough to prevent him flying again.

Military air arms also became aware (albeit slowly) of the detrimental effects of 'altitude sickness' on aircrew performance, but after the outbreak of war in August 1914 this had a low priority, as the practical operating height for many types of aircraft was generally lower than 10,000ft (3,048m). The maximum attainable ceiling for Service biplanes and even monoplanes such as the Fokker E.III and E.V/D.VIII, the Morane Saulnier L and N and the Junkers D.1 and CL.1 was significantly below the 29,530ft (9,000m) attained by the Aviatik C.VIII, one of the last reconnaissance types built by Germany before the Armistice.

The operational ceiling of the majority of First World War combat aircraft remained in the 10,000 - 15,000ft (3,000 – 4,600m) band for the duration of the conflict, which nevertheless recorded some refinements in military aircraft, including the provision of the first oxygen systems. From the single, small cylinder provided, the

Balloon lead

The balloon, one of the simplest of air vehicles, was the pioneer of high-altitude flight. Jean Francois Pilatre de Rozier and the Marquis d'Arlandes ascended over Paris on 21 November 1783 to an altitude of 3,000ft (914m). This incredible, courageous achievement in a fire balloon was soon copied, and the first flight by a balloon filled with hydrogen gas and carrying a person took place on 1 December that same year. Unexpectedly, the sole occupant of the gondola, physicist Jacques Alexander Cesar Charles, found himself uncontrollably rising to 10,000ft (3,000m). The balloon carried only a few pounds of ballast, insufficient to check its ascent, and Charles wrote of the ill effects of aero-otitis media, manifest as a violent pain in the ears and jaw. He never flew again.

FASTER, FURTHER, HIGHER

pilot could take a few gulps of life-sustaining gas before seeking a lower altitude, preferably as fast as possible. The oxygen contained in the cylinders lasted only a few minutes.

By 1918 the Sopwith 7F.1 Snipe, a fighter capable of undertaking combat patrols at 25,000ft (7,620m) was equipped with oxygen bottles and a mask for the pilot. Under the impetus of war it had proven to be relatively easy to attain high altitudes in the lightweight aeroplanes of the period powered by uprated engines. The well-being of the pilot was not such an advanced science, and although physical fitness was monitored through medical examinations, advice was routinely ignored as individuals sallied forth to do battle without adequate protection. But there was a price to pay; aircrew reported sluggish reactions to simple control functions owing to oxygen starvation and extreme cold. Only so much could be achieved by wearing layers of clothing, boots and gloves, but that remained the normal remedy for some considerable time.

Such counterproductive impediments to operations were being better understood by the end of the war in November 1918, although oxygen systems where only slowly introduced as standard equipment in Service aircraft. There was a tendency among Service pilots to flout regulations governing equipment which was widely viewed as unnecessary, heavy and cumbersome, an attitude that extended to parachutes as well as oxygen apparatus. Rudimentary tests to determine a person's ability to function without oxygen were abandoned when it was confirmed that oxygen was required for all personnel flying above 12,000ft (3,660m).

100,000 feet high

Balloons continued to set altitude records, and in 1935 two Americans, Captains O A Anderson and A W Stevens, boosted the figure to 73,395ft (22,370m) by ascending in their craft *Explorer II* over Rapid City, South Dakota, on 11 November. Assisted by the National Geographic Society, the flight was recognised by the Fédération Aéronautique Internationale (FAI) as a record for aircraft irrespective of class, and stood until 19/20 August 1957, when Major David G Simons reached 101,516ft (30,942m) at Crosby, Minnesota.

Westland test pilot Harald Penrose in the Houston-Westland PV3, one of the two aeroplanes that made the first flight over Mount Everest (29,028ft/13,167m) on 3 April 1933. The crew had heated flying clothing and an oxygen system, but the pilot sat in an open cockpit. Earlier, the PV3's supercharged 630hp Bristol Pegasus IS3 nine-cylinder radial engine enabled Penrose to attain 35,000ft (10,668m) during a test flight.

THE UPPER REACHES

Slow acceptance

While it was relatively easy to provide military aircrew with oxygen masks to wear in the event of extreme altitude flight, during the 1920s and early 1930s many sorties continued to be flown at less than 20,000ft (6,100m). The technical advances made with military aircraft during the period after the First World War were relatively slow, in that the 'traditional' biplane configuration was widely retained. Although the importance of oxygen was recognised, aircrews continued to accept the discomfort inherent in high-altitude flight by believing in the option of diving to below 'oxygen height', which brought about a recovery.

Clearly, however, issuing individual oxygen masks to passengers in civil aircraft was not the answer for the commercial airline operator. In order for aircraft to operate above 12,000ft (3,660m) and thereby enjoy more economical performance and smoother flight in the calmer upper air streams, the oxygen provided for passengers to breathe inside the cabin had ideally to equal that at ground level and remain constant. Air temperature decreases by about 3.5 degrees per 1,000ft (300m), so that at 35,000ft (10,670m) it drops to minus 65 degrees Farenheit. To maintain this vital comfort for its human cargo, the cabin section of the aircraft had to be completely sealed off from outside temperatures and under pressure, preferably the same as that prevailing at ground level.

Early pressurisation experiments encountered several challenges. By their very nature, pressure cabins had to be 'double skinned' and fully leak-proof, which meant they were inevitably heavy. Effective sealing of the compartment was also found to be difficult, and if it was required to accommodate multiple passengers in a special fuselage section, the overall size of the aircraft would have to increase. This in turn meant an inevitable rise in weight and the need to use more powerful engines, with higher fuel costs and some operating restrictions, using only the larger airports.

On 28 September 1936 Squadron Leader F R D Swain, seen here in his somewhat primitive pioneering pressure suit, set a new world altitude record of 49,944ft (15,223m) in the Bristol 138A high-altitude monoplane.

In routine flying in the Services, things were less advanced. Here, an RAF pilot prepares to make a meteorological flight from Duxford Airfield, Cambridge, in the UK, in an Armstrong Whitworth Siskin fighter in the mid-1930s.

FASTER, FURTHER, HIGHER

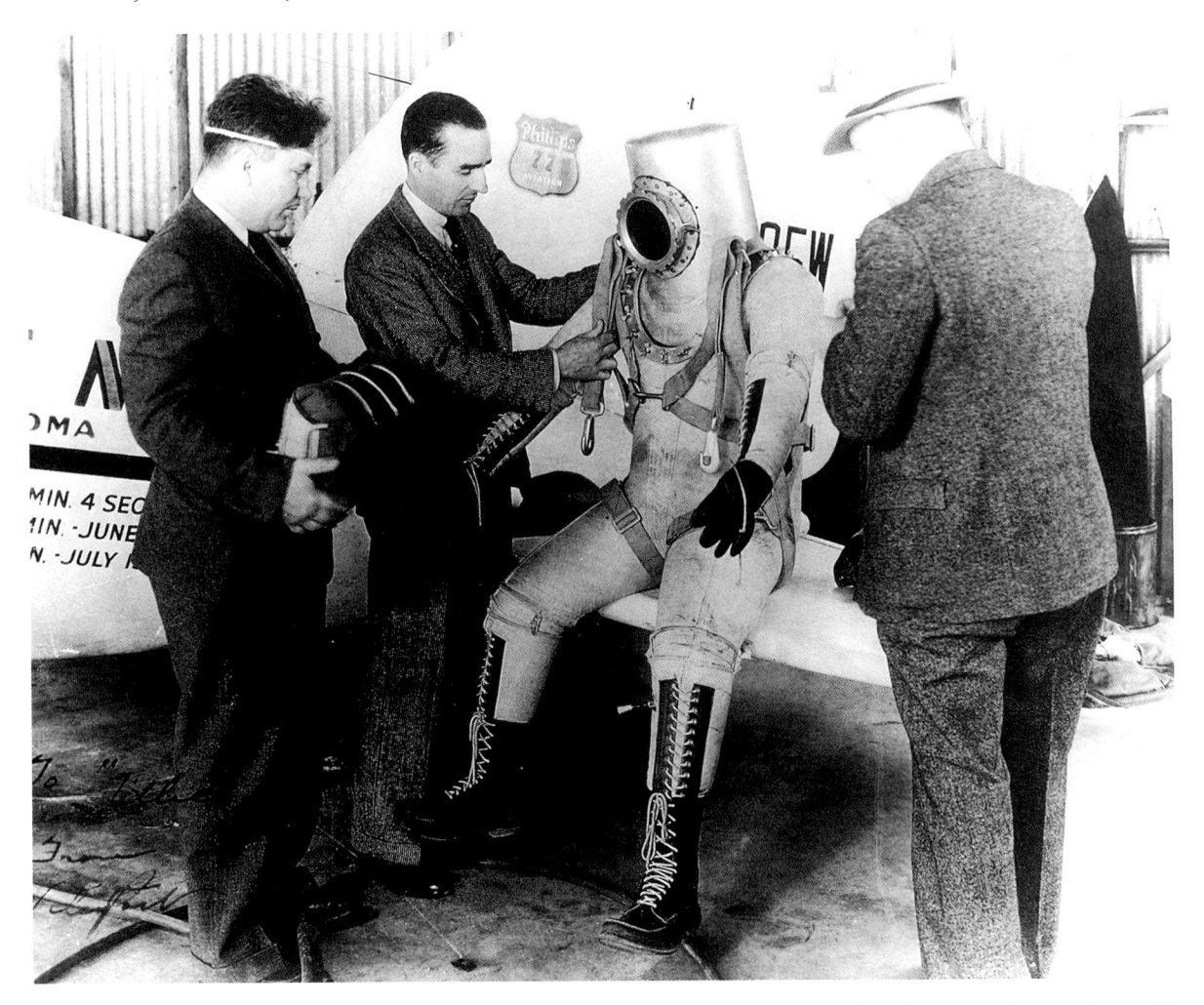

American record-breaking pilot Wiley Post (left) and Billy Parker of Phillips Petroleum adjust the straps of one of the world's first pressure suits in December 1934. Wearing this suit and flying his famous Lockheed Vega 5 Winnie Mae, *Post made an attempt on the world altitude record in December 1934 and conducted a number of sub-stratosphere transcontinental flights in 1935.*

Progress towards sustained higher-altitude flying by standard civil and military aircraft was boosted by such events as the flight over Mount Everest by two Westland biplanes, the P.V. 3 and P.V. 6. Flown by mixed RAF and civilian crews, these aircraft cleared the 29,028ft (13,167m) summit on 3 April 1933. Three years later the RAF ordered the Bristol 138A for an attempt on the world altitude record. The aircraft beat the existing record of 48,698ft (22,087m) in September 1936 by attaining 49,967ft (22,665m), then repeated the feat in June 1937 by achieving 53,987ft (24,488m), which took the pilot, Flight Lieutenant M J Adam, 2¼hr. This flight secured the record for Britain for only a few months, as it was lost to Italy the following year. Such record attempts were generally curtailed by the outbreak of the Second World War in Europe, but further progress towards pressurised passenger aircraft was made shortly beforehand in Britain and the USA.

Research into practical pressure cabins involved a number of experimental aircraft, those in Britain including the GAL 41, a pressurised General Aircraft Monospar. Employing an auxiliary engine to maintain cabin air at a differential pressure of 7lb/in^2 (0.49kg/m^2), the aircraft first flew in May 1939, two years or so after the Lockheed XC-35, a US Army-sponsored modification of the Electra. Also used to test engine superchargers in high altitude operation, the XC-35 flew for the first time on 7 May 1937. Able to reach 31,500ft (14,290m), it differed from the standard transport by incorporating a completely circular-section fuselage without the standard windows. This latter modification was carried out in order not to compromise pressurisation, which posed

THE UPPER REACHES

some challenges in regard to completely sealing the cabin. After a false start using cloth strips soaked in glue, a good seal was successfully achieved with Du Pont neoprene tape.

Despite the technical challenges, Boeing designers and aeronautical engineers fitted the first pressure cabin for commercial airline use in the Model 307 Stratoliner of 1937. Capable of attaining 26,500ft (8,077m), the Stratoliner was an offshoot of the B-17 Flying Fortress bomber, with a new fuselage accommodating up to 35 passengers. En route to providing passengers a significant new service with pressurised comfort (at a differential of 2.5lb/in^2), Boeing found that ventilation and temperature control could vary depending on the ambient temperature, and where and when the aircraft might be required to operate. In tropical climates and at lower altitudes during the summer the temperature control showed a significant variation. Pressure was equal to 8,000ft (2,440m) inside and outside the cabin up to an altitude of 15,000ft (4,570m), but there was then a gradual drop, and at 20,000ft (6,100m) the pressure was the equivalent of that prevailing at 12,000ft (3,660m).

As was widely recognised, the advantage of pres-

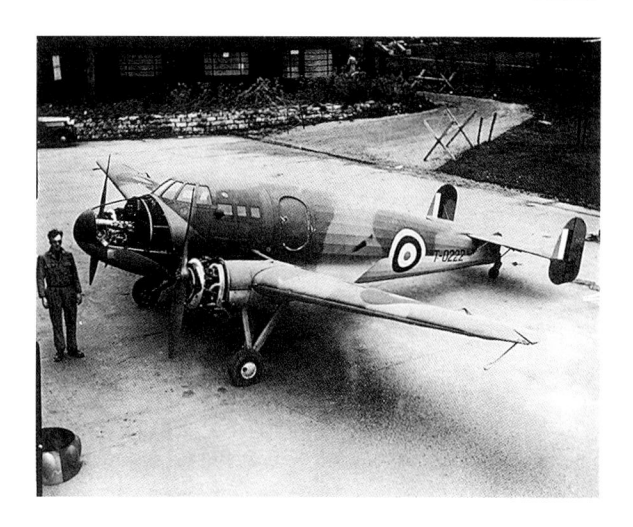

In 1939-40 experiments were made in the UK with this General Aircraft GAL 41 Monospar, a special version of the Monospar series with a complete pressure cabin in which pressure was maintained by an auxiliary engine mounted in the aircraft's nose. The fabric-covered rear fuselage was retained.

The Lockheed XC-35 was a pressurised version of the Electra. Sponsored by the US Army and first flown in 1937, it had a fuselage of completely circular cross-section, with much smaller windows than usual.

FASTER, FURTHER, HIGHER

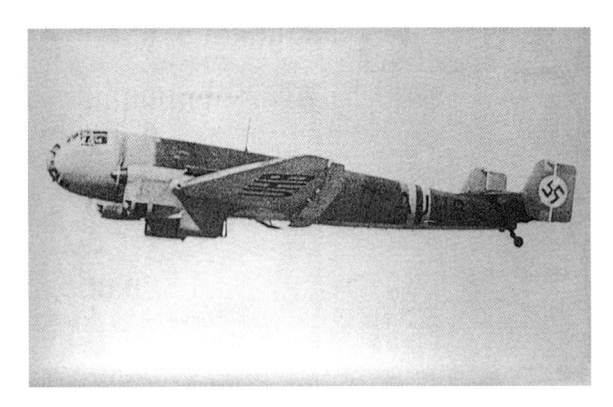

Early in the Second World War the obsolete Junkers Ju 86 bomber was adapted to produce a highly effective aircraft for high-altitude surveillance, the Ju 86P. Air bled from one of the engine superchargers maintained fuselage pressure.

Throughout the late 1920s and 1930s there was a gradual all-round increase in technical reliability of aircraft in many categories. By 1936 the Douglas DC-3 was offering airline passengers new standards of comfort, albeit at low altitudes.

To fight the war, Britain developed a range of unpressurised heavy bombers with generally modest operating ceilings, whereas, by bombing in daylight, the Americans theoretically raised the maximum operational altitude, though a lower height band was generally adopted by crews flying the unpressurised Boeing B-17 and Consolidated B-24 Liberator. These bombers were actually designed to reach 37,500ft (11,430m) and 32,000ft (9,750m) respectively. German interceptor fighters, especially the Messerschmitt Bf 109, were also able to fight at the bombers' altitudes and above (up to 38,500ft/11,734m). At the same time it was recognised that extreme altitude was the safest option for the unarmed reconnaissance aircraft that had become vital for checking target damage and surveying enemy activity. The actual deployment of pressurised aircraft for this special purpose varied among the combatants; the RAF used special high-altitude versions of the Supermarine Spitfire and de Havilland Mosquito with cockpit pressurisation, which were used operationally in conjunction

surised airliners lay in their ability to climb high enough to avoid flying through the rougher, more turbulent air at lower levels, which had hitherto been the lot of all commercial aircraft. In 1938 Boeing responded to a requirement for a pressurised version of the B-17. This emerged as the B-29 Superfortress, which made its first flight in September 1942.

In Italy, experiments in stratospheric flight were carried out during 1939-40 with the Piaggio P.111, which had a pressure cabin and was powered by specially supercharged P.XII RC engines, designated 100/2V and delivering 1,000hp.

The fuselage of Boeing's B-29 Superfortress bomber contained three pressurised compartments. The forward compartment, housing the pilot, copilot, bombardier, engineer, navigator and radio operator, was linked by a tunnel to the rear compartment, just forward of the fin strake, which housed the top gunner and side gunners, who operated the barbettes via remote control, and also had folding berths for two relief crewmen. The tail gunner had a separate pressurised compartment.

with unpressurised machines to cover targets at lower altitudes.

In Germany, considerable research into pressurisation resulted in new versions of standard Service bombers, although very few were actually deployed by the Luftwaffe purely in the intended reconnaissance role. They included the Dornier Do 217P (with a maximum ceiling of 49,215ft/15,000m), the Junkers Ju 388L (42,640ft/13,000m) and the revolutionary Ju 287 (39,360ft/12,000m), all of which showed much technical promise. Apart from the Ju 388L, which entered service in small numbers shortly before the war ended, few aircrew were able to fly these and other interesting prototypes, the Luftwaffe high command instead opting for dual-purpose machines.

The singular exception was the Junkers Ju 86P, a remarkably successful adaptation of an obsolete bomber type, used to fly the majority of the Luftwaffe's high-altitude surveillance sorties at heights up to 39,360ft (12,000m). Junkers had gained much experience in the design of pressure cabins with the Ju 49 of 1931 and the Ju (EF) 61, which appeared in 1936. Pressurisation was maintained by bleeding air from one of the engine superchargers.

Later in the conflict the pressurised Arado Ar 243 jet bomber pointed the way to the future by photographing potential targets in Italy and England while virtually immune from interception at an altitude of 39,300ft (11,980m).

Few technical advances happen in isolation, and other nations explored pressurisation for photo-reconnaissance purposes during the war years. Japan's Kawasaki Ki-108 and Italy's Piaggio P.111 were practical examples of the research carried out in this field, though neither type achieved operational status.

Crews of the vast fleets of unpressurised bombers that fought over Europe were provided with demand oxygen systems, only the B-29 incorporating fully-pressurised fuselage sections in its original design. The Superfortress represented a pinnacle of wartime progress in terms of a conventional front-line bomber powered by reciprocating engines. Produced in quantity, the B-29 – having been exclusively deployed against Japan, usually at altitudes much lower than its design ceiling of 31,850ft (9,700m) – became a useful postwar 'mother ship' for research aircraft.

With the coming of peace in 1945 the ponderous progress of the prewar period was replaced in the West by an international race for better air weapons. Western nations had achieved much with aircraft that in the main were of conventional construction and had reciprocating engines. By contrast, German scientists and technicians

FASTER, FURTHER, HIGHER

had brought forward the next generation of aeronautics with new airframes and powerplants including turbojets, rocket motors and ramjets. While these innovative designs had not proved to be decisive weapons, they had in several instances been used operationally, and their failure was partly attributable to technical shortcomings such as the enforced use of poor-quality materials in areas such as turbine blade manufacture. With her military position growing increasing dire, Germany had lost vast quantities of 'quality' raw materials. High-speed flight had been achieved, but reliability had suffered to a significant extent.

Jet propulsion had not, of course, been exclusive to the Germans. Allied efforts had almost caught up with fighters such as the Messerschmitt Me 262, which remained the most advanced front-line fighter in the world in 1945.

With the war over, defence budgets were trimmed, but they picked up fairly rapidly, particularly in the USA, where capital was made available to explore numerous new approaches to aeronautical propulsion, some of it incorporating the results of German research. Broadly speaking, turbojet power and wing sweep was where the future lay in most of the important civil and military aircraft projects for the future. Fortuitously, the USA grasped this by redesigning the North American F-86 jet fighter to incorporate a swept rather than 'straight' wing. By its bold gamble in ordering the Sabre, the USA could field a world-class combat aircraft when it had to in 1950. When war broke out in Korea barely five years after the end of the Second World War, it appeared to have wider implications for world peace, and US air power was deployed under the flag of the United Nations. Against the F-86 the North Koreans deployed the MiG-15, which led to air combat between the two most modern fighters in the world.

In Britain, numerous new types were able swell the programme at the annual industry showcase at Farnbough, among them (in 1957) the promising Saunders-Roe SR.53, a high-performance fighter prototype that derived its power from both a Spectre rocket and Gyron turbojet. Developed as the SR 177, it might well have been a success as a naval interceptor. It was a sign of the times that it was far from unique in being cancelled before its potential could be realised.

Project X

By contrast, the USA seemed prolific in its desire to extend the frontiers of flying, to capture records and generally modernise its armed forces across the board, taking advantage of new technology. To achieve that end, a far-reaching programme of testing with a series of fascinating aircraft known collectively as the 'X-planes' began. The National Advisory Committee for Aeronautics (NACA; renamed National Aeronautics and Space Administration (NASA) in 1958) and the US Services

The prototype of a mixed-powerplant high-altitude interceptor, the Saunders-Roe SR.53 of the late 1950s had an Armstrong Siddeley Viper turbojet supplemented by a de Havilland Spectre rocket motor. It could reach 50,000ft (15,240m) in 2 minutes 12 seconds from brake release.

jointly funded programmes which invariably employed modified B-29 and B-50 Superfortresses to carry research vehicles designed to probe the 'upper limits' to great heights before they were launched. The veteran bombers saved the weight of the fuel that the tiny research aircraft would otherwise have had to carry for ground take-offs.

A primary goal for Bell and the USAF was for America to be the first country to attain Mach One in level flight. That race was won by the X-1 on 14 October 1947, when Chuck Yeager attained Mach 1.06 (700mph/1,126km/h) at an altitude of 45,000ft (20,412m) on the fiftieth flight of the X-1. Yeager bettered this speed the following March by flying the original aircraft to Mach 1.45 (957mph/1,540km/h) at 50,000ft (22,680 m).

An attack on the altitude record was subsequently made on several occasions with the Bell X-1, and unofficially the figure was raised to 90,440ft (27,5660m) by the X-1A in August 1954, and to 125,907ft (38,376m) by Iven Kincheloe on 7 September 1956. The swept-wing X-2 had also captured an unofficial world speed record on 23 July that year, Frank Everest pushing the aircraft to Mach 2.8706 (1,900.34 mph/3,057.6km/h) after he had flown the aircraft supersonically for the first time the previous April.

Of all these dramatic flights at hitherto unattainable speeds and altitudes, none came as more of a relief in aviation circles than the cracking of the mysterious 'sound barrier'. Penetrating the dangerous wall of air that built up as aircraft flew faster had long seemed all but impossible, but at last it had been breached without undue harm to pilot or aircraft. Pilots of the early X series relied on what was then becoming an old-fashioned demand oxygen system, but it saved the weight of a pressure cabin in aircraft where every pound counted.

THE UPPER REACHES

New clothing

During the latter stages of the Second World War the stresses imposed on the body during high-speed aerial manoeuvring had brought about the need for a pressure or 'G-suit' to maintain blood supply to the pilot's brain and prevent blacking-out. Germany experimented with both the G-suit and ejector seat, and American suits were manufactured and used in combat before the end of hostilities. Both water and air were used to inflate abdominal and arm/leg joint sections of the suits selectively, and the oufits were quite successful within their limitations. But, as aircraft speeds increased, more sophisticated suits with instantaneous response to high-speed manoeuvres were required. Another postwar invention was designed to protect the pilot's head under all flight attitudes and in the event of accidents; the so-called 'bone dome' became standard equipment for flying at any altitude. This included ascents into the upper reaches by balloons, which, as Major Simons' 1957 feat showed, were still capable of leaving conventional aircraft behind in terms of the great heights they could attain.

That research flying could hold unforeseen hazards for test pilots was obvious, but engineers could do little more than make their designs as safe as possible and modify them as and when necessary. A small army of Service and civilian test pilots meanwhile accepted the risks. With courage and outstanding airmanship, individuals such as Chuck Yeager, Scott Crossfield and Frank Everest gradually 'pushed the envelope' to achieve the next goal, that of exceeding Mach 2 in level flight.

Few of the military and civilian teams working for NACA or the Royal Aircraft Establishment (RAE) at Farnborough in the UK and engaged on high-speed research could totally insure against mishaps as man strove to fly ever faster and higher. There were many incidents and accidents, some of them fatal. During the 1950s and 60s a succession of pilots also achieved (albeit unintentionally) several aviation 'milestones', including the first successful supersonic bale-out. That feat was claimed by North American Aviation test pilot George F Smith on 26 February 1955, when his F-100 Super Sabre became uncontrollable during a dive. Although battered by the tremendous forces imposed as he abandoned the plunging fighter heading for the earth at 1,140ft (517m) per second, Smith made a quick recovery and was medically cleared to fly again by 23 August. By then, coincidentally, Flying Officer Hedley Molland had undergone a similar experience when, on 3 August 1955, his No 263 Squadron Hawker Hunter refused to respond to the controls during a high-speed dive. Molland baled out lower than Smith, at about 25,000ft (11,340m), where the speed of sound is higher, but also survived without lasting ill-effects.

Few pilots would choose to bale out into the teeth of a raging electric storm, but Lieutenant Colonel William Rankin, US Marine Corps, had no choice when his Chance Vought F-8U Crusader suffered an engine seizure at 40,000ft (12,192m) on 26 July 1959. Although the external temperature was minus 70 degrees, Rankin was at that height specifically to avoid the thunderstorm brewing below him. Now, without power, he had to eject. Suffering terrible pain from alternate heat and decompression, Rankin tumbled, at the mercy of the elements. Breathing oxygen when denser air allowed him to move his arms and use the flapping mask, the terrified Marine waited for his parachute to open at 10,000ft (3,000m). At that point he fell into the thunderstorm, was nearly drowned by rain, half-blinded by lightning and deafened

Pilots of the rocket-powered Bell X-1A and X-1B had simple demand oxygen systems to eliminate the need for weighty pressure cabins.

Britain's English Electric Canberra bomber gained a reputation for its exceptional high-altitude performance. In this aircraft, Canberra B Mk 2 WD952, a testbed powered by a pair of Bristol Olympus turbojets, Wing Commander Walter F Gibb set two records for absolute height, reaching 63,668ft (19,406m) on 29 August 1955 and 65,889ft (20,083m) on 28 August 1957.

FASTER, FURTHER, HIGHER

by thunderclaps. Fearing that his parachute would never stand this strain, Rankin finally did float free of the violent air before crashing into tree-tops and slipping painfully to the ground. Like other pilots who had experienced bale-outs under extreme conditions, he recovered fully to fly again within weeks.

Developed during the Second World War, ejection seats have saved hundreds of lives, and pilots have resorted to them under all kinds of flight emergency, including riding careering jet fighters off the decks of carriers. Such entirely involuntary 'tests' have demonstrated the 'zero-zero' qualities of the seats at little more than ground level. During the 1950 and 1960s numerous hazards associated with unpredictable flight characteristics became better understood en route to significantly increasing aircraft capability in range, speed, altitude – and safety.

For years the world speed record rose only by small increments, and it was not until 10 March 1956 that a manned aircraft achieved over 1,000mph (1,609km/h) in level flight. Peter Twiss, flying the Fairey F.D.2, put 1,132mph (1,821km/h) on the clock and the record was confirmed by the FAI on 1 June. It did not stand for long (and it was the last that Britain claimed), and, by the turn of the twenty-first century, flights that took aircraft past Mach 3, 4 and beyond, and above 100,000ft (30,480m) made such records in the 'lower air' arguably less significant than they once were. In an era in which Western nations designed a whole series of interceptors to climb fast and destroy the enemy bombers that Cold War scenarios predicted, altitude attainment remained important. These same projections of future conflict also highlighted the vulnerability of installations such as runways, and vertical/short take-off and landing (VTOL) appeared to offer many tactical advantages.

In Britain that challenge was taken up when the world's first V/STOL jet fighter first took wing as the Hawker P.1127 in October 1960. Behind this prototype

A pilot in pressure suit and 'bone dome' poses with the prototype Lockheed F-104 Starfighter. As the capabilities of the pilots' mounts improved, so too did the clothing required to ensure that its wearers could withstand the increasing physical demands.

lay a lengthy programme of experimentation with VTO, data having been gathered by such vehicles as the Rolls-Royce twin Nene-powered Thrust Measuring Rig, the so-called Flying Bedstead, and the more conventional Short SC 1 of 1957. Developed as the Harrier and McDonnell Douglas AV-8 series, Hawker's Kestrel led to a highly successful V/STOL programme that remained unique until it was finally duplicated by the Russians with the shipboard Yakovlev Yak-38 of 1971. Currently, the VTOL Lockheed-Martin F-22/X-35C Joint Strike Fighter represents a highly significant programme that will provide the latest generation of jet fighters to equip the US air forces and the RAF, replacing the Harrier among other types and bringing the wheel full circle, forty-plus years after the first flight of the P.1127.

Commercial expansion

The search for sustained operating altitudes for civil airline flying was far from a flippant goal. Smoother airstreams above 12,000ft (3,650m) meant that, if aircraft could adopt such heights as their standard cruising altitude, their passengers would enjoy a far smoother ride well 'above the weather'. The first airline operator to offer such a standard of comfort would naturally secure a sizeable share of the market. Lockheed took that lead with its pressurised Model 049 Constellation, which found itself in immediate competition with the unpressurised Douglas DC-4. Douglas soon passed on to the pressurised DC-6 and Boeing's Stratocruiser appeared as a derivative of the KC-97 tanker, which was also pressurised. These fine aircraft, each powered by four reciprocating engines, firmly established new long-distance air routes that would in their turn be flown by jetliners.

Lockheed's prewar work for Pan-American Airways, the launch customer for the Model 049, had to be suspended for the duration of the conflict, but thereafter production of civil Constellations, as against military models, picked up. Lockheed designed the Model 049 with a cabin pressure differential of 4.18lb/in^2, very close to the 4.16lb/in2 (2.92kg/m^2) of the rival DC-6. Both aircraft cruised at 310mph (498km/h) and carried 54-56 passengers at typical operating heights of 18,000ft (5,490m) to 21,000ft (6,400m).

The limited endurance of rocket and ramjet engines meant that they clearly had few immediate postwar civil applications, but turbojets had huge advantages for intercontinental airliners. In the twenty-first century mixed-power-source engines are back in vogue, as many of the seemingly insurmountable drawbacks of the 1950s have been overcome in the intervening decades. There were, by contrast, few doubts about the turbojet; that power source completely revolutionised high-altitude flying for commercial purposes, and all postwar airliner designs concurrently incorporated pressure cabins as standard. Varying degrees of wing sweep were incorporated for improved performance.

Commercial aviation's rapid progress in the field of turbojet airliners was perhaps too rapid in the case of the de Havilland Comet, which entered service with British Overseas Airways Corporation on 2 May 1952. With a cabin pressurised to 8.2lb/in^2, the Comet set new standards and represented a very proud British milestone when it inaugurated the world's first long-range jet airliner service. The Comet's lead was tragically lost after a series of fatal crashes in 1953 and 1954. Investigation proved that faulty cabin widow design, coupled with metal fatigue, caused the fuselage to crack under pressure and create a chain reaction that led to explosive cabin decompression. Belatedly returned to service in 1958, the Comet 4 was able to carry 60 passengers nonstop across the Atlantic, but by then the lead had passed to the USA.

As the effects of violent decompression became known, in that anything not secured is sucked out of any rupture exposed to the outside air, attention focussed on the design of airliner doors and windows to prevent a similar occurence. It rarely happened, but when the door of a Turkish Airlines McDonnell Douglas DC-10 came loose on 3 March 1974 with similar disastrous consequences, the accident (then the world's worst in terms of loss of life) showed that there was no room for complacency in attention to such fundamental design details.

In the USA the Boeing company secured a lucrative order for a new military jet transport which led to the KC-135 and the outstanding Model 707, the world's first international jet airliner in sustained commercial terms. After 18 June 1957, the day that the USAF took delivery of its first KC-135A, Boeing hardly looked back.

Using the well-proven layout of four underslung turbojets, the Boeing 707, which flew for the first time on 15 July 1954, was designed to fly 132 passengers at a speed of Mach 0.82 at 35,000 to 41,000ft (10,670 to 12,500m). With a cabin pressure differential of 8.6lb/in^2, the aircraft maintained a sea-level pressure up to 22,500ft (6,360m) and 7,000ft-altitude pressure (2,133m) at 40,000ft (12,190m).

Jetliners quickly established an enviable safety record, although here, as in other areas of aviation, there were lessons to be learned. Among these was rule that true air speed decreased and stalling speed increased with an increase in altitude. At their near-maximum subsonic speed of Mach 0.9, airliners faced the prospect of quickly going supersonic if they stalled at altitude or were obliged to lose height quickly. Catastrophic airframe failure would inevitably follow. Airliners were not stressed to overcome supersonic velocity, and steps were therefore taken for crews to limit the commercial operating ceiling to below 40,000ft (12,190m) and to monitor airspeed and stall-warning devices closely.

Such feedback helped when the turbojet airliners first encountered clear-air turbulence (CAT), which could

FASTER, FURTHER, HIGHER

strike without warning, high speed winds battering the airliner and causing fatal stalls and airframe failure in a number of instances. Although CAT is survivable, normal weather warning radar cannot detect what is, in effect, not there. Modern wing design includes the fitting of spoilers and other drag-inducing devices to help prevent catastrophe in the event of unplanned manoeuvres due to rough weather conditions. Despite such dangers, early transcontinental jet air travel proved to be extremely safe.

The 707 inevitably had its US rivals. Convair produced the 880 series and Douglas the DC-8, the first generation of long-range jet airliners that revolutionised air travel for millions. Flight tests with the single Boeing Model 367-80 prototype proved that cruising at 25,000ft (11,340m) would become typical on regular transatlantic crossings. The following Model 707-100, which first flew on 20 December 1957, had similar operating figures, with a ceiling of 37,500ft (17,000m).

While the Russians made progress with multi-engined airliners such as the Tupolev Tu-104, which initially at least owed their design origins to long-range bombers, the West brought American. Britain and the rest of Europe tended to concentrate on smaller airliners designed for shorter-length, lower-altitude routes. These included the Sud-Est Caravelle, the BAC One-Eleven and successful turboprop types such as the Vickers Viscount and Bristol Britannia. The Vickers VC10, which entered British airline service during the mid-1960s, represented virtually the sole (and relatively brief) European challenge to US domination of the big-jet market.

America's enviable position as world leader did not entirely overshadow the competition, as in Europe, but Boeing in particular was working towards a new generation of airliners to cater for a huge anticipated increase in demand for intercontinental air travel in the 1970s and 1980s.

This demand was primarily met by the giant 363- to 490-seat Boeing 747. Dubbed the 'Jumbo Jet' by the press of the day, the 747 proved to be one of the world's most technically outstanding and commercially successful aircraft of all time. Stressed (as most otherwise sedate airliners are) way beyond the flight profiles to which it complies for 99 % of its working life, the 747 was thrown all over the sky in test flights that proved the resilience of the airframe. The first of a 'family' of 747 derivatives, the Dash 100 Series, first flew on 8 February 1969. Capable of reaching 45,000ft (20,412m) the 747-100 normally cruised at 30,000ft (13,600m) at a speed of 595mph (957km/h).

The original Boeing 747 series was joined by the Special Performance (SP) version in 1976. With a similar ceiling and speed range as its larger counterpart, the short-fuselage 747 offered reduced passenger-mile costs on long-range routes. Given its success, it was unlikely that Boeing's widebody would not to be copied, and to some extent it was by the three-engined DC-10 and Lockheed TriStar, which began to capture sectors of a market expanding so quickly that all players were assured of a slice of the cake. Airlines regularly operated their widebodies at altitudes 30,000 to 35,000ft (9,114–10,668m). Once this optimum operating height had been established for the big jets there was actually little need to vary it; the main concern of the airlines was to increase speed where possible, extend range without compromising the number of seats offered by a given type on a typical high-density route and, as ever, reduce fuel costs. Manufacturers responded by developing turbofan engines, which gave significantly higher efficiency compared with turbojets, with a more economical fuel burn over the entire speed range. In addition, turbofans can replace older engines without major modifications. Physically much larger than the older powerplants, turbofans have been fitted retrospectively to many civil and military aircraft that fortuitously adopted the 'wing pod' engine configuration.

In common with the rest of the world, European carriers tended to buy American equipment for use on long-range routes, which became more economical as aircraft sales skyrocketed and passenger numbers steadily rose. But manufacturers such as British Aerospace and Aérospatiale of France were meanwhile collaborating on an exciting possibility; air travel across the globe at supersonic speeds. Boeing had begun research towards a similar goal as early as 1952, but at that time the prohibitive cost of building a very large aircraft appeared to be beyond the resources of a single firm. Each Boeing supersonic transport (SST) carried a projected price tag of $16,000,000, which was believed to be prohibitively high. A government-led design competition into an SST was let in 1963, and Boeing, which was pronounced the winner in 1966, opted for a variable-wing design, the Model 2707. Estimated to have a speed of Mach 2.7 (1,800mph/2,900km/h) and cruise at 64,000ft (19,507m), the aircraft would have carried up to 350 passengers, a far more realistic number in commercial terms than Concorde. In the event, the project was cancelled on 24 March 1971, Boeing having in the interim replaced the variable wing planform with a fixed delta.

Flights by large aircraft at Mach 2 were hardly common in the early 1970s, and relatively little data was available to the designers of the first commercial SSTs. Two fundamental challenges they faced were the reduction of material strength through surface heating and distortion due to thermal differentials. To keep costs within limits it was decided to opt for an aluminium alloy, in this case Hiduminium RR.58 under British nomenclature and AU2GN to the French. Flight and static tests showed that, to avoid overheating and the resultant 'stretching' of the skin, the speed of Concorde would be critical. It was determined that a limit of Mach 2.05, a reduction from the original 'never-exceed' speed of Mach 2.2, should be adhered to. A white paint finish reduced heat absorption by 10%, an important saving when the aircraft was intended to enter an atmosphere exceeding boiling point at 60,000ft (18,288m). As it is the skin heat limit is set at 400 degrees K (127 degrees C), RR.58 proving well able to stand up to the condition of sustained flight at Mach 2.2-plus.

Europe pressed ahead with its SST, and the fruits of Anglo-French collaboration produced a relatively small airliner powered by a quartet of unsuppressed Rolls-Royce RB.211 military engines (the most powerful available at the time). The prototype flew on 2 March 1969. An elegant, unmistakable new shape in the sky, it had an impressive performance including acceleration of 0 to 201mph (0 to 323km/h) in 24 seconds. But Concorde soon ran into hastily-concocted restrictions for its supposed detrimental impact on the environment. British Airways and Air France, the only long-term operators, circumvented these by agreeing to restrict dashes at supersonic speed over areas of ocean, but otherwise offered a unique service at speeds of Mach 2 (1,336mph/2,150km/h), linking London and Paris with New York in record time.

Challenged briefly by the Russians with their similar but ill-fated Tupolev Tu-144, which first flew on 31 December 1968, Concorde went on alone to carve its own niche in world air travel, setting new standards for speed. By 2001 Concorde's airframe life had been estimated at another ten to fifteen years, one reason being that less corrosion is evident when operating at an altitude of 55,000ft (16,765m), where the atmosphere is dry.

High-flying spies

During the Cold War, surveillance by Western aircraft of areas of the Soviet Union proliferated and a range of high-flying aeroplanes such as Lockheed's ingenious jet sailplane, the U-2, were developed. Optimised to reach 65,000-70,000ft (19,812 to 21,336m), the U-2 and the derivative TR-1 represented a weight-saving programme that resulted in an 80ft (24.38m) wing weighing only 3lb/ft² (14.6kg/m²), a third of that of a conventional jet-aircraft wing. Pressurisation was dispensed with, the U-2 pilot using instead a full-pressure suit.

The General Dynamics adaptation of the English Electric Canberra, reworked as the RB-57F, was capable of operating at 64,000ft (19,507m). Lockheed then put its considerable experience into the awesomely sophisticated Lockheed SR-71, which enabled the USAF to maintain its lead in this field. Capable of reaching altitudes of 80,000ft (24,384m)-plus at speeds exceeding Mach 3 (2,200mph/3,540km/h), the Blackbird blurred

Designed specifically for high-altitude surveillance, the Lockheed U-2 and its derivative, the TR-1, seen here, had very light wing structures and, like the Bell X-1s, were unpressurised, their pilots wearing full-pressure suits instead.

the technological division between the conventional aircraft and the space vehicle. Sealed into a cockpit that resembled a space capsule (it was not actually an escape capsule, but had a conventional ejection-seat system), the two-man crew of an SR-71 regularly viewed the outside world as hues of deep blue and black as their altitude increased. They also they routinely witnessed the skin of their machine glowing red-hot as kinetic energy built up. In military terms the SR-71 proved all but immune from interception by earthbound defence systems, despite the existence of the Soviet MiG-25 'Foxbat', which was designed specifically to counter such extreme-altitude overflights of Soviet territories.

From the English Electric Canberra, via the licence-built Martin B-57, General Dynamics in the USA evolved the RB-57F for high-altitude reconnaissance. Power was provided by two 18,000lb-thrust Pratt & Whitney TF33-P-11 turbofans in the nacelles, plus two 3,300lb-thrust J60-P-9 turbojets in underwing pods. The RB-57F entered service in 1963.

FASTER, FURTHER, HIGHER

A reconnaissance aeroplane with an exceptional all-round performance, Lockheed's SR-71 could exceed Mach 3 and attain altitudes in excess of 80,000ft (24,384m).

Looking more like astronauts than airmen in their high-altitude flying suits, Major Noel F Widdifield, reconnaissance systems officer (left), and Major James V Sullivan, pilot, pose in front of their SR-71.

While fixed-wing aircraft were making headlines with ever more impressive performance figures, the helicopter had steadily been creating new altitude and endurance records. Relatively small machines were built in the West, but in the vast reaches of Russia far larger rotorcraft were required. Size has never actually been a problem in building aircraft, and 12 February 1969 saw the enormous Mil Mi-12 (V-12) helicopter lift off for the first time. Still the largest helicopter ever built by a considerable margin, it was an interesting design exercise but one rejected by the Russians themselves in favour of more conventional layouts. Mil, however, has yet to be surpassed in the sheer dimensions of its rotorcraft.

Helicopters have their own categories under the FAI regulations governing records, and although they clearly have limitations in terms of altitude, this is perhaps their least important attribute in comparison to payload. Height records nevertheless continued to be set. On 23 July 1981 Charles Praether, flying an Augusta A.109A in Pennsylvania, set a new helicopter altitude record with a height of 20,000ft (6,096m). This was subsequently recognised under the FAI's sub-class E-I-d.

As high as aircraft get?

The immense challenges offered by practical manned space travel were taken up by several counties after the Second World War, but only the USA and Russia succeeded in funding entire launch sytems, from booster rockets to space vehicles. Less obviously, other countries contributed significantly to the 'space race' by a high degree of technical input as well as actual hardware. In the meantime, numerous test programmes were undertaken to explore the feasibility of such endeavours, taking into account the physiological effects on the human body and the necessary protection needed for both astronaut and machine. The race to place the first human on the Moon took precedence, and when the USA won, in July 1969, other, arguably much greater prizes still beckoned. Steps towards achieving some of them had been taken by the North American X-15, ten years before the first moon landings.

The contract awarded to North American Aviation stressed that the resulting aircraft was to be advanced enough to provide realistic flight data right on the threshold of space. The X-15 would be a pure research vehicle to explore aerodynamics, the effects of extreme stresses on structures, and those physiological aspects that would determine the equipment required for a reliable life-support system for future astronauts.

The first flight of the X-15 took place on 8 June 1959. As before, the aircraft was air-launched to conserve fuel burn, the mother ship being a Boeing B-52 Stratofortress adapted with a special wing rack to accommodate the X-15's gross weight of 33,000lb (14,968kg). Powered by a Reaction Motors XLR99 liquid-propellent rocket motor providing 57,000lb (25,855kg) of thrust at sea level, the wedge-section-winged X-15 was fitted with a rudimentary undercarriage consisting of a twin-wheel nose oleo and two rear-fuselage skids. Three X-15s were built, and a regular flight programme was established. By 9 November 1961 the goal of the original design speed of Mach 6.04 (4,093mph/6,585km/h) was reached for the first time. This preceded the aircraft attaining its design altitude goal of 246,700ft (75,190m), on 30 April 1962. An unofficial world absolute speed record of Mach 5.92 was achieved on 27 June 1962, but on the 17th of the following month Captain Robert M White took the first X-15 up to 314, 750ft (142,770m), which was recognised by the FAI as a world absolute record for the aircraft's class.

Such extreme altitude records were set during the X-15 programme that the question arose as to where conventional flight ended and space travel began. The Americans neatly solved that conundrum by awarding astronaut's wings to any pilot who exceeded heights of 50 miles (80km) while piloting the X-15. In total, 199 flights were made by the original trio of X-15s built, these representing one of the most successful research programmes in US history. A subsequent rebuild and modification of the second airframe after a November 1962 crash created the X-15A-2. Benefiting from remedial work, a larger XLR99 engine and a fuel load boosted by two drop tanks, the shortlived A-2 programme took flight test altitudes to 354,000ft (107,900m).

While the X-15 was in a class of its own, the Russians continued the development of military prototypes which would ultimately provide data for new aircraft designed to equal and, wherever possible, surpass those built in the West. In traditional Russian fashion these were often tested to the maximum and flown into the record books. On 25 July 1973 the Mikoyan Ye-266 established a new world altitude record by flying to a height of 118,898ft (36,240m).

A similar 'flight to the limit' pre-production evaluation programme for new combat aircraft was in evidence in the West. Like McDonnell's superlative F-4 Phantom before it, the F-15 Eagle went after a raft of world records, and in late January 1975 a specially modified A model aircraft named Streak Eagle set eight time-to-height records. At the end of a sixteen-day programme the aircraft took just 3 minutes 27 seconds to reach 98,425ft (30,000m) from a runway standing start. The Eagle, a fine all-round fighter and the mainstay of the USAF in recent decades, has been the subject of an ambitious development programme that has inevitably turned the aircraft into a 'mud mover' gound-attack type. But other fields of research went very much in the opposite direction. During the late 1970s it was proposed that the F-15A/B could use the highly-capable LTV ASTM-135 anti-satellite missile (ASAT) to shoot down orbiting

Altitudes of 354,000ft (107,900m) were reached by the North American X-15A-2, a substantial rebuild of the second X-15 airframe. Note the two drop tanks, which greatly increased fuel capacity.

probes. To achieve such a kill the F-15 made a zoom climb to 80,000ft (24,384m) and launched the two-stage ASAT which then accelerated to 11,000mph (17,703 km/h) to impact a satellite travelling at 17,000mph (27,359km/h) in an orbit about 320 miles (514km) above the Earth. The idea was the subject of a practical test on 13 September 1985, and the satellite was duly destroyed. Anti-satellite missiles were not entirely new, and the USA had followed a system developed by the Soviets in the mid-1970s. But along with several other programmes along similar lines, the USA decided to shelve ASAT, fearing that a similar weapon could be used against its own satellites in the event of war.

With stealth technology taking tangible form in the Lockheed F-117, which made its first flight as the YF-117A on 18 June 1981, extreme altitude has been somewhat traded for a very low radar signature or radar cross-section (RCS) at less extreme maximum operating heights. The Nighthawk's ceiling is a relatively modest 52,000ft (15,849 m), actually less than that of the B-52, which is explained when one considers that the aircraft is intended to be part of an integrated strike force operating with older, less-sophisticated types. A similar ceiling figure is quoted for the other stealth aircraft in current US service, the Northrop-Grumman B-2. Operating at just 50,000ft (15,850m) the 'Spirit' (first flown on 17 July 1989) also relies on a low detection rate from hostile radars to perform its mission, which is planned around several hi-lo-hi or hi-hi-hi bombing options.

Mention of current military aircraft at the beginning of the twenty-first century brings up a number of projects that have taken on an air of mystery under US Government denials. One is the much-discussed Aurora (which is almost certainly a cover name), a hypersonic spaceplane that has undoubtedly flown and has probably

FASTER, FURTHER, HIGHER

This specially modified McDonnell Douglas F-15, named Streak Eagle, *set eight time-to-height records in January 1975.*

Space Shuttle Orbiter Columbia *was the first of its kind to go into space, being launched from Kennedy Space Center on 12 April 1981 and landing as a winged aircraft at Edwards Air Force Base after a mission lasting 2 days, 6 hours, 20 minutes and 52 seconds.*

been well funded for considerable future development. Any speed and altitude barriers to such aircraft have largely been conquered by extensive research cloaked in the kind of secrecy that kept the F-117 out of the headlines for years. There have, of course, been strong hints despite the official denials. As long ago as 1988 the US press reported that work was proceeding on a stealthy reconnaissance aircraft with a speed of Mach 6 (around 4,000mph (6,440km/h).

There need be little doubt that aircraft like the Aurora are also designed to operate at altitudes far in excess of those achieved by the SR-71. One hint as to the actual existence of highly-capable new aircraft was the almost meek attitude of the USAF to the retirement of the SR-71, long before the end of its useful life. Officially, the story was that the Blackbirds were to be replaced by satellites, but this statement did not hold up under closer scrutiny. If data gathering is to remain classified, a satellite with its highly visible launch, low operating life and easily tracked orbit is hardly the answer. On the other hand, overflights of sensitive areas by piloted aircraft can be planned to be highly unpredictable to defences. As it was also realised that satellites would be a prime target in the event of a nuclear war, the new generation of surveillance aircraft surely exists today. Yesterday's altitude and speed limitations are no longer valid.

There have been numerous historic pointers to the practicality of hypersonic flight, much of it accumulated by the X-20 Dyna-Soar and X-15. The former's value was all but eclipsed by the outstanding performance of the X-15 – and, as everything has to be paid for, by the budget constraints of the early 1960s. In addition, international treaties to limit the military uses of space connived to kill the programme. Such research also ran into the belief that unpiloted vehicles and missiles could do the job of aircraft in a future conflict. However, when more realistic views were analysed, the studies were taken off the shelf, dusted down and reinstated, and with the passage of time some dramatic advances had been made. Among the proposals put forward for spaceplanes powered by rocket and ramjet engines were a number based on modified shuttles.

During the visionary 1960s a class of aircraft studied under the general heading of Transatmospheric Vehicles (TAV) envisaged flight profiles for spaceplanes that went into orbit and could descend to 250,000-350,000ft (76,200–106,680m) for reconnaissance purposes before returning to orbit (at speeds of Mach 26-28). Such performance figures have since been proved to be quite practical, mainly through the results obtained by the space shuttles.

Flying into space
That it was actually possible to blast a winged aircraft beyond the mighty pull of the Earth's gravity field – and, most importantly, to return in it – remained the stuff of science fiction until 12 April 1981. Until then, although the concept was sound, tangible evidence was still lacking. That changed when *Columbia*, the Rockwell-designed Space Shuttle Orbiter, made its maiden flight. Following some important test flights with *Enterprise*, the original vehicle, which was secured to the top of a Boeing 747 acting as the lower component of the 'world's largest biplane' to furnish essential flight data, the programme went into space for the first time.

Columbia and its subsquent counterparts: *Challenger*, *Discovery*, *Atlantis* and *Endeavour*, proved that not only could such a vehicle hitch a ride into outer space strapped to a pair of rocket motors, but that it could be flown back into the Earth's atmosphere and make a conventional runway landing. This opened up immense possibilities for research into an alien environment, promising far-reaching programmes entailing civilian scientists landing on the Moon and working in space laboratories, initially in an orbiting shuttle and later in a fully equipped space station.

Evidence that it was possible to literally fly an aircraft through space, rather than resort to cumbersome re-entry vehicles of the type used in the early Moon landings, was highly significant. And it was also discovered decades ago, in the 1950s. What was also significant was that attaining extreme altitudes of up to 100,000ft (30,480m) and speeds in excess of Mach 5 were theortically feasible, but much of the practical hardware had to await the development of new materials able to withstand the extreme heat levels generated by flights on the 'threshold of space'. Funding such programmes was also difficult; even in the USA, where the availability of billions of dollars has rarely been a problem if a 'national security' justification for a programme can be proven, the budget has not always stretched to funding purely civil programmes. Not that the military has always had an easy ride; changes in the US administration during the extended gestation period of modern airborne systems inevitably bring policy decisions that may be at variance with those of the last political party, and the cancellation of certain projects on the grounds of expense, obsolecence or, conversely, being too futuristic and unproven. As a result NASA and the US military have sometimes been obliged to postpone quasi-space-vehicle research which, although innovative, might no longer be perceived as having any immediacy in the list of government priorities. The somewhat cloudy division between military and civil programmes in the USA has been the subject of speculation and accusation that numerous programmes have been kept secret from the taxpaying public, sometimes to the point of paranoia.

The positive thread running through such speculation, misinformation and doubt is that aeronautical programmes which show merit, and may be awaiting new technology to make them work, are rarely abandoned completely. Numerous advanced programme concepts of the 1950s and 1960s are only now being proven to be sound. The right type of airframe materials, new fuels and greatly increased engine power have become available in the intervening decades.

Protected by a shell of heat-resistant and replaceable tiles to withstand the enormous surface temperatures generated during re-entry, the American shuttle ostensibly carries a useful 55,250lb (170,000kg) payload and can accommodate up to seven crew members. Rockwell's orbiter vehicle offers handling qualities not far removed from that of a conventional airliner once the pilot gets used to the difference in size. Much of the early research into the most practical configuration for a reusable space vehicle was done between 1963 and 1974, when NASA and the USAF embarked on separate programmes involving both unpiloted and piloted lifting bodies carried to altitude slung beneath the wing of a B-52, which has a useful ceiling, in 'G' model form, of 55,000ft (16,764m).

Piloted flights began in April 1969 with the Martin Marietta X-24A. Released at an altitude of 42,653ft (13,000m), the astronaut/pilot had the option of firing a rocket motor or simply gliding back to Earth. The motor enabled speeds of up to 994mph (1,600km/h) to be attained at an altitude of 70,540ft (21,500m). After 2.5 minutes, with its rocket fuel exhausted, the X-24A became a glider.

These tests and a subsequent NASA series with the Northrop ML-F2 and HL-10 were highly successful, but in the event a less-complex double-delta planform was chosen for the space shuttle, one that was about the size of DC-9 airliner — and one that the Russians almost exactly duplicated for their Buran (Snowstorm) Shuttle.

After a false start in 1969, Buran, built by the Molniya design bureau of Moscow, was launched for its first unmanned test on 15 November 1988. Unlike *Columbia*, Buran featured a fully automated landing system but had

The first piloted lifting-body aircraft tested by NASA and the USAF was the Martin Marietta X-24A, which made its first piloted flight in April 1969 after air-launch from a Boeing B-52.

THE UPPER REACHES

no reusable rocket engines. It was carried into orbit by huge Energia rocket boosters, but no piloted flights are known to have taken place before the collapse of the Soviet Union. Similarly to NASA's early 747 Shuttle composite tests, the Russians mounted the Buran on top of a six-engined Antonov An-225, then the heaviest and most powerful aircraft in the world. Another victim of Russia's budget-restricted space programme, only one An-225 was completed, but it gained many world records following its first flight in November 1988. Among these was a cruising altitude of 40,485ft (12,340m), recorded during a flight in which the aircraft captured no fewer than 106 records. The An-225 began flights with Buran aboard on 13 May 1989, the year that the combination appeared at the Paris Air Show. In common with other interesting Russian aviation programmes, Buran fell victim to financial restrictions when the old Soviet Union fragmented.

Without delving too far into the realms of pure space travel, which is really beyond the scope of this volume, it is worth observing that the Space Shuttle has in effect kept piloted flight, most of the history of which has been conducted within the Earth's atmosphere, very much on course right out into space. Rather then render one of humanity's most remarkable and demanding skills totally obsolete, super shuttles and the Hotol and Aurora space-planes mean that flying, as most people understand it, remains very much a part of cutting-edge technology. This means that the human body, with all its shortcomings and frailties, is not, as was long believed, quite the weakest link in the chain. Flying may no longer be quite 'by the seat of the pants', but neither has it been totally replaced by banks of computers. That in itself is surely a comfort to a great many people.

The process need not stop with even the next generation of space shuttle-type vehicles. A hypersonic 'scramjet' powerplant, which operates on the ramjet principle, using hydrogen fuel ignited by oxygen forced into the engine at supersonic speeds, has already been tested. An aircraft so powered will, it is confidently predicted, be perfectly able to carry passengers at speeds exceeding Mach 7.6. To many, flying at eight times the speed of sound will at last bring air travel within reasonable bounds in terms of time saving. It is interesting to speculate exactly where such machines might be going when they enter service! In short, the advances of recent years have demonstrated that the barriers to high-altitude, high-speed flight (which often go together) have gradually been overcome, and that in the twenty-first century there is really no limit, technically speaking, to how high piloted vehicles that remain similar to conventional aeroplanes will be able to fly.

The spin-off of an arms race that stopped short of a Third World War was a huge leap forward in technology, materials and techniques, speeded by a conflict that never was. During the Cold War a man walked on the Moon and remotely piloted space vehicles reached out for distant planets and stars, the race to be first and to stay technically ahead extending far outside the Earth's atmosphere. The end of running the wrong race brought about the right prize; the satisfactory results we see in all aspects of aviation today.

Bibliography

Bowers, P M, *Boeing Aircraft since 1916* (Putnam, London, 1989). A comprehensive history of the aeroplanes produced by this famous US manufacturer.

Davies, J K, *Space Exploration* (W & R Chambers, Edinburgh, 1992). An historical survey of humankind's endeavours to explore beyond the Earth's atmosphere.

Francillon, R J, *Lockheed Aircraft since 1913* (Putnam, London, 1982). The products of this innovative aircraft manufacturer covered in detail.

Gunston, B, *The Osprey Encyclopedia of Russian Aircraft* (Osprey, London, 1995). An unprecedented English-language survey of aircraft built in Russia, from the earliest times to the present day.

Miller, J, *The X-Planes: X-1 to X-29* (Specialty Press, Minnesota, 1983). Complete histories of the USA's famous X-series of experimental aeroplanes; recently comprehensively revised.

Pearcy, A, *Flying the Frontiers: NACA and NASA Experimental Aircraft* (Airlife, Shrewsbury, 1993). Describes the extraordinary variety of aircraft, both purpose-built and specially modified, used in this institution's wide-ranging research work.

Robinson, D H, *The Dangerous Sky* (Foulis, Oxfordshire, 1973). A comprehensive and readable history of aviation medicine, related in layman's terms.

Taylor, J H & Mondey, D, *Milestones of Flight* (Jane's, London, 1983). A useful chronological record of significant worldwide aviation developments and records from 863 BC onwards.

Thetford, O, *Aircraft of the Royal Air Force since 1918* (Putnam, London, 1995). Brief development and Service histories of all the aircraft that have served in

5
Taking the Strain
Ray Whitford

The need for thin wings

German research during the 1940s had shown a clear appreciation of the value of reduced wing thickness for high-speed flight. Thinner aerofoil sections not only have lower drag coefficients, but also delay the onset of drag rise due to shock wave formation. The latter was the main argument in favour of thin sections, which generally, together with sweepback, have been successfully employed on nearly all subsequent high-speed aircraft. Following the introduction of essentially German ideas on sweepback for high-subsonic fighters such as the North American F-86 Sabre, MiG-15 and Saab 29, thickness ratios were held at around 12 per cent because it was thought that the delay in the drag rise did not warrant the structural weight penalty of thinner wings. In addition, a wing of conventional thickness, permitted by the use of sweepback, maintained reasonably high maximum lift and provided room for both leading- and trailing-edge lift devices, together with useful fuel volume and stowage for the undercarriage. At higher speeds such thick sections are impractical because the excessive airflow acceleration over the surface produces velocities that exceed the local sonic velocity and leads to premature shock wave formation. This was delayed by more sweepback, since 30 degree sweep plus 10% thickness is roughly equivalent at subsonic speeds to 40 degree sweep plus 12% thickness.

By the late 1940s it had been conclusively established that thinner sections were essential for still higher Mach numbers. Since the shockwave drag due to thickness is proportional to thickness ratio squared, the demand for aircraft capable of fully penetrating the supersonic flight

North American's F-86 Sabre, represented here by an F-86F, had light-alloy and titanium structure. Its swept wing had a 12% thickness:chord ratio.

TAKING THE STRAIN

The F-86's successor, from the same stable, the F-100 Super Sabre, used titanium in the aft fuselage. Inboard ailerons were employed to avoid aileron reversal, as the wing was insufficiently stiff in torsion

regime caused thickness ratios to tumble even further, being stopped short of structural limits only by increases in sweepback. Thus, with the combination of sweepback, the benefits of thinness were to delay and reduce subsonic drag rise, reduce transonic trim changes, give gradual variation in lift coefficient and extend buffet boundaries and reduce supersonic drag.

The F-86's successor was the supersonic F-100 Super Sabre, having a wing of 45 degrees sweep and 6% thickness and powered by an afterburning Pratt & Whitney J57 turbojet. Its thin wing, however necessary for supersonic flight, was inadequately stiff in torsion and the aircraft required a new form of roll control, which will be examined later in this chapter. The design philosophy of the Mach 2 XF-104 Starfighter (called, by Lockheed's publicity department, 'missile with a man in it') was altogether different. Somewhat similar in layout to the earlier Douglas X-3 in having had stubby unswept wings, the F-104 required a thickness ratio of only 3.36%, giving a maximum thickness at the root of 4.2in (10.7cm) and a leading-edge radius so small as to be sharp. In contrast, other contemporary aircraft of similar top speed, such as the English Electric Lightning, had a wing of 60 degrees sweep and 5% thickness. An alternative configuration was the delta wing, which combines sweep with higher physical wing thickness by exploiting the planform's inherent structural advantage of a low thickness ratio. This was used by Convair for its F-102A and F-106A interceptors, and B-58 supersonic bomber, though the company most associated with the delta is Dassault, whose Mirage series of aircraft has sold worldwide.

Structural penalties of sweep

A swept wing has a greater structural span than a straight one of the same area and aspect ratio. This, together with the greater stiffness needed to counter the bigger air loads generated by the higher speeds, led to a growth in wing weight, to maintain adequate bending strength and stiffness. Demands for yet higher speeds soon began to cause wing thickness ratios to drop and, if no resort to the delta was made, even heavier wings.

Another effect of fundamental importance was intensified by the introduction of sweepback. All wings, straight or swept, are subject to torsion under aerodynamic loads. Sweepback introduced additional and severe torsion because the applied loads on the wing act significantly aft on the wing root. This means that the loads in the rear spar are also increased, since the shear force and torsion reactions are additive, whereas at the front spar they are subtractive. The opposite is the case with the forward-swept wing, in which the front spar is more highly loaded though such a configuration is subject to other, more extreme problems, which are covered later in this chapter. The increased torsion introduced by sweep required more structural material, leading to a further increase in wing weight. In addition, the wing-root torsion box contained within the fuselage is also heavier, since it has to carry the kink loads imposed when the wing loads are transferred to the fuselage. These torsion loads are reduced by the delta planform, in which the rear spar is much less swept.

Wing structural form

The compressive and tensile loading in wing skin panels is proportional to bending moment and section thickness, and inversely proportional to the wing section's second moment of area (itself proportional to thickness cubed). For thin wings there is a special need to keep local centroids of the skin panels as near as possible to the skin planes (i.e. as far as possible from the neutral axis of the wing). Furthermore, torsional stiffness is of primary importance, so it is necessary to place as much material as far away as possible from the axis of twist in order to enclose as big an area within the wing box as possible. In other words, it is better to have the material within the skin than in the stringers. This furthered the trend towards thicker skins.

The fundamental problem of wing skins is to make them carry shear, torsion and bending loads. This was achieved with previous 'low-speed' sections by attaching stiffeners to the skin. But a large amount of riveting reduces structural efficiency: fastener holes raise stress and also reduce the effective area of the material in tension on the lower skin. Furthermore, riveted construction spoils external smoothness. As loads grew, riveting became impractical with the required thicker skins and stringers. As compressive axial loading (or end load) increased further, the trend was towards very thick skins support by vertical shear webs. To use the optimum stringer depth and thickness would have led to impractically deep stringers (they would have got too close to the wing's neutral axis), so an arbitrary limit of 2in (5cm) was set on stringer depth and extra material was put into the skins. Beyond about 5 ton/in (2 tonne/cm) end load there was a radical change in the relative geometry between skins and stringers with a greater proportion of end load being carried by the skin

A very thin, multi-spar straight wing with a thickness:chord ratio of 3.36% enabled Lockheed's F-104 Starfighter to attain Mach 2.

TAKING THE STRAIN

itself. The effect on the rib pitch above 5ton/in was a sudden change from an increasing pitch to a decreasing one, to resist the compressive buckling loads.

Integral construction

Aircraft surface structures must conform to contour, shape, and clearance requirements while supporting specified loads. Generally, these loads include normal pressure loads; overall bending (wing and fuselage bending), which causes axial stresses in surface elements; and overall torsion which puts shear in the skin. Conventional (1940s) stressed-skin wing structure met these requirements with high efficiency in terms of weight. Essentially such structures consist of: (1) the external skin; (2) a system of stiffeners (stringers) that make the skin rigid in places and, together with the skin, carry the major surface stresses; (3) a system of regularly spaced internal supports (ribs) that establish the column length of the skin/stringer combination and also transfer shear load; and (4) at the heart of the wing, the main spars, whose purpose is chiefly to preserve the shape and carry beam shear. Generally there is taper in size and strength of all elements along the span of the wing.

Up to the end of the Second World War external structures were mainly fabricated by wrapping comparatively thin skins over an internal framework and allowing them to spring to shape. The separate elements were joined by riveting to form a built-up structure (skin-stringer panels). By the mid-1950s, with speeds and wing loads rapidly increasing, much thicker skins were needed. These were difficult to wrap and had to be pre-formed by rolling and later by stretch forming. Very high powered rollers were needed for curving the skins, and machining was necessary to provide the required surface finish.

The need for increased structural efficiency and reduced production costs led, in the early 1950s, to the development of large sections of integrally stiffened structures in place of the usual multi-piece skin/stringer construction. As the name implies, integral construction means that complicated sections are formed in one piece, thereby cutting out all joints between them. There are several methods: machining from a slab or billet, forging or rolling as a sheet product, extruding or casting. The immediate advantages of such a structure are the reduction in the number of parts and attachments and the reduction in the number of handling expenses and

Integral machining, using numerically-controlled machine tools, played a prominent part in the construction of the BAC TSR.2 in the 1960s. Small quantities of aluminium-lithium (Al-Li) alloys were also used.

assembly tooling. Other advantages are also obtained: reduced weight, improved surface smoothness and simplified sealing.

Integral construction was required because thin-winged, high-speed aircraft needed very thick skins with internal webs rather than stiffeners. The alternative form of construction involved no new design principles and it had several advantages: inherent stability in compression (of chief interest) due to intrinsic solidity of the single machined component and the consequently thicker sections matched the requirements for increased stiffness. It also allowed more efficient use of material, since skin thickness was more easily tapered to coincide, for example, with the spanwise decrease in bending loads. An additional advantage of integral construction was the ability to combine many functions into a single complex part to fit within the available space. The considerable reduction in number of parts, joints, and fasteners saved cost and time in organisation and paperwork, including stress-approved drawings. The reduced number of parts also gave a reduction in production control problems and reduced cost of administration. The lack of joints, seams, and holes provided good resistance to fatigue, reduced the sources of leaks in, and a good structural and volume-efficient form for integral fuel tanks. It also yielded appreciable weight savings of up to 5% of structure weight, and the high surface finish, free from manufacturing distortions, was beneficial for drag especially at high speed, both of which improved aircraft performance.

That said, integral construction had some disadvantages. It needed a large capital outlay on special equipment, hence overall economics needed careful scrutiny, although, as is often the case, once installed there was a tendency to justify capital expenditure by using a new facility wherever possible, regardless of merit. There were also manufacturing difficulties; huge forgings and extrusion presses were needed, together with complex and expensive close-tolerance skin milling tools. Moreover, double curvatures were difficult to produce, and even single curvature was not easy. Furthermore, the larger and more complex the component, the greater the chance of production errors. It was expensive to scrap large components which in some cases represented only 5% of a large slab of expensive material, the rest having been machined away.

One of the first large-scale uses of integral construction was for the delta-wing Convair F-102A supersonic interceptor. The wing skins started as billets of Alcoa 2024 aluminium alloy. The skins were continuous over each fuel tank bay, those for the rear bay consisting of 8ft x 20.8ft (2.43m by 6.35m) sheets. In view of the fact that the whole of each wing box was a fuel tank, quality control was especially rigorous. All interior surfaces were smeared with a sealant and joints made with Scotchweld, a thermosetting adhesive applied in sheet form. Each tank was then baked at 160°C (320°F) to bond the Scotchweld, after which the access panels in the bottom skin were locked in place and the complete assembly pressure tested to 0.5 atmospheres. The wing was found to be extremely fuel-tight.

For very thin wings with small torsion box areas there was a need to develop structures capable of carrying high compressive loadings and to provide thick shear material, principally to meet aeroelastic requirements which we shall examine later in this chapter. Multi-web construction became superior because the deeper stabilising webs provide adequate support for the skin in compression. The added advantage is that since the webs are spanwise they carried shear loads. The transition from integral construction of skin-stringer panels to multi-web was very dependent on wing thickness. A very thin wing is suitable in multi-web form for loadings as low as 8-10ton/in (3.2-4tonne/cm). From a torsion/shear thickness viewpoint, the multi-web form is best with the section comprising a number of closed cells. The thick sheet of multi-web construction is fairly easily coupled with web tapering because machining is simplified by the absence of stringers. For straight wings (e.g. Lockheed F-104 Starfighter), flutter requirements are more critical, and increased spar boom material was necessary because of the small wing depth.

Aeroelasticity

Higher speeds, thinner wings, stronger materials and more refined design caused a change in concepts that once prevailed on the subject of stiffness. In the past, when an aircraft structure was designed to meet a specific strength requirement, the resultant stiffness was often adequate. Sometimes it was necessary to add some material for more stiffness, to correct flutter or resonant vibration problems, but until the modern aircraft arrived, strength not stiffness was the major concern.

A desirable feature is a wing very stiff in torsion, and

The F-104's multi-spar wing needed more spar boom material to compensate for its thinness and prevent flutter problems.

TAKING THE STRAIN

The wing skins and integral tanks of the Convair F-102A were made from 2000-series alloy, the lack of joints, seams and holes minimising leakages. This was an early major use of integral construction.

this gives great advantage to the stressed-skin wing, but often at the cost of more weight than necessary from the pure strength point of view. There is another consideration which demands great torsional stiffness, and that is the efficiency of the aileron control. If a wing is too flexible in torsion, the application of the aileron twists the wing to such an extent that the rolling moment due to the aileron is largely offset by a moment of opposite sign due to the twist of the wing. In the extreme case the rolling moment will be of the opposite sign so that the aircraft rolls in the opposite direction to that intended. Wing torsion in response to control deflection decreased the available rolling moment as a function of dynamic pressure (inversely proportional to altitude and directly proportional to speed squared), resulting in an aileron reversal speed. This had been a critical design problem for Second World War fighters, but was exacerbated by the advent of much faster jet aircraft with their thin swept wings. For these, the coupling of bending and torsion meant aileron reversal requirements were much more difficult to meet due to line-of-flight twist. Much of wing design is controlled by the requirement for torsional rigidity.

This was precisely the case with the Boeing XB-47. The aircraft had been designed with 35 degree sweep-back and an aspect ratio of 9.4 with a constant wing thickness of 12%. Originally, Boeing had proposed engines within the fuselage but the United States Air Force (USAF) refused to accept this on fire safety grounds. George Schairer, in charge of the design, had the bright idea of using the aircraft's large span to space the six engines in pods on slender pylons well outboard. This satisfied the USAF's concern and reduced aerodynamic interference between wing/pylon/pod, and gained the structural benefits of inertia relief.

During windtunnel tests on around 50 different pod positions, to establish which gave the best solution, it was found that favourable aeroelastic effects were produced with the outer pods near the tips of the wings. Under certain conditions, the weight and aerodynamics of the pod tended to dampen the tendency to pitch-up that other locations produced. But little was known about building such wings when design of the B-47 began in 1945. Indeed, no structural designer had never dealt with such a wing before. Not only was it swept, but it had concentrated engine masses hung below, forward and widely spaced. That it would be very flexible was known, indeed it was intentionally designed to flex with the aerodynamic loads and those coming from the engines. It was by far

Favourable aeroelastic effects resulted from the positioning of the engine pods along the Boeing B-47 Stratojet's wing, which was designed to flex under aerodynamic loads and those imposed by the engines.

the most flexible wing that Boeing had ever attempted.

The problem confronting the wing structure design engineers was that the torsional stiffness requirements needed to be established so that construction of the aircraft could begin. However, there was no previous experience from which to extrapolate. Windtunnel techniques had yet to be evolved to handle the elastic problem, and a complete theoretical solution had yet to be developed. Without a computer there was no hope that solutions to the complicated equations could be solved in a reasonable time. Faced with this dilemma, the project engineer performed a strip integration solution, computing the wing torsion in response to aileron deflection. With these results an estimate was made of the torsional rigidity requirements to keep the aileron reversal speed above the design limit speed. Unfortunately the solution did not account for the fact that, when the ailerons deflected, the wings bent, and when they do, aft-swept wings twist: leading-edge down for the wing bent upwards, and vice versa.

A third approach was undertaken by a few Boeing experimentalists who put together a makeshift steel-sparred balsa wood dynamically-scaled model (similar to that used previously for the Bristol Brabazon) that was set on a spindle in the windtunnel with ailerons deflected and permitted freedom to roll. The tunnel speed was increased until the model's rate of roll started to decrease and then reversed: this was the model's aileron reversal speed. It came quite close to predicting that at full scale, but the test was too crude to be taken seriously, and the theoretical solution came too late to influence the design of the B-47. As a result, the torsional stiffness requirements were underestimated and the actual aileron reversal speed was too low and became a problem for the B-47. During structural testing it was found that the wingtip deflected 35ft (10.7m) between maximum positive and negative loads, and it was realised by its designers that this would have a very serious effect on stability and control of the aircraft. This is an understatement.

Spoilers

What was needed, as also for the prevention of flutter, was a type of control which applies a rolling moment without twisting the wing, in other words a control that applies a change of lift near the centre of lift of the wing instead of near the trailing edge, namely a spoiler. The next Boeing bomber, the B-52, used both ailerons and spoilers for roll control (through the A-F models) and eventually dispensed with ailerons for the G and H versions.

Designers of the F-100 were well aware of the XB-47's roll control problem, so used inboard ailerons that left so little room for flaps on the relatively short-span wing that none were fitted to the earliest aircraft. One of the first fighters to be equipped with spoilers was the Republic F-105 Thunderchief, which had a four-segment system for use at high speed and conventional ailerons for low speeds. Nowadays, both commercial and military aircraft make wide use of spoilers for high-speed lateral control, and conventional ailerons (plus the spoilers) at low speed.

Materials

The history of the development of aluminium alloys and steels, which for some decades were the dominant metallic materials used in aircraft structures, is a catalogue of improvement of those material properties of first concern to the designer. Ultimate strength in particular has been vastly improved over the years, and a value that is one per cent of the elastic modulus (the stress to double the length of the material) is now commonplace in steels and titanium while a similar value is in sight for aluminium alloy. Of course, while materials of high strength will continue to be much sought after, the environment for an aircraft structure is often such that other properties of the material are of equal or even greater concern. Suffice it to say that alloys of steel and aluminium that became the mainstay of aircraft structures in the first half century of flight have far from run their course, and will continue to find a place for some decades to come. However, while recognising a continuing future for the 'conventional' materials, some thought must also be given to the other materials that helped to expand the flight envelope of manned aircraft greatly.

Before any true comparison of the various candidate materials can be made, it is necessary to define the operating envelope of the aircraft. In the 1950s a whole range of aircraft were designed whose performance was to far outstrip their fastest contemporaries. The most significant aspect that distinguishes the design of the structure and choice of materials of supersonic aircraft from those of lower-performance types is the high temperature envi-

Ten per cent of the McDonnell Douglas F/A-18's structure is carbonfibre composite. Aeroelastic wing twist at high dynamic pressure required leading-edge deflection to roll the aircraft. Visible on top of the wing-root leading-edge extension, just aft of the cockpit, is a small fence to reduce buffeting of the twin vertical tail surfaces.

ronment caused by kinetic heating in sustained supersonic flight. Leading edges of all components experience stagnation temperatures (which is a function of the square of the Mach number) such that flying at Mach 2.2 at 35,000ft (10.7km) will cause a leading-edge temperature of around 135°C (275°F). So far as metallic materials are concerned, the advent of commercially-available alloys of titanium made big inroads for use in supersonic aircraft of all kinds.

Aluminium alloys

The first supersonic aircraft, the rocket-powered Bell X-1, was stressed for ultimate loads of +18g/-10g and was built primarily of 75-ST aluminium alloy. It achieved Mach 1.04 in October 1947. However, the flights of the aircraft were so short (measured in a few minutes) and the Mach numbers so low that no problems were encountered with kinetic heating. However, when it came to the supersonic transport (SST), required to cruise for three hours or more at Mach 2, infinitely more care had to be given to materials.

The Royal Aircraft Establishment (RAE) at Farnborough was charged with organising the UK's SST programme, and under its deputy director, Sir Morien Morgan, it formed the Supersonic Transport Aircraft Committee (STAC). After prolonged study of several alternatives, the STAC Final Report of March 1959 recommended an SST to cruise at Mach 2.2 or 1,450mph (2,330km/h), with transatlantic range. Any speed higher than this would have resulted in structural temperatures too high for long airframe life using the well-tried aluminium alloys. The alternatives of steel and titanium posed great problems due to inexperience, cost, weight and, in all probability, a long and troublesome programme. Nevertheless, while the design was evolving the use of titanium or steel continued to be investigated, but aluminium alloy was chosen for three main reasons:

1 It enabled the wealth of knowledge of the properties and fabrication of aluminium alloys to be used.

2 At the time, the manufacturing problems of steel and titanium were substantial.

3 On a lightly loaded structure, like a delta wing, the thinner sheet materials resulting from the use of steel or titanium would probably have necessitated additional stiffening to restore compression stability, and at least partly cancelled out their weight saving.

Even so, the major problem was to find an aluminium alloy that would withstand the sustained kinetic heating. At the time there was a scarcity of specific design data available to the designer to allow assessment of the most suitable structural aluminium alloy for a high-speed civil transport, that would meet temperatures between 100-200°C (212-392°F) for an airframe life of 30,000hr. The

The Bell X-1A attained Mach 1.04 with conventional aluminium alloys. Fortunately its flights were so brief and the speeds sufficiently low that kinetic heating was never a problem.

problem was essentially due to the relatively low range of elevated temperatures of interest coupled with the very long life requirements. Consideration of materials for engines were not on the basis of temperatures much below 200°C (392°F), and engine designers were satisfied if the test periods were up to 1,000hr. On the other hand, the designer of subsonic aircraft, for which there is no kinetic heating problem, was quite satisfied with a knowledge of the room-temperature behaviour of various aluminium alloys.

The alloy finally chosen for the Concorde, as it became known, is referred to as Hiduminium RR58 in the UK (in France as AU2GN, similar to AA 2168). Although not the strongest aluminium alloy, it was chosen because of its superior long-term creep strength at around 120°C (250°F) for 0.1% total plastic deformation in 20,000hr. An Aluminium-Copper-Magnesium-Nickel-Iron-Titanium alloy, Hiduminium RR58, was evolved from an alloy originally developed to withstand the rigours of use in engines. It had been specifically developed for gas turbine engine applications operating between 200-250°C (392-482°F) and was available in many forms and had, by the late 1950s, given satisfactory service as gas turbine impellers, spacer rings, compressor blades and aero engine pistons. Steel, titanium (1.5% of structural weight) or nickel alloys are used in the few areas subjected to very high temperatures.

Metallic honeycomb materials, developed to give light, strong and stiff structures, were used extensively during the 1950s and since. The Convair B-58 Hustler bomber had a multi-spar delta wing with 60° leading-edge sweep, an aspect ratio of 2.09 and aerofoil sections varying in thickness ratio from 3.46 at the root to 4.08 at the tip. The required high fineness ratio elements of this aircraft, which was capable of Mach 2, made extensive use of aluminium alloy honeycomb panels. Most of the outer covering of the aircraft consisted of such panels, having outer and inner alloy skins bonded to a honeycomb of aluminium and glassfibre. In addition to light weight, this type of structure has a smooth exterior surface.

Even today, the majority of the world's Mach 2+ combat aircraft are made primarily from aluminium alloys having a specific gravity of about 2.8 and are dominated by:

1 Damage-tolerant Aluminium-Copper AA 2000 series used in fatigue-sensitive areas such as pressure cabins and lower wing skins, two areas prone to fatigue due to the long-continued application and relaxation of tensile stresses. The standard material is 2024-T3.

2 High-strength Aluminium-Zinc AA 7000 series have been used for internal ribs, frames and undercarriage components. For upper wing skins, which have mainly to resist mainly compression as the wing flexes upward in flight, 7075-T6 (with Zinc and Chromium introduced) was used widely on fighters and commercial aircraft.

It is expected that the use of the AA 2000 series in future fighters will be minimal (though both the General Dynamics F-16 and McDonnell Douglas F-15 used it, in thick plate form, for upper wing skins). The AA 7000 series still seems to have a future, especially with the introduction of the powder metallurgy process which gives the material a much finer structure, though this appears to depend on the success of the Aluminium-Lithium alloys (see later).

Titanium
Titanium was discovered in 1790 by an English cleric, William Gregory, during his studies of black beach sands in Cornwall. Then, in 1932, Dr Wilhelm Kroll, a native of Luxembourg, devised a method of extracting titanium which in 1947 was developed into a commercial production process in the USA. By the early 1950s titanium was regarded as the most radically different new material for aircraft. It offered strengths slightly less than the stainless steel suitable for airframes, but at about 40 per cent of the weight. Furthermore, titanium is a suitable alternative to aluminium alloy when prolonged operating temperatures are greater than the 150°C (300°F) that can be withstood by aluminium without excessive deformation due to creep, or when greater strength is required, albeit with a weight penalty. When compared with aluminium alloys it has higher strength/weight, longer fatigue life, much higher temperature strength retention, does not corrode under typical aircraft usage and has better crack growth resistance.

Offsetting its attractive properties is an extreme chemical reactivity, and there were considerable difficulties to be overcome in refining the metal from its ore and forming it into forgings and castings. Moreover, titanium alloys were significantly more (10-20 times) expensive than those of aluminium, and more difficult to machine, drill and cut. These problems delayed its introduction until the demands of aviation bulldozed them aside. Its original use was in the compressor blades and discs of jet engines, though early alloys were used in the aft fuselage of the F-100 of the 1950s. The 'workhorse' Ti-6Al-4V was developed in 1956 in America, and is still the most commonly used form.

Titanium was used extensively for the primary load-bearing substructure on the North American X-15 hypersonic (Mach 5+) research aircraft. The whole exterior skin of this Mach 6.8 aircraft was made from a high nickel-chromium alloy called Inconel X, because skin temperatures could reach up to 650°C (1,200°F). The first production aircraft to be made largely of titanium was Lockheed's YF-12/SR-71 Blackbird high-altitude reconnaissance aircraft, capable of 2,200mph (3,540km/h), at over 80,000ft (25km), where skin temperatures of up to 320°C (600°F) were encountered. The Blackbird's chief designer, 'Kelly' Johnson, recalled: 'Of

FASTER, FURTHER, HIGHER

the first 6,000 pieces of Beta-120 titanium we lost 95 per cent. With the help of Titanium Metals Inc we attacked the problems vigorously, investigating such factors as hydrogen embrittlement, heat treatment procedures, forming methods and design for production. We solved the problems but at considerable cost.' At first B-120 was extremely difficult to cut, and drills had to be reground after every 10 holes; but by the end of the Blackbird's development the alloy was being cut 10 times faster and drill bits were lasting 12 times as long.

Within the space of two decades, titanium was developed from the status of a relatively exotic material to one in which production was measured in tens of thousands of tons per annum. By 1967 nearly 90 per cent of the titanium used in UK went into aeronautical applications, and one third of the weight of current jet engines is titanium. The initially limited use of titanium was due to the high first cost of the material, together with the very high fabrication and forming costs when conventional manufacturing techniques were employed. This explains why, for example, General Dynamics' YF-16 technology demonstrator had virtually no titanium. Hot forming, laser cutting and isothermal forging all helped to reduce cost, but this came after the F-16. The cost of machining titanium parts had been halved by the early 1980s.

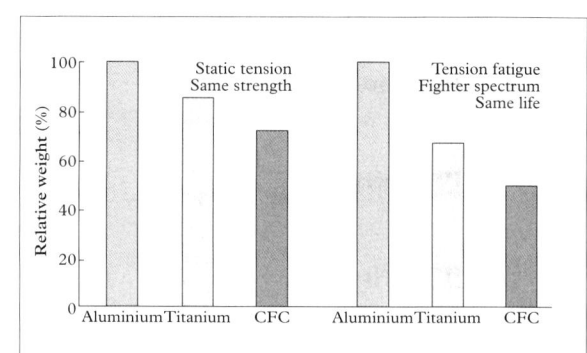

Figure 1. *Composites are structurally efficient in terms of weight. Comparison of aluminium alloy 7075-T73, titanium 6Al-4V and a carbon/epoxy laminate.*

Titanium alloys have subsequently found use in commercial aircraft as high as 10 per cent of structural weight, but their most extensive use has been in combat

Convair's B-58 Hustler bomber, which was capable of Mach 2, had a multi-spar wing with panels made from aluminium honeycomb, and stainless steel honeycomb engine pods. The wing structure represented 14% of the aircraft's take-off weight.

102

Kinetic heating necessitated the use of a titanium primary load-bearing structure and Inconel X skinning on the North American X-15 hypersonic research aircraft, which could comfortably exceed Mach 5.

aircraft. For example, 24.4% by weight of the structure of the Grumman F-14 Tomcat swing-wing fighter is of titanium, used for the wing carry-through structure and upper and lower wing skins. The McDonnell Douglas F-15 fighter and ground-attack aircraft used 26% for lower wing skins and its aft fuselage, and the Rockwell B-1B bomber (21%) used a titanium forging weighing 5,315lb (5.4 tonnes) for its wing box and wing-pivot lugs. Although the use of titanium has been to a lesser extent on European aircraft, the variable-sweep Panavia Tornado has a high titanium content (17.5% of structural mass), including its electron beam-welded wing carry-through box.

Although, on both strength/weight and fatigue/strength bases, titanium has the edge on aluminium alloys and steel, there was no prospect that it would completely displace aluminium alloys (except for high-temperature applications), because strength is not the only criterion. Rigidity is frequently essential, particularly where thin members act in compression, and a reasonable bulk such as is provided by low-density light alloys is necessary to resist buckling.

Super plastic forming and diffusion bonding (SPF/DB)

Many metals (including titanium and aluminium alloys) can be made to deform superplastically at specific temperatures (Ti-4Al-6V will do this at 930°C [1,700°F], about half its melting temperature), when elongations of up to 15 times have be achieved. Complex and deep drawn shapes can be produced in one operation, using relatively low mechanical forces. Although SPF has been well known for over 80 years, the relatively new combined process of superplastic forming and diffusion bonding, a method pioneered by Rockwell using mechanical forming in the 1980s, enables complex shapes to be formed and fastened, giving lighter structures and cheaper fabrication. British Aerospace further developed the technique by using inert gas (nitrogen) pressure of about 20 atmospheres to form thin sheets, in effect by a hot stretching process, whereby material is forced into or over one-piece dies. The monolithic structure of the formed component also adds to its strength.

Diffusion bonding effectively took the monolithic concept further. Under certain conditions of temperature and pressure some clean metallic materials of the same composition will bond themselves (by absorbing their own oxide layers) as a result of diffusion across the mating surfaces held in intimate contact. The process is such that no trace of a joint can be seen on sectioning. The process has been used with Ti-6Al-4V, and the real commercial breakthrough came when it was established that the DB environment for this alloy was identical with that needed to form it superplastically. The results from work in this area have shown that in some applications savings of up to 50 per cent in cost and 40 per cent in weight are attainable. While SPF/DB tools are relatively expensive, the cost is not as great as that of matched tools for hot forming. So far as SPF/DB is concerned, reduction in manufacturing costs is substantially achieved through a reduction in personnel hours that can approach 45%. Reduced weight comes from elimination of fasteners,

FASTER, FURTHER, HIGHER

No less than 21% of the Rockwell B-1B's structural weight is titanium, including its wing box and the pivot lugs for the variable-sweep wings. It also features extensive use of carbon fibre composites – in the tail, engine nacelles and flaps, for example.

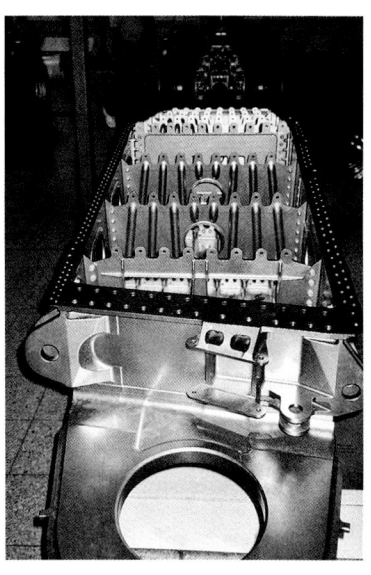

The Panavia Tornado has an electron-beam-welded titanium wing carry-through box.

joint overlaps and the ability to produce more efficient structural forms.

Steels

Where tensile strength greater than that offered by titanium alloys is required, the high-strength steels are still supreme. A range of maraging, high-strength Nickel-Chrome, precipitation-hardened steels are used in critical areas such as undercarriage units and other compact but highly loaded fittings, their use often being dictated not by weight considerations but by the lack of space available.

Low-carbon precipitation-hardened stainless steels such as 18/8 (18% Chromium-8% Nickel) were used in airframes where high strength and excellent corrosion resistance were needed and where high temperatures exist. Such steels have a negligible loss of mechanical properties at 120°C (250°F) and have been used for some supersonic aircraft, though their main disadvantage is a high specific gravity, about three times that of aluminium alloys.

One of the first research aircraft to exploit the high-strength, high-temperature performance of stainless steel was the Bell X-2, designed in 1946 to explore kinetic heating at Mach 3. Compared with the aluminium alloys used in the Bell X-1, the X-2's materials were quite exotic. Stainless steel was used for the wings and tail unit, with tapered skins and stringers to give maximum strength at minimum weight. Control surfaces incorporated spot-welded assemblies where practical. The fuselage was made of K-Monel, an alloy of nickel and copper with high thermal conductivity and an ultimate tensile strength twice that of 18/8 stainless steel (but heavier), to handle the high temperatures caused by kinetic heating. K-Monel could stand up to 400°C (750°F), whereas the best aluminium alloys of the day begin to soften and lose strength before 150°C (300°F).

The Bristol Type 188 (named after the 18/8 stainless steel) was designed in the late 1950s as a research vehicle for a possible Mach 3 interceptor, at a time when America had the monopoly on titanium technology. Firth Vickers supplied the steel for the skins and produced the forged engine mounting rings that were to be machined out to the required thickness between internal ring frames corresponding with multiple inner and outer wing attachment lugs. The external structure followed the normal pattern of thin skins, with local buckling at high loads between the attached stiffeners. This was no different from the early forms of aluminium alloy structures, and the design analysis was much the same. A very successful method of joining sheet to stiffener and sheet to sheet, as an alternative to riveting, was developed at Bristol. This was a technique of welding by controlled local fusion of the materials to be joined beneath an arc struck from an electrode surrounded by an inert atmosphere of argon gas. The first of the two prototypes was rolled out in April 1961, though by this time the company's Bloodhound Mk II surface-to-air missile was performing so well in firing trials that doubts were raised regarding the need for a Mach 3 interceptor and the project was cancelled.

Another largely stainless-steel aircraft was the North American XB-70 Valkyrie Mach 3 strategic bomber, which first flew in September 1964. The majority of the XB-70's structure had to carry flight loads at skin temperatures of about 250°C (475°F), while some areas of the inlet duct and leading edges reached temperatures as high as 330°C (630°F). Steel honeycomb sandwich construction met the XB-70 requirement for low structural weight, high structural efficiency, insulation and excellent surface smoothness. The most widely used structural material in the XB-70 (68% by weight) was PH15-7Mo stainless steel honeycomb, used for most of the wing and lower fuselage as well as a large portion of the upper fuselage. Titanium alloys, of which 12,000lb (5,400kg) was used in the XB-70, formed most of the non-fuel-containing structure. The entire 61ft (18.6m) long forward fuselage (including the foreplanes), the aft fuselage in the engine compartment area and, to a lesser extent, the internal structure of wings and control surfaces, were of titanium alloys.

Composites

The demand for low structural weight and high stiffness requirements, especially for fighters, have often exceeded the capability of conventional aluminium alloys, requiring resort to titanium ones. To achieve better specific properties (strength/weight and stiffness/weight) the aircraft industry has been, since the mid-1940s, the spearhead for the development of composite structures employing laminates of very fine aligned fibres of glass, Kevlar (an aramid fibre), or as discussed below, boron and carbon. Composites bring enormous opportunities because of their low density, combined with high strength and stiffness, and excellent fatigue properties. The small-diameter, high-strength, high-modulus (stiffness) fibres provide the basic strength while the matrix, typically of resin, stabilises the tiny fibres and acts to redistribute the load in shear between fibres. But the application of advanced composites required new design philosophies and manufacturing technology compared with traditional metal structures.

The Grumman Corporation in the USA was early in the field with boron filaments, using them with great success for the skins of the F-14 tailplane in the 1970s. Boron however, had the disadvantage of high cost and, at 0.0038in (100mm) diameter, was too thick to form sharp corners. By 1966 Johnson, Phillips and Watt at RAE Farnborough had perfected a technique for the production of carbon fibres with highly satisfactory characteristics and 0.0003in (8mm) diameter.

Composites give the aircraft designer use of the essentially anisotropic nature of the reinforcing filaments (whose properties are concentrated along their length) that could be made, with a bonding epoxy resin, into laminates with the filaments laid in varying direction (the most common being orientations of 0°, 45° and 90°) to produce the panels required to withstand varying stresses from whichever direction they occurred. The directional properties of these composite panels can be used to provide specified deformations for lifting surfaces under air load, so-called aeroelastic tailoring, to improve aircraft performance. Tailoring could be used to obtain a lower weight design that satisfied all of the applicable design constraints such as strength, flutter and divergence. It could also be used to obtain specified wing twist and camber and so have beneficial effects on aerodynamic drag, control effectiveness and air load distribution, leading to increase in payload or range.

Sponsored by the Defense Advanced Research Projects Agency, the forward-swept-wing (FSW) Grumman X-29 was built to test the theory that tailoring the directional stiffness properties of composite wing skins could achieve the required torsional stiffness at much lower weight than was possible with conventional metal construction. The aerodynamic advantages of the FSW were claimed to be considerable. But, whereas aft-swept wings wash-out (decrease angle of attack with upward bending) so loading is reduced with no structural divergence, the opposite is true of FSW. Divergence is a very serious problem with a FSW built from metal, since upward bending under load induces wash-in with conse-

Designed to explore kinetic heating at Mach 2, the rocket powered Bell X-2 here displays the effect of extreme temperatures on its paint finish after a flight in which it reached Mach 2.87. The canopy's fully-tempered glass could withstand 500 degrees Centigrade.

quent load increase, leading to structural failure unless a severe weight penalty is accepted. With carbon fibre composite (CFC) skins varying in thickness from 0.4-0.2in (1.1-0.5cm), the substructure of the X-29's wing was a mixture of titanium and aluminium spars and ribs. Ground testing of the wing to 100% design load within the full 8g envelope showed good agreement between predicted and actual strains. Furthermore, the flight test programme was very successful and verified the optimum composite skin thickness distributions, ply orientations and bend/twist coupling technology for the aero-elastically tailored wing.

Compared with 2000 and 7000 series aluminium alloys, CFCs offer mass savings of 20-30% due to their remarkable specific properties. However, the resulting structures have been much more expensive than their metal counterparts, due in part to the expensive raw materials, and the fact that the major emphasis was on maximum weight reduction. To accomplish this objective the design approaches have concentrated on structural simplification, reduced part count and the elimination of costly design features. Composite materials are ideal for structural applications where high strength/weight and stiffness/weight ratios are needed, though the benefits depend very much on detail design. A further advantage is the ability to mould complex shapes, which also reduces waste material, machining costs, and the number of parts in complex structures by a factor of about three, so reducing joining costs.

The development of the Harrier into the AV-8B Harrier II by McDonnell Douglas and British Aerospace gave the first major production opportunity to use CFC. The aircraft, which first flew in 1983, had 25% of the airframe structure weight and 50% of its wetted area made of composite materials. A major use was in the forward fuselage, the whole of the one-piece wing and the tailplane. The cockpit had 88 separate parts and 2,450 fasteners, compared with 237 parts and 6,440 fasteners in the metal AV-8A.

Since the 1980s CFCs have been increasingly accepted, and airliners have used a growing amount of them (up to 16% structure weight), though this has lagged behind trend-setting fighters. The Eurofighter Typhoon, for example, has 40% structure weight from CFC. This is due to cost being a more important consid-

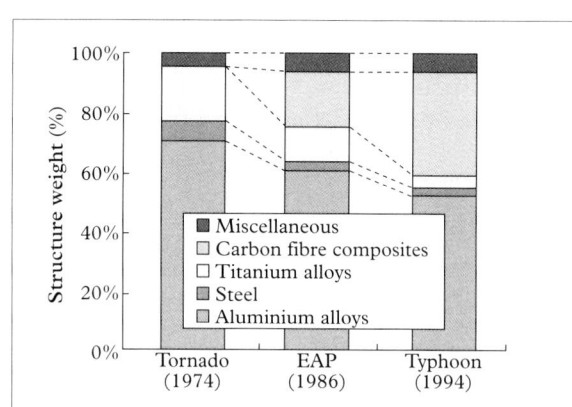

Figure 2. *Material usage of three European aircraft shows the relatively large use of titanium in the Tornado variable-sweep aircraft, the declining use of aluminium alloys and increased use of carbon fibre composites, with year of first flight.*

Two hypersonic North American aircraft with the US National Aeronautics and Space Administration at Edwards Air Force Base in the 1960s, serving as research aircraft. The XB-70 Valkyrie Mach 3 strategic bomber employed a welded stainless steel and titanium honeycomb structure; the X-15A-2 has an ablative coating and a pair of large external fuel tanks.

TAKING THE STRAIN

eration for airlines, and a general conservatism deriving from greater concern for safety and preparedness to adopt a more wait-and-see policy. Damage tolerance is still proving a problem for all composites, however. The vulnerability of brittle materials to impact and to stress concentration is not a new problem, and it occasionally catches designers unawares, even for metals, most recently Aluminium-Lithium (see page 108). Low-velocity impact may be invisible from the outside, that is, it will be either internal delamination or 'back face' tensile failure. Stress concentrations are also extremely dangerous for materials whose through-thickness strength may be only 1/40 of its in-plane values.

In the context of 'Faster, Further, Higher', while the advent of aluminium alloys significantly extended the speeds that aircraft could potentially attain (compared with wooden structures), CFC has not reset the limits, since the maximum temperature is currently similar to that of aluminium. However, it is true that the introduction of CFC has resulted in significantly lighter structures and cheaper (at least theoretically) to produce aerodynamically optimised shapes. This means they can fly further on the same amount of fuel and carry a greater load in proportion to their structural weight. Therefore, though the adoption of CFC has led to a revolution in the way aircraft are made, it has not exactly revolutionised their performance. Furthermore, despite the increasing use of CFC, especially for thin fighter wings where stiffness is the prime consideration, many parts of future

The carbon fibre composite skin and spars in the Eurofighter Typhoon's multi-web wing.

The dark grey areas on this unpainted McDonnell Douglas AV-8B Harrier II reveal the use of carbon fibre composites in the wings, tail surfaces, forward fuselage and wing fences.

FASTER, FURTHER, HIGHER

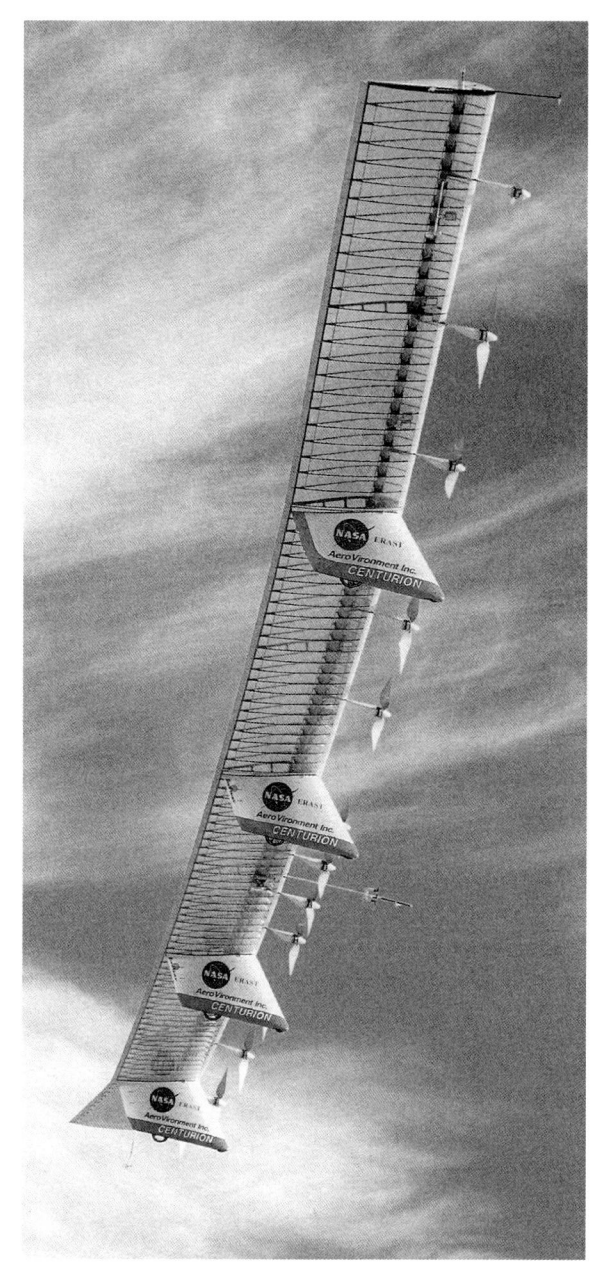

The extraordinary remotely piloted, solar-powered Centurion, designed, built and operated by AeroVironment Inc of California for NASA's Environmental Research Aircraft and Sensor Technology programme, makes extensive use of carbon fibre composites in its light yet stiff structure. It is designed to operate at altitudes between 90,000 and 100,000ft (27,400 - 30,500m).

The first major use of aluminium-lithium in the UK was in the BAe Experimental Aircraft Programme (EAP) demonstrator, seen here during handling trials in mid-1987. Carbon fibre composites were used for the wing.

fighters will be made from metallic materials. This is especially the case where loading is locally intense, for substructures, regions experiencing high temperatures, and for components where the inherent toughness and ability to withstand damage of metals is needed (e.g. leading edges).

Aluminium-lithium alloys

The most efficient means of reducing structural mass is to reduce the density of the material, rather than increasing strength, stiffness or toughness, though these measures would allow further weight reduction. A material density reduction of 10% gives an equal weight saving. This far outweighs the saving gained by similar proportional increase in mechanical properties: a 10% increase in strength gives only a 3% weight saving. The great attraction of Aluminium-lithium (Al-Li) alloys is its reduced density, about 10% less than that of typical AA 2000 series alloys, whereas it has about the same strength and is 10% stiffer. Al-Li alloys are not entirely new. Indeed, the North American A3J Vigilante of the 1950s and the BAC TSR.2 of the 1960s used such alloys in small amounts. There were problems in developing the alloys, particularly with scatter in short transverse fracture toughness and with unexpected crack growth directions, but these have been largely overcome.

During the late 1980s a number of Al-Li based alloys were developed which met these targets to varying degrees and were manufactured on a semi-commercial basis. Considerable quantities were supplied in various wrought forms to numerous aerospace companies for evaluation. However, for various reasons only two of these, the general-purpose and damage-tolerant AA 8090 and AA 2090, survived to reach commercial status and incorporation in various aerospace products.

In 1986 AA 8090 flew almost simultaneously on both sides of the Atlantic. British Aerospace used the alloy to replace AA 2014 for the flaperon skins and undercar-

The upper wing skin panel of Boeing's X-32 contender for the Joint Strike Fighter (JSF) competition comprised hundreds of layers of composite tape. The JSF wing skins were the largest thermoplastic composite parts ever produced.

riage doors for its Experimental Aircraft Programme (EAP). Al-Li can be specially processed to sheet possessing the necessary properties to allow superplastic deformation, and the EAP's undercarriage doors were produced in this way. The finished components were part of the stressed skin and therefore a Class 1 component, but were 20% lighter, half due to AA 8090 and half due to SPF. The number of parts was reduced from 96 to 11 and the number of fasteners from 1,466 to 540, giving a saving in weight, manufacturing cost and complexity. It was planned to use the AA 8090 alloy as a replacement for the AA 2014 in the major wing carry-through frames of the Eurofighter Typhoon, though this has not occurred to date. The first American aircraft incorporating AA 8090 to fly was the McDonnell Douglas F-15 STOL/M demonstrator. By using AA 8090 to replace AA 2014 for the upper wing skins, component weight was reduced by 9%, strength increased by 5%, and the fatigue crack growth rate was reduced by a factor of 2-3 times.

Primarily because of the high cost of lithium, safety precautions in casting, the need for scrap segregation and handling, and closer control of processing, product costs are more than three times those of conventional aluminium alloys. But the advantage is that manufacturers can use existing equipment to machine it and workers need no special training. Airbus already uses about 1,100lb (500kg) of Al-Li on the wing leading edge of each A330 and A340, saving about 110lb (50kg) per aircraft. Furthermore, on the A340 some two tonnes of the material with a weight saving potential to grow from the initial 440lb to 2,400lb (200kg to 1,100kg) if all applications materialised, would mean that Al-Li use would match the use of composites within the A340's structure. It is known that Airbus is giving serious attention to Al-Li in the A380 and A400M.

It was also found that lower levels of lithium, when present with specific levels of other elements, can have beneficial effects on certain other properties. This has led to the development of another new Al-Li alloy, AA 2195, in which the inclusion of lithium was not made primarily for weight saving, and any inherent density reduction was secondary. Alloy AA 2195 is based on findings that lithium additions to aluminium impart cryogenic properties, and with the additional presence of copper and silver, can lead to very high strength. This particular alloy was used for the manufacture of new external cryogenic fuel tanks for the US Space Shuttle and replaces the conventional alloy AA 2219. Furthermore, AA 2195 can be welded, a feature not commonly found in high-strength aluminium alloys.

Bibliography

Dorey, G, 'Materials development for future aerospace structures', *Aerospace* May 1989.

Erickson, R A, 'Flight characteristics of the B-58 Mach 2 bomber', *Journal of the Royal Aeronautical Society*, November 1962.

Gordon, J E, *The New Science of Strong Materials* (Penguin, London, 1986).

Heath, W G, 'The structural effects of kinetic heating', *Journal of the Royal Aeronautical Society*, November 1959.

Heppe, R R, Melcon, M A, and Stauffer, W A, 'Structural design philosophy for the supersonic transport' ICAS-64-585.

James, D, 'The use of high strength aluminium alloys', *Journal of the Royal Aeronautical Society*, August 1966.

—— 'Kinetic heating in relation to structural integrity', *Journal of the Royal Aeronautical Society*, February 1961.

Johnson, C L, 'Some development aspects of the YF-12A interceptor aircraft', AIAA-69-757.

Kennedy, A J, 'Looking ahead in aeronautics-14 The prospect for materials', *The Aeronautical Journal*, January 1969.

Megson, N J L, 'Materials--Metallic and non-metallic', *Journal of the Royal Aeronautical Society*, January 1961.

Middleton, D H (Ed), *Composite Materials in Aircraft Structures* (Longman, 1990).

Niu, M C Y, *Airframe Structural Design* (Conmilit Press, 1991).

Peel, G J, 'Advanced materials for aerospace', *The Aeronautical Journal*, December 1996.

Raha, J E, 'Grumman's forward swept wing feasibility studies and X-29A technology demonstrator', SAE-82-1454.

Ripley, E L, 'Structural tests for the supersonic transport aircraft', 11th Anglo-American Aeronautical Conference, September 1969.

Ross, J W, & Rogerson, D B, 'XB-70 technology advancements', AIAA-83-1048.

Sandorff, P, and Papen, G W, 'Integrally stiffened structures', *Aeronautical Engineering Review*, February 1950.

Satre, P, 'The supersonic air transport-True problems and misconceptions', AIAA-69-759.

Schairer, G S, 'Pod mounting of jet engines', Fourth Anglo-US Aeronautical Conference, 1953.

Schleicher, R L, 'Structural design of the X-15', *Journal of the Royal Aeronautical Society*, October 1963.

Shaw, H W, 'Recent developments in titanium', *Journal of the Royal Aeronautical Society*, August 1966.

Smith, A F, 'The current status and applications of Aluminium-Lithium alloys', Airworthiness Aspects of New Technologies, RAeS Seminar, 1996.

Strang, W J, & McKinley, R M, 'Concorde in service', *The Aeronautical Journal*, February 1979.

Whitford, R, 'Bell X-2', *Air International*, August & September 1996.

6
Propulsion
Bruce Astridge

The end of the Second World War came at a time of transition in the aerospace industry. During the war, aerodynamically advanced airframes had achieved performance levels at which compressibility problems were encountered, and the ensuing loss of control had at times resulted in fatal accidents. Since the most unexplained incidents arose when aircraft were totally destroyed by diving into the ground, specific controlled tests were carried out with Supermarine Spitfires, Lockheed P-38 Lightnings and North American P-51 Mustangs to investigate the phenomenon and determine the problem to be addressed, if performance boundaries were to be extended. Since propulsion powers did not exist to carry out such tests in sustained level flight, the programme had to entail both diving of aircraft from altitude to defined or possible recovery levels and various tests of models, both aircraft mounted and free-fall.

The power limitation resulted from both propeller (and hence propulsive) efficiency reduction at high speeds, and from piston-engine performance limits set by size and complexity. However, by the end of the war new alternative propulsion systems in the form of both turbojet and rocket engines were becoming available, without associated propeller and propulsive efficiency deficiencies. The ability to achieve operating speeds into the transonic regime, and the capability to investigate fundamentally the whole problem, was imminent.

A further legacy of the period was that of a large and developing transport fleet. It was here that the propeller and the piston engine would continue to occupy the 300-400mph (480-645 km/h) speed range, to make a valuable and extended contribution to the establishment of a modern air transport industry.

The last large piston engines

With the cessation of hostilities in 1945, the piston engine continued in military service, but in ever more specialised areas such that the main influence on engine requirements was the transport and civil industry. The piston engines that developed into, and dominated, this era of peacetime use were typified by those summarised in Table 1.

These powerplants included variously the main technological advances proven during the war.

The Bristol Centaurus featured the replacement of the poppet valve with the sleeve alternative, and so relieved the engine of the hot cylinder head of the more conventional configuration. The prize realised from the consequent reduction in pre-ignition effects was the ability to

The 2,820hp Bristol Centaurus typifies the final development of the air-cooled radial engine in the complexity of its cooling-air ducting. It is untypical in its incorporation of sleeve valves, which confer the neat cylinder heads.

Table 1: Large piston engines

	Bristol Centaurus 661	*Pratt & Whitney Wasp Major R-4360*	*Wright Turbo-Compound R-3350-30*
Configuration	Two-row, 18-cylinder radial	Four-row 28-cylinder radial	Two-row, 18-cylinder radial, compound
Capacity, litres	53.6	71.5	54.9
Valve system	Sleeve	Poppet	Poppet
Max. rating, SL, hp	2,820	3,520	3,250
Weight, lb (kg)	3,170 (1,437)	3,520 (1,597)	3,300 (1,497)
Applications	Airspeed Ambassador, Blackburn Beverley, (Sea Fury)	Boeing Stratocruiser, C-97 Stratofreighter, (Boeing B-50)	Lockheed Super Constellation, Douglas DC-7

FASTER, FURTHER, HIGHER

The Lockheed Constellation installation of the 3,250hp Wright turbo-compound, exhibiting the cleanest possible installation of a complicated engine. The exhaust turbines inherent to the compounding of this engine discharge their exhaust air rearward of the cooling flaps at the 12, 4 and 8 o'clock positions.

Lockheed's L.1649A Constellation was the aerodynamic and aesthetic epitome of the post-Second World War piston-engine airliner, and the principal exponent of the Wright Aeronautical Corporation's compound engine.

increase pressure ratio for power and economy. Features such as the deletion of continual tappet adjustments made the engine easier to maintain.

The Pratt & Whitney Wasp Major was notable for its large capacity, more than 71 litres, but also for its configuration by which its frontal area was restrained, 28 cylinders in a four-row configuration.

The Wright Cyclone R-3350-30 featured an ambitious application of compounding, by the use of exhaust-driven blow-down turbines which, via fluid drives, supplied power back to the crankshaft. A very significant improvement of about 20% in engine fuel consumption relevant to the base engine was realised from this compounding. The complexity of the engine was responsible, initially, for some poor overhaul times and lack of reliability, but considerable improvements were made before the engine was displaced from first-line operation by the jets.

It is notable that all of these engines were air-cooled. The liquid-cooled engine (such as the commercial ver-

PROPULSION

The First Whittle engine, the WU, in its original configuration, with a double-sided centrifugal compressor, a single-stage axial turbine (with short connecting shaft) and single combustion chamber in a large volute.

sion of the famous Rolls-Royce Merlin) failed to make the transition successfully from wartime to postwar commercial or long-range use. Other complex – or monstrous – liquid-cooled engines were producing remarkable powers, but none of them competed ultimately with either the air-cooled radials or the turbojet. Engines in this category included the impressive Napier Sabre H-24, which was yielding more than 4,000hp, the Rolls-Royce Eagle of the same configuration of 3,500hp and, typically of the monstrous unlikely, the Lycoming XR-7555 of 36 cylinders, 127 litres capacity and 5,000hp – with a target of 7,000hp. Applications for such projects disappeared as the turbojet took over for high-speed research and military applications.

The genesis of the jet engine

It is impossible to consider the origins of the jet engine without reference to the foresight and efforts of (Sir) Frank Whittle in the UK. As a young RAF officer he patented an early engine design in 1930. He was convinced that it was possible to design and develop components (particularly compressors) of high enough efficiency for an operable engine, and he also recognised that such a powerplant could be superior to any propeller unit at speeds above approximately 500mph (800 km/h).

This can be understood fundamentally in consideration of the efficiencies of the process. Any powerplant produces thrust by converting the heating value of fuel burnt into an increase in kinetic energy, and the efficiency of this process is termed thermal (or internal) efficiency. As with all heat engines, the level of this efficiency is dependent on cycle pressure ratio and combustion temperature, as well as on component efficiencies. The efficiency of conversion of kinetic energy to propulsive work is termed the propulsive (or external) efficiency, and this is affected by the amount of kinetic energy wasted by the propelling mechanism. This waste can be expressed as $W(V_J-V_P)^2/2g$, where W is the mass flow passing through the engine, V_J is the jet velocity at the propelling nozzle and V_P the airspeed of the aircraft, and the velocity to which the jet dissipates externally. It can be seen that this loss is reduced if the jet velocity tends to the aircraft velocity. At low aircraft speeds, therefore, the use of a propeller for low jet velocity maximises propulsive efficiency, but at higher speeds, where the aircraft speed increases towards the jet velocity, the simple jet waste velocity relevant to ambient reduces, giving propulsive efficiency closure between the processes. In practice the propeller itself has its own losses at high airspeeds, where velocities across its aerofoils become transonic, and this compounds the closure.

The equations of the process can be reduced to the following simplified and familiar form which expresses the principle clearly:

$$\eta_p = \frac{2}{1 + V_r/V_p}$$

In summary: at low airspeeds the propeller accelerates a large mass of air to low velocity, and is optimum for best propulsive efficiency, but at higher speeds (500mph plus) the simple jet exhaust can be optimum, due to its own improving propulsive efficiency and the avoidance of propeller inefficiencies.

The use of a fan to absorb some of a turbojet's kinetic energy to increase airflow and reduce jet velocity results similarly in an improvement in propulsive efficiency. Hence, the turbofan or bypass engine.

The early jet engines

The most significant programmes in the definition of early jet engines were those in Great Britain, Germany and that burgeoning in the USA.

In Britain, Sir Frank Whittle favoured the centrifugal compressor over the axial in his early work, despite his original patent showing an axial/centrifugal configuration. The next five years were very frustrating, in that no interest from industry could be raised and the ministry remained indifferent, to the point of refusing even the tiny funding for patent renewal at the end of its five-year term. Nevertheless, Whittle persisted and achieved some backing for the formation of a company, Power Jets Ltd, to build a test engine, the WU (Whittle Unit), which featured a double-sided centrifugal compressor, a single large combustion chamber curling round the engine, and a single-stage axial turbine. After damage to the rotor caused by a casing foul, the engine was run successfully in April 1937. Failures and rebuilds followed until the engine, the appearance of which came to signify the classical Whittle configuration, demonstrated a more practical layout with multiple (=10) reverse-flow combustion units.

FASTER, FURTHER, HIGHER

An order for a flight engine was gained and, following appropriate development, this finally flew successfully in the Gloster E.28/39 on 15 May 1941. By this time the official attitude to Whittle's engine had changed, and an order was placed for eighty Gloster F.9/40 fighters with W.2B Whittle powerplants. The scenario that followed was somewhat tortured, with the Ministry of Aircraft Production (MAP) first appointing Rover to build Power Jets engines, and later (1943) sanctioning a Rolls-Royce takeover of an effort that was by then faltering. The Whittle engine was then developed to a robust and very satisfactory family of centrifugal engines.

Whittle had favoured the centrifugal compressor for robustness and ease of development to adequate efficiency, and the success of his programme helped cement the centrifugal engine as standard for early British jet aircraft. However, the axial compressor was not neglected in Britain to the extent that this visible preponderance suggested. At the Royal Aircraft Establishment (RAE) Dr A A Griffiths had progressed from his work on axial turbine blading to propose, in 1929, the development of a very complex gas-turbine-driven propeller unit, a turboprop. This proposal featured both contrarotation and contra-flows, and was rejected at this time, but with sanction for some more conventional axial compressor testing and some combustion research.

Metropolitan-Vickers in Manchester had a background in axial steam turbines, and was brought in to build the components of an industrial gas turbine power unit which ran by late 1940. At this point Metropolitan-Vickers became aware of Sir Frank Whittle's work, and elected to redirect effort to what became the F.2 turbojet. This engine flew in the Gloster F.9/40 Meteor, and was developed into a very satisfactory 4000lb-thrust engine, the F.2/4 Beryl for the Saunders-Roe SRA.1 flying-boat fighter. Metropolitan-Vickers also used the F.2 as a basis for test-stand work on a ducted-fan version, almost doubling its thrust to 4,600lb (2,090kg) and making a major cut in specific fuel consumption of up to 60%. An open-rotor version of the ducted fan (called a thrust augmenter in those early days) was also tested. Both options were ahead of their time and were not pursued.

However, when Metropolitan-Vickers was persuaded to leave the industry in 1948 as part of industry rationalisation, it was able to pass on to Armstrong Siddeley Motors the initial design of what became the Armstrong Siddeley Sapphire. This was an eminently successful turbojet engine with a very acceptable axial compressor.

The third build of the Whittle WU. Although the configuration of the basic rotating machinery was unchanged, the engine's appearance was transformed by the adoption of multiple reverse-flow combustion chambers. The thrust capability of the WU was typified by the only measurement made on the similar second build of the engine; 480lb (218kg).

Prime German jet engine production in the Second World War centred on the Junkers Jumo 004 and this engine, the BMW 003A, both of which were of axial configuration and of modern appearance. They had thrust ratings of 1,980lb and 1,760lb.

In Germany, a pioneer of Sir Frank Whittle's stature managed to gain initial support of which Sir Frank could only dream. Hans Pabst von Ohain had studied physics at the University of Göttingen, and at the end of 1935 he patented his design for a turbojet. This design featured a centrifugal compressor, a combustion chamber and a radial inflow turbine. After some experimentation with a model, von Ohain wrote to and obtained substantial backing from Ernst Heinkel, the aircraft manufacturer, and by 1937 he was able to run a thrust-producing rig. By August 1939 a flight engine, although producing only 380kg (838lb) of installed thrust, flew in the He 178 research aircraft. A limited flight programme was completed, which gave sufficient confidence in the new form of propulsion for the design of a fighter, the He 280, to begin. It also encouraged some very diverse German projects from Heinkel and, eventually, from the established engine companies BMW, Junkers and Daimler-Benz. It is perhaps a denial of this diversity of approach that the most significant German engines that were to appear in service late in the war were the BMW 003 and the Junkers 004. Both were what could be described as classic simple axial-compressor engines of quite modern configuration and appearance. Each had axial compressors, a cannular or annular combustor, an axial turbine and a variable nozzle to assist engine matching and handling throughout the flight envelope.

The German designers had chosen to accept the challenge of the more difficult development of efficient axial compressor, to pursue the benefits of smaller frontal area and the potential of high efficiency. Both engines were in service at the end of the war, particularly in the Messerschmitt Me 262 and Heinkel He 162 fighters and the Arado Ar 234 reconnaissance bomber. Overhaul lives and thrust development were limited by the shortage of

Powered by Jumo 004 engines, the Arado Ar 234B Blitz entered service in 1944. Prototypes of the planned Ar 234C flew later with four BMW 003s. The slim configuration of both engine types enabled the use of aerodynamically clean nacelle installations with straightforward pitot intakes.

nickel, and hence high-temperature materials, in Germany, but the axial engine had arrived alongside the centrifugal in Great Britain.

Industry in the USA was slow to start, but made great strides once the significance of the new propulsion systems was realised. In 1941 the chief of the Army Air Staff, General Hap Arnold, recommended a NACA study of jet propulsion and also made a fact-finding visit to the UK. This resulted in a licence agreement for access to Whittle's work and permission to use it in military and civil aircraft. Also, the Power Jets W1.X bench engine was sent to the General Electric Company in Massachusetts, where it supplied the datum for the J33, the 4,000lb (1,815kg) design that the Allison Engine Company took responsibility for in 1945.

The axial engine was not neglected. Nathan Price at Lockheed was working on an impossibly complex two-shaft axial-compressor engine, but it was the realisation by GE of its greater future potential that lead to the I-40 engine, identified as the J35 in its Army/Navy guise. The engine developed 4,000lb (1,815kg) of take-off thrust initially, and featured an eleven-stage axial compressor, tubular combustion chambers and a single-stage turbine.

Thus, by 1945, the centrifugal compressor engine was established as a robust basis for the gas turbine industry in Great Britain and the USA, and the axial engine was proven in Germany and its potential recognised in all counties.

Early high-speed projects
The onset of compressibility effects on Second World War high-performance fighters, and some accidents caused by buffet and control loss, suggested by 1943 that something needed to be done to understand and overcome the adverse effects of the phenomenon. It was also recognised that the transonic region (Mach 0.75-1.3) was difficult to study or visualise in conventional windtunnels of the time, owing to the most radical and complex physical changes taking place in this region. In fact, shock waves forming from the model reflect from the walls back to it (a phenomenon called choking) and make data recording and interpretation very difficult.

Nevertheless, Germany engaged in a research programme of extensive windtunnel work and development of wing shapes (e.g. the swept configuration). As a result, a multiplicity of advanced aircraft projects appeared in the later war years. In the USA and Great Britain recognition of the windtunnel problem resulted in the acceptance of the necessity for some full-scale investigations to supplement such research information as could be learned.

In Great Britain this resulted, as early as 1943, in the issue by the MAP of specification E.24/43. This somewhat uncharacteristic, revolutionary specification called for an aircraft capable of flying through Mach 1.0 to

Initial flight testing of the projected Miles M.52 was to be made using a 2,000lb-thrust Power Jets W.2/700, but the ultimate powerplant for this high-speed aircraft would have been an early application of the ducted fan or bypass engine, with reheat in the fan duct from built-in combustion chambers.

1,000mph (1,610km/h, or approximately Mach 1.5) at 36,000ft (10,970m). The Miles Aircraft Company was favoured to build the aircraft as the company's chairman and managing director, F G Miles, had already demonstrated an impressive record for lateral thinking. Frank Whittle, inevitably for the time and also because of his early advocacy of such a programme, was selected with Power Jets to produce a suitable powerplant. The RAE and National Physical Laboratory (NPL) were to supply support.

Miles evolved, from ballistic studies, a bullet-shaped aircraft with an elliptical planform straight wing with clipped tips. The wing thickness:chord ratio was 7.5% at the roots and only 4.5% at the tips, and it was designed to operate supersonically within the V-shaped shock wave from the aircraft nose. For early subsonic trials it was planned to use the Power Jets W.2/700 centrifugal engine. It was recognised (in 1943) that its 2,000lb (908kg) of thrust would result in a very limited flight envelope, but to extend it the engine was to be fitted later with an aft fan thrust augmentor, and combustion cans in the duct for afterburning, or reheat. Progress with the project was steady, with slow-speed testing of the wing on a Miles Falcon light aircraft and windtunnel investigations. By early 1946 90% of the detail design was complete, assembly jigs were ready and the component assembly programme well under way. The augmentor fan had also been built. Then, in February, the bombshell struck. The Director-General of Scientific Research at the MAP, Sir Ben Lockspeiser, sent a note to F G Miles, telling him that it had been decided to stop all work on the project. The project was still secret at this time and for some months to come, and the reasons for the cancellation given over this period were variable and unconvincing.

PROPULSION

The only technical excuse arose from a belated recognition of the research in Germany being made available from intelligence surveys of the industry after the war. This was interpreted, mistakenly, as implying the need for wing sweep for supersonic flight. Propulsion concerns do not seem to have figured, especially as back-up powerplants in the form of both jets and rockets were becoming available for different versions of the aircraft. Lack of belief in any imminent supersonic flight, economy and concerns about pilot safety were all expressed and the result was terminal. All tools were scrapped, the mock-up broken up and an opportunity to investigate supersonic flight, along with early experience of a ducted fan engine and reheat, was lost.

The American parallel programme began with an aerodynamically similar vehicle and grew to a massive and extensive research programme running to this day. In 1935 the International Fifth Volta Congress on High Speeds in Aviation had been held in Italy. Theodor von Kármán was one of the attending aerodynamicists whose belief in supersonic flight was reinforced by this conference, as was also his belief that the USA should initiate a broad programme on the subject to achieve a forward leap in aircraft performance. Several other prominent engineers at the US Army's Wright Field (Ezra Kotcher) and NACA (John Stack) contributed to the definition of a programme and a vehicle so that, by 1943, with new propulsion systems becoming available, the whole concept was beginning to look practical. It was suggested that the programme should be initialised with the military supplying the funding, the industry carrying out the development and NACA doing the flight testing, the basic definition for the massive on-going X-Plane programme. The industry was already involved, with Bell Aircraft Corporation contributing to the debate and decisions. The culmination of all this activity was a contract, on 16 March 1945, for the X-1 aircraft. By this time propulsion studies had decided that, with the state of current gas turbines and the lightness of rocket units, the greatest probability of success would be achieved by use of a rocket engine, the Reaction Motors XLR11. This was a four-chamber liquid-oxygen and ethyl alcohol engine with 1,500lb (680kg) thrust available from each chamber and the ability for individual firing of each. There was no other throttle control. Three X-1s were built, and the first aircraft went supersonic, with Chuck Yeager at the controls, on 14 October 1947. Although the third, the most advanced, was destroyed before it could make a powered flight, the first two completed an extensive programme to Mach Numbers above 1.2. The rocket system conferred very limited periods of powered flight and therefore, except for one specific take-off test, which was successful, the aircraft had to be launched from a Boeing B-29 or B-50. It did, nevertheless, prove that a manned aircraft could fly safely above sonic velocity, and with powers that were now becoming available from gas turbines as well as rocket engines.

Other high-speed programmes were being pursued around the world, particularly in the USSR (where there was a strong tradition of rocket research) and France, where one of the more unusual aircraft configurations was pursued in the search for speed, using a ramjet as its primary propulsion unit. The ramjet is a mechanically simple (and aerodynamically complex) continuous-duct jet propulsion unit with few moving parts. Air compression is by diffusion of the inlet air in an intake duct to subsonic velocity for combustion of the fuel, and expansion to ambient conditions for thrust through a nozzle. The ramjet, therefore, produces no thrust statically and must be accelerated to the order of 300mph (480km/h) to start operating. Thus any vehicle so powered must have auxiliary, or supplementary, means of propulsion, or be air-launched. It is then satisfactory to approximately Mach 6.0, where compression (in decelerating the air to subsonic velocities for combustion) is so great that the pressure level becomes unmanageable and dissociation of oxygen molecules begins to take place, with massive heat losses. However, initial programmes in France with piloted aircraft were not so ambitious, with simple exceeding of the speed of sound only the initial target. An early pioneer of the use of the ramjet was René Leduc, who designed his air-launched 0.10 aircraft in 1935-37 and flew it after the war (1947), when it reached only 500mph (805km/h). The 0.21 and 0.22 followed, the latter including an ATAR turbojet for take-off and acceleration, but this was not until the late 1950s, when alternative aircraft were well established with the Mach 2.0 performance target. No Leduc ramjet managed to exceed Mach 1.0. The Leduc aircraft looked very futuristic, with fuselages forming a giant duct around the centrebody housing the pilot. No production or really significant research aircraft emerged from the programme, but it was an interesting exploration into the third option of gas propulsion.

The second generation – the turbojet

The postwar gas turbine industry was founded on much of the pioneering work done in Great Britain and, indeed, Pratt & Whitney, the most dominant manufacturer from the late 1950s, was given its head start by means of licence production of the Rolls-Royce Nene and Tay centrifugal engines.

It was realised, however, on both sides of the Atlantic, that the progress to more efficient engines would realise the benefits of potential leaps in range and speed. The axial compressor was recognised as the better prospect for installation, on account of its lower frontal area; but also because it had the best prospects for component efficiency. Also, it was recognised – it was only too obvious – that improved overall performance would be enhanced

FASTER, FURTHER, HIGHER

by improvement in cycle efficiency as well, and it is fundamental that this would mean increases in overall pressure ratio and combustion temperatures. Given a solution to handling and starting problems unrelated to the propulsion unit, the axial compressor would be capable of the increased pressure ratios, simply by adding stages. Multiple centrifugal compressors to the same end were clumsy, with losses in their convoluted flow paths. (Against this, the Dart turboprop has been successful for many years; but other factors including design for both performance in other areas of the engine, and for reliability, have dominated).

In 1945 the most significant engines in Great Britain were of centrifugal configuration. Rolls-Royce, however, (despite the success of the Nene and Derwent) was well into the design of an axial Avon engine, or, as it was known at the time, the AJ65. The compressor handling was initially poor, and required a great deal of effort and some help from Armstrong Siddeley, the inheritor of Metropolitan-Vickers experience. During the programme, in 1949, a compressor with five stages of variable stators was built as the first demonstration in the world of this technology; but it was not used in production, where variable inlet stators and bleed valves proved adequate. The Avon went on to power many research aircraft and to a long life in both military and civil service. The first production engine, the RA.3, entered service at a take-off thrust of 6,500lb (2,950kg) 1950. By 1956 the RA.24 with a 15-stage axial compressor and air-cooled turbine blades was available at 11,250lb (5,100kg) thrust. The commercial RA.29 Avon for the de Havilland Comet and the Sud-Aviation Caravelle, with a further 'O' stage on the compressor and a third turbine stage, gave a take-off thrust of 12,600lb (5,720kg). Its compressor was operating at a pressure of more than 8.0.

In the meantime, at Bristol, after some initial work on a turboprop called the Theseus, the Engine Division of the Bristol Aeroplane Company decided to initiate design work on a large turbojet for a future bomber. The brief for the engine included specification of a compound, or

The second major redesign of the by now very successful Rolls-Royce Avon engine running on a test stand. Its basic take-off thrust was 9,500lb, and for the early flight testing of the Fairey F.D.2 it was fitted with reheat and a simple two-position eyelid-type nozzle.

two-spool, arrangement, with an axial low-pressure (LP) compressor and a centrifugal high-pressure (HP) unit. A target thrust of 9,000lb (4,085kg) was established and an overall compression ratio of 9.0 specified, in an attempt to achieve economy and range. This pressure ratio was well above current world experience. The centrifugal HP compressor proved incompatible with the axial LP compressor of six stages and with associated turbine dimensions, and an axial HP compressor of eight stages was therefore adopted. This took place after some delay due to the demise of the bomber project, and the first engine ran on 6 May 1950. It demonstrated very good aerodynamics and some less-desirable mechanical characteristics. Nevertheless, this was the first two-spool engine to run, and its basic premise was soon proven. Development of the engine to a very good mechanical standard was rapid, and the engine proved to have remarkable potential, albeit with considerable design changes, for development to more than three times its original design thrust. The ultimate Olympus powers the Concorde SST and produces 29,300lb (13,300kg) of take-off thrust dry. It has also seen very extensive development and use as power for electricity generation sets and marine propulsion.

Pratt & Whitney had carried out some turboprop design work, and production of the J30 from Westinghouse, at the end of the war. Then, however, the US Navy bought the Rolls-Royce Nene, which was Americanised and produced under licence by Pratt & Whitney as the J42 by October 1948. The Rolls-Royce Tay, as the J48, followed logically. Pratt & Whitney was looking concurrently for the next forward step of technology, and from abortive high-pressure single-spool designs progressed to the JT3 (or J57 military) engine.

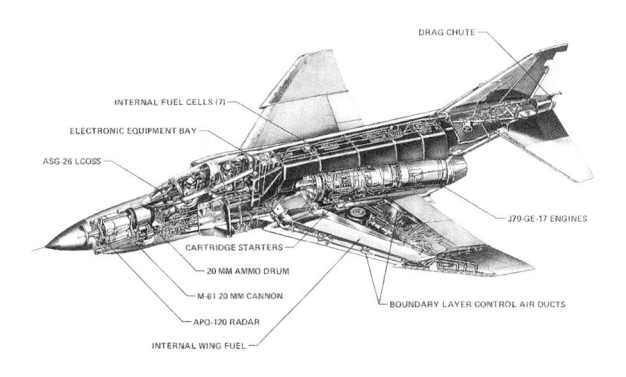

The original 1940s target thrust for the Bristol Olympus was 9,000lb. The first production Mk 100s entered service in the Avro Vulcan at 11,000lb thrust, and the next major development, the Mk 200 series, powered the Vulcan B.2 at 17,000lb. The ultimate Olympus, the 593, seen here, powers the Anglo-French Concorde supersonic airliner. Its thrust of 29,300lb without reheat has been achieved by extensive redesign and development, but its basic two-shaft turbojet configuration remains unchanged.

Several redesigns were demanded to realise the benefits of this engine, but it emerged finally in a two-spool configuration which gave a high overall pressure ratio of 12.5. During 1951 it powered Boeing's YB-52 (which its advent had saved from being powered by turboprops) and, with reheat, the North American YF-100A. It went on to power the initial versions of the Boeing 707 and Douglas DC-8. This engine was probably the most significant of its generation, and powered many types of aircraft, military and civil, in both its original turbojet guise and later as a turbofan, of which it formed the basis.

General Electric (GE) pursued the same goal in a different way, and with a different but equally successful result. In this period GE was seeking to build on the very large production success of its axial single-spool J47 engine. The objectives included fuel economy at cruise conditions (assumed to be subsonic) and high thrust at supersonic conditions. High pressure ratio was therefore relevant, and GE decided to take a serious look at a compressor with sufficient stages for the target pressure ratio, and with variable stators to make it operable over its working range. It was quickly concluded that this was preferable to the two-spool alternative, in that it would be lighter in weight, and the decision was made in late 1952 to go for a seventeen-stage compressor with seven stages of variable stators. The new engine flew under a North American B-45 Tornado bomber in May 1955, in the Douglas XF-4D Skyray in late 1955, and in a supersonic installation for the Convair XB-58 bomber in November

The successful alternative to the two-shaft configuration for higher compression ratios, the variable-stator compressor, was proven by General Electric in its highly successful J79 turbojet engine, seen here in the ubiquitous McDonnell Douglas Phantom II. Seven of its seventeen compressor stages featured variable stators to enable handling at all engine speeds.

Fairey F.D.2

One Rolls-Royce R.A.5 Avon turbojet with afterburner
Span, 26ft 10in; length with typical probe, 51ft 7 1/2in; height to tip of fin with main undercarriage oleos under load, 11ft 0in; track, 91.6in; gross wing area, 360sq ft.

This cutaway, drawn by *Flight* artist Leonard Clow in 1958, portrays the essential features of the Fairey F.D.2's structure and primary systems. The small inset diagram just above the cockpit gives an indication of the shock-wave pattern at one of the main engine intakes at a Mach number of about 1.4. The brake parachute system, which was streamed from the container numbered 33 on the main drawing, is shown at a slightly reduced scale compared with the cutaway.

Most highly-swept or delta-wing aircraft adopt a marked nose-up attitude at low speeds. This fact, coupled with the F.D.2's long nose and shallow canopy, led the Fairey designers to provide the aircraft with a downward-hinging nose. The mechanism for this is depicted in the adjacent drawing.

The operating jack A turns a transverse torque-tube upon which are mounted two cranks B; these are attached via a pair of ties C to the nose section D, which is hinged at E. In the 'up' position the nose is locked by latch-pins at F. The nose undercarriage leg G is retracted to the rear by the jack H, which pulls up the cross-member J; the doors are operated by the two jacks K, which are also energised when the nose is moved to its 'down' position.

Key

1. Compass detector unit
2. Gaseous oxygen
3. Communications radio
4. Whip aerial
5. Oxygen valve and charging point
6. Access to radio, undercarriage selector, brake control valve and oxygen valve
7. Access to flying controls and air thermometer 7A
8. Cabin pressure ground socket
9. Pressure head
10. Canopy jettison (external rescue)
11. Canopy release latch
12. Brake accumulator
13. Flying-control rods and drooping-nose linkage
14. Hydraulic header tank
15. Micronic filters
16. Fuel filler
17. Access panels to equipment on top decking
18. Access to main leg hinges
19. Access to flying controls, rudder bar and trim actuators
20. Access to rudder pedal hinges
21. Electric starter
22. Accessory gearbox drive via universal-jointed shaft
23. Cabin air-conditioning equipment (see system diagram)
24. Cabin cooling-air outlet (fed from lower boundary-layer duct)
25. Access panels to engine mounting
26. Variable-ratio gearboxes (port, aileron; starboard, elevator) governed by electric actuators 26A, and incorporating aileron load-relieving unit 26B
27. Engine rear mounting
28. Elevator power unit (port and starboard)
29. Rudder power unit
30. Aileron power unit
31. Main undercarriage jack
32. Main undercarriage shock strut
33. Braking parachute container
34. Skin stiffener over wheel bay
35. Wing fence
36. Boundary-layer bleed
37. Flying-control rods
38. Variable-area propelling nozzle
39. Afterburner retaining rails
40. Tail bumper
41. Differential lever assembly
42. Electric actuator for airbrake servo valves
43. Airbrake hydraulic jack (four)
44. Transfer-pipe joint
45. Engine-bay ventilation intakes
46. Fireproof bulkhead
47. Horn balance
48. Braking parachute assembly

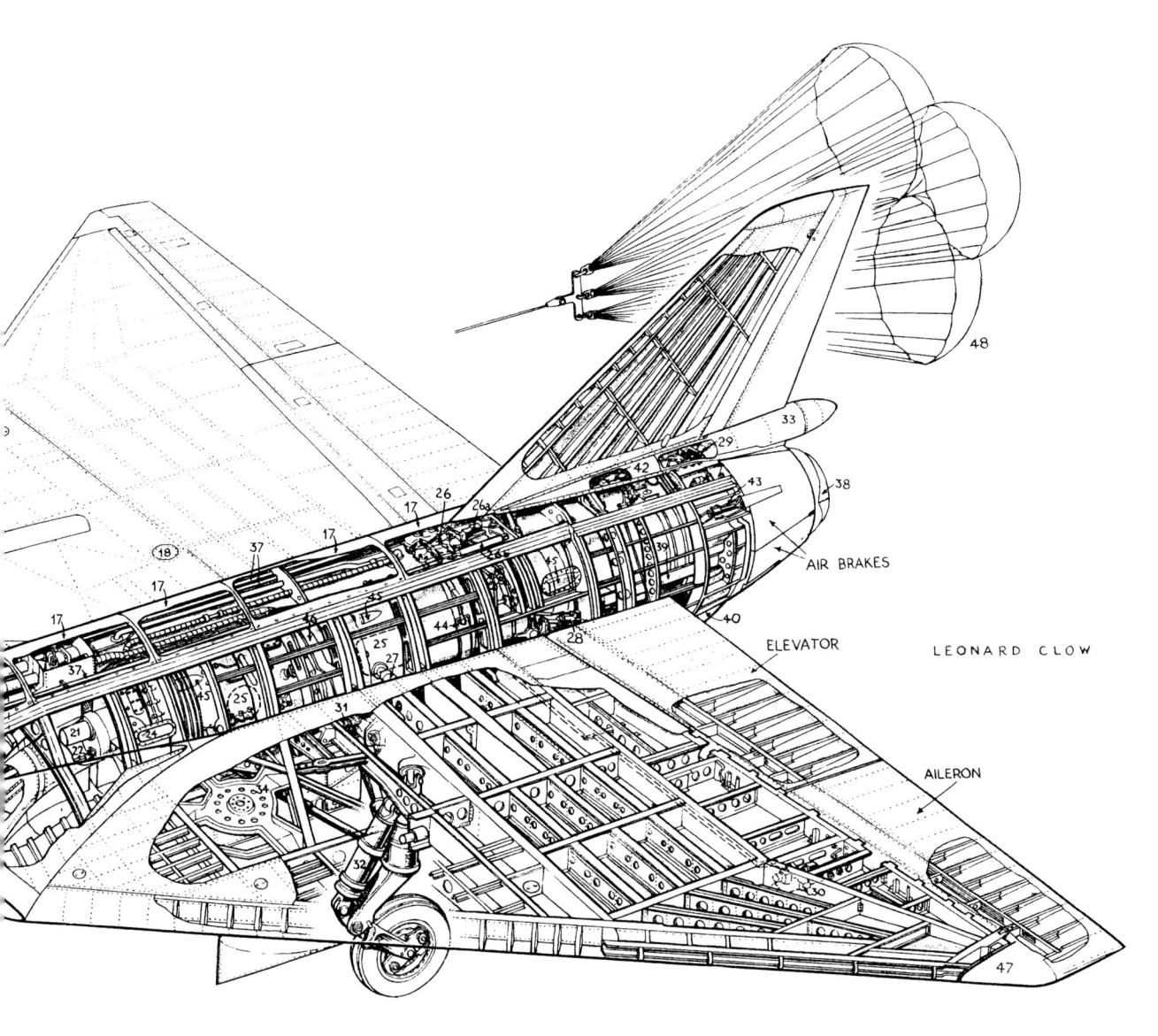

The diagram below shows the disposition of the main units of the air-conditioning system.

1	canopy seal exhaust	10	temperature controller
2	windscreen sandwich ducts	11	gate valve
3	demisting vents	12	venturi
4	punkah louvre	13	cold-air unit
5	thermostat	14	compressor exhaust
6	canopy seal	15	flow augmenter
7	water extractor	16	static vent
8	pressure-ratio controller	17	pressure controller
9	pre-coolers	18	foot distributor vents

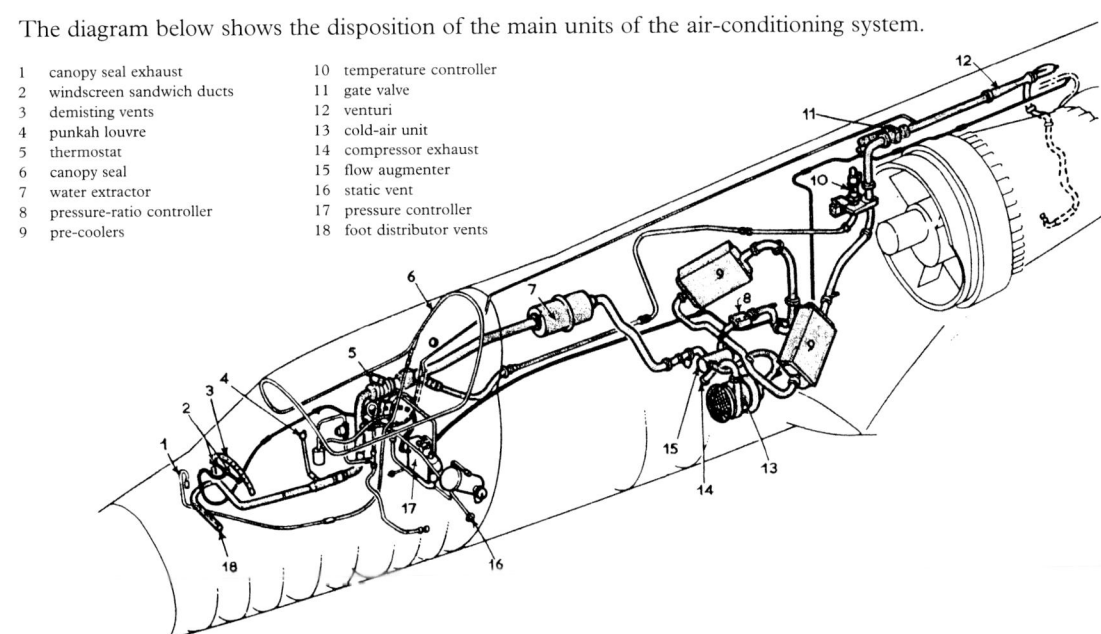

1956. Its success can be measured in its production run of more than 17,000, and its many installations.

Thus, by the mid-1950s, engine technology had progressed to the successful adoption of practical alternatives for high pressure ratios (two-spool and/or variable stators), to the use of reheat for thrust boost and, in fact concurrently, to the adoption of the turbofan or bypass configuration. Without going into detail, it should be noted that parallel developments were being made in other parts of the world, most notably in the USSR, Canada and France. In the first-mentioned it might have been at the cost of engine reliability and engine life (some of it due to different philosophies), but the same technologies were being proved and used.

Supersonic turbojets
Penetration of the so-called sound barrier or, more properly, achievement of sonic velocity, had been achieved by an airframe powered by rocket engines, its thrust:weight ratio being much better than any available turbojet. But the practicality of supersonic speeds was now proven, and gas turbine powerplants of much better thrust:weight ratio were becoming available very quickly. Designers and aerodynamicists were soon to start considering the implications inherent in their engine installations.

Early thinking on gas turbines had recognised the potential of jet-pipe reheat for thrust boost, but getting maximum boost at best fuel consumption required optimisation. Engines for reheat featured a turbine exhaust diffuser to reduce fluid velocity for burning the boost fuel, some form of flame stabilisation frame (gutters) and an increased-sized jet pipe to accommodate the increased temperature and velocity of the exhaust fluid. The propelling nozzle had to be variable so that, when reheat was selected, it opened to give an area suitable for the increase in volume of the gas stream and did not give increased back pressure on the engine to compromise its operation. The variability had to be progressive (usually by multiple flaps with an operating sleeve) if the system included throttle capability, or could be two-position (such as by a pair of eyelid-type shutters) if the reheat was a simple on-or-off system.

The pressure ratio of a turbojet nozzle at high thrust is high and normally choked – i.e. gas velocity at the throat is sonic and a shockwave normal to the gas flow forms across the throat. The nozzle is operated with its exit static pressure higher than ambient, and both momentum and pressure terms contribute to the thrust. The most efficient use of gas energy is to continue the expansion to ambient and increase velocity, and a way of achieving some of this is to use a divergent nozzle section downstream of the convergent throat; i.e. a con-di nozzle. Higher supersonic nozzle-pressure ratios made the use of such a nozzle even more relevant.

The other prime consideration at supersonic flight conditions was air-intake treatment. Compressors require inlet air velocity at entry to be no more than mach 0.4-0.5, and at subsonic conditions this could be accomplished reasonably simply by a rounded lip intake and diffuser. At supersonic speeds a powerful shockwave, normal to the airflow, forms across the inlet and generates losses in pressure. By flight speeds around Mach 1.5 these losses become prohibitive, and it becomes necessary to achieve more efficient airspeed deceleration by the generation of smaller, multiple shockwaves. The loss across the shockwave is a function of the Mach numbers ahead of and behind the shock, and smaller multiple shocks total less loss than that of a large single shock. The prime ways of achieving this are by a shaped centrebody, translatable or not, depending on speed range and control of the losses required, or some form of ramp structure to vary areas in the intake as required for shock control. Further complexity arises in the supply of auxiliary intake doors for supplementary and dump air-flows to match intake to engine requirements at different powers.

Supersonic speeds were, therefore, driving considerations of the powerplant and airframe in a much more integrated manner. To reinforce the need for this, it is worth noting that compression in the intake at Mach 2.0 can be higher than that by the engine compressor. With these considerations in mind, a plethora of high-speed research programmes into the development of turbojets emerged in all countries with major airframe and engine industries.

In Great Britain, the abrupt cancellation of the M.52 was followed, eventually, by the establishment of a new research programme. The programme specification was originally for research studies of transonic flight with swept wings, but it was quickly realised that, to be of value, supersonic flight should be included. Fairey Aviation and English Electric responded with their different proposals and, after some delay, contracts for two aircraft each were awarded to both in 1950.

The Fairey aircraft, the FD.2, was a delta of minimum size for its task, and was powered by a Rolls-Royce Avon with a basic on/off reheat system, including an eyelid-configuration, two-piece variable nozzle. It also featured an innovation that would reappear in later supersonic aircraft, the 'drop-snoot' nose. This was necessary to ensure reasonable pilot visibility at the high angles of attack during landing inherent in the tailless delta-wing configuration. In the case of the FD.2, the droop structure included the cockpit as well as the long nose. The first aircraft flew on 6 October 1954, but, after some further flights, power was lost at 30,000ft (9,150m) due to a system fault and fuel starvation, and test pilot Peter Twiss performed a masterly deadstick landing at Boscombe Down. Damage was minor, but continuation of the flight testing was delayed until August 1955, and the first supersonic flight was achieved in October. By November

Mach 1.56 had been demonstrated, equivalent to more than 1,000mph at 30,000ft (1,609km/h at 9,150m), and thoughts of the world air speed record (at the time held by an F-100 at 822mph (1,323km/h)) started to emerge. On 10 March 1955 the first record of over 1,000mph (1,128mph at 38,000ft (1,815km/h at 11,580m)) was achieved with a Rolls-Royce Avon RA.28 giving 13,100lb (5,950kg) thrust. Meanwhile, the second aircraft had flown, and an extensive flight programme ensued with both aircraft. Supersonic flight was explored down to low altitudes and a programme was flown (in Norway) to investigate intake effects on axial compressors, particularly stage one blade stresses. The first aircraft was eventually fitted with an ogival planform wing and used in configuration research for Concorde. Fairey made several proposals for fighter developments of the FD.2, but the 1957 government postulation of the end of manned fighters for the RAF aborted the whole process. Dassault, in France, was left to show the potential of the delta-wing configuration in its highly successful Mirage III, with a closely equivalent reheated turbojet, the SNECMA Atar 09C.

Meanwhile, the promise of the English Electric proposal had resulted, in 1952, in the conversion of the contract to one including fighter versions of the aircraft. The research aircraft were designated P.1A and the fighters P.1B. The first of the former flew on 4 August 1954, and later in the month achieved supersonic speed, the first British aircraft to do so in level flight. The engines were Armstrong Siddeley AS.SA.5 Sapphires of 8,100lb (3,677kg) thrust and, initially, no reheat. The engines were superimposed and staggered in the fuselage to produce minimum frontal area, and the intake was a fixed pitot type in the nose. The engines were later fitted with reheat and Mach 1.56 was achieved during the test programme. However, the P.1As became less useful when the P.1B fighter version prototype emerged in April 1957,

English Electric's P.1A emerged from the same research programme as the Fairey FD.2, and in this prototype form, with a fixed-pitot-type engine inlet and Armstrong Siddeley Sapphire engines, achieved similar speeds. The design evolved into the highly impressive P.1B Lightning fighter.

as this version had Rolls-Royce Avons with full reheat (with appropriate multi-petal variable nozzles) and a pitot inlet with shaped centrebody. Its power to weight ratio neared 1.0, which conferred it with impressive performance and a long and very satisfactory Service career. However, it was not blessed with large fuel capacity, or turbofan engines, and its range was not impressive.

The other turbojet-powered supersonic aircraft in Great Britain that should be mentioned is the BAC TSR.2. Rather than for research, this aircraft was a fully integrated weapon system and the subject of extensive industry studies. The contract went to British Aircraft Corporation in October 1960. Power for Mach 0.9-1.1 at low altitude and Mach 2.0 at medium altitudes, was to be supplied by two Bristol Olympus 22R engines, the latest version of the two-spool engine that had proved capable of modification and development to 19,600lb (8,896kg) dry, 30,200lb (13,708kg) reheated take-off thrust. The first aircraft flew on 27 September 1954 and performed a short and successful flight programme until 5 April 1965, when the whole programme was cancelled. It can be argued that all this turbojet research and experience was not lost, as the eventual Concorde SST was powered by an even more developed Olympus engine, with fully integrated variable intakes and nozzles with electronic controls.

While Mach 1.0 had been exceeded in the USA with a short-duration rocket-powered aircraft, the necessity for test of sustained supersonic flying had not been ignored. A USAF request for proposal to Douglas as early as 30 December 1943 asked for transonic speed capability for 30min. A contract followed eventually in May 1945. It was specified that the vehicle should take off under its own power, but any combination of turbojets, ramjets or rockets would be acceptable for propulsion. It was not until mid 1949 that the final propulsion system (the J34 turbojet with reheat) was selected. The protracted time was largely driven by propulsion studies owing to inadequate thrust from the various chosen turbojets that were really the only option to meet the sustained-duration requirement. The aircraft featured fixed engine inlets, although the effect of variability was one of the studies made in pursuit of thrust.

The aircraft, the X-3 Stiletto, was as sharp looking and as thrusting as its name implied, and it flew first on 15 October 1952, but no serious testing started before April 1953. It went supersonic in a dive only and failed to contribute significantly to the supersonic research, as the mantle was by then being seized by the rapidly moving fighter programmes, and the designers had confidence that even Mach 2.0 was available with the right technologies.

In contrast, however, two very significant and advanced turbojet programmes began to emerge in the USA at this time, leading to the GE J93 and the Pratt &

FASTER, FURTHER, HIGHER

The Douglas X-3 Stiletto was a most striking aircraft. However, it was thwarted in its intent to explore sustained supersonic flight by delays in development of the requisite small-diameter axial jet engines – and the lack of room for the installation of alternatives.

Pratt & Whitney's J58 powered the Lockheed A-12 and SR-71 Blackbird. Its demanding role dictated the provision of variable geometry and other advanced features such as turbine blade cooling and the incorporation of exotic materials.

Whitney J58 engines. The former eventuated from the October 1954 USAF General Operational Requirement (GOR) No. 38 for an Intercontinental Bombardment Weapon System Piloted Bomber, with a performance to include Mach 3.0 dash capability. Various detail changes were made to the GOR, but by November 1955 design contracts were awarded to Boeing and North American, and in 1956 GE and the Allison Division of General Motors were asked to submit engine proposals. After a fierce competition GE succeeded, and in September 1957 the USAF gave approval to start development of the J93-GE-3. The aircraft requirements then stood at a speed of Mach 3-3.2, a target altitude of 70,000-75,000ft (21,330-22,860m), a range of 6,500 to 10,500 miles (10,460 to 16,900km) and a gross weight of 475,000-490,000lb (215,000-222,260kg).

The first engine sent to the test cell was a J93-GE-1 in September 1958. In March 1958 GE received the go-ahead for the developed J93-GE-3 version, and in July 1958 Air Material Command approved a further growth version, the J93-GE-5. This engine would burn high-energy fuel. The J93-GE-3 went to test in July 1959, and became the first operational engine in the USA to run with air-cooled turbine blades. Originally, GE had wanted to use nickel-plated molybdenum blades, but the USAF ruled this out because of the fear that any hole in the surface would cause the molybdenum blade to oxidise and fail catastrophically. For the air-cooled blades GE developed a process for drilling long, radial cooling holes. The high-energy fuel (HEF) programme was soon cancelled, and eventually (in April 1961) it was decided that there would be no flight programme beyond the prototype North American XB-70. The first flight of the aircraft with six J93 engines was on 21 September 1964, and the first Mach 3.0 flight, with J93-GE-3 engines on 24 October 1965. The second XB-70 crashed after an F-104 chase-plane crashed into its fin, and the remaining aircraft was taken into a joint USAF/NASA programme in March 1967, flying last on 4 February 1969.

The engine was designed as a single-spool turbojet, and the high engine face temperature at Mach 3.0 was recognised by the use of titanium for the front-stage blading and steel in the latter. Two independent groups of variable stators were included for engine handling over the wide range of flight conditions and powers. A forward set was applied in the first three stages to ensure stall margin, inlet-distortion tolerance and low idle thrust capability, and a rear stage set allowed maximisation of airflow for Mach 3.0. The combustor was fully annular for maximum heat release, and the turbine blades used Udimet 700 material and featured the small-hole air cooling and long shanks to remove the turbine disk from the hot stream. The installation featured a fully modulating reheat system with a con-di exhaust nozzle, and the inlet of two-dimensional (rectangular) section featured ramp adjustment for shock control. Although the XB-70 programme was severely truncated, the engine design and development was of inestimable value to GE.

The other impressive US turbojet development was instigated in the late 1950s by the need to replace the

PROPULSION

Lockheed U-2 spyplane with another that would be more survivable in the face of the growing Russian ground-to-air threat. This meant flying higher, faster and further, and the Pentagon was thinking in terms of 100,000ft (30,480m), 2,000mph (3,220km/h) and over 4,000 miles (6,440km). Clarence L 'Kelly' Johnson of Lockheed was confident that an aircraft with such capabilities was possible, and a USAF contract with Lockheed for the A-12 unarmed reconnaissance aircraft followed. North American X-15 experience illustrated the design problems to be surmounted, and also many of their solutions. A new, high-temperature titanium alloy was also available for the main airframe structure, leaving the major unsolved problem as the engine – which had to be air breathing for the sustained flight implicit in the range.

The A-12 flew first with a Pratt & Whitney J75 on 26 April 1962, while the new engine, the JT11/J58, was available from the end of the year, and for service in 1963. Among the new problems to be solved for the engine were the high, and continuous, temperatures inherent in the speed: for fuel at inlet, for lubricants, for turbine blading. Obviously weight was critical, and dictated cycle optimisation for the most important flight conditions, take-off and extended cruise. For take-off, the lightest solution was probably a reheated moderate-pressure-ratio, single-spool reheated turbojet. However, at high-speed cruise the high engine inlet temperature in relation to the turbine inlet temperature would dictate poor cycle efficiency, and the complete bypass of the engine by the airflow was a possible option in order to use the jet pipe as a ramjet. Studies of the possible ways of doing this showed that the bypass ducting and blocking mechanisms necessary to shut off the engine would make for an unacceptably heavy solution. The turbojet therefore had to be retained, with some optimisation for a compromise between the possible modes. But with the engine still operating in the supersonic mode the compressor would be operating at low corrected speeds and would be seriously short of, or completely lack, a margin for surge (the gross breakdown of airflow through the compressor). Variable inlet-guide vanes and stators could alleviate this problem, but would not enable very large thrust increases.

The engineers therefore accepted that some variable geometry was necessary, and evolved the compromise of, at cruise, bypassing air from a mid-compressor stage (stage four) through six bypass pipes both to restore compressor surge margin and boost air into the jet pipe for extra fuel burn and boost. The configuration that had emerged was therefore a turbojet with variable geometry and some very advanced features. The engine turbine demanded new thinking from Pratt & Whitney, and thus the company's previous uncooled, shrouded and tip damped blade was supplanted by an unshrouded convection-cooled blade with root dampers.

The aircraft was highly successful, and gave thirty years of unique service to the USAF. It is noteworthy, however, that the engine was the last turbojet built by Pratt & Whitney. Many high-speed service aircraft still use turbojets, but from now on the turbofan (in its various guises) was to be the dominant configuration in both subsonic and supersonic regimes, and in military and civil usage.

Turbofans

The adoption of the turbofan, or bypass, engine was driven in the civil world by ever-increasing range and fuel efficiency requirements. In the military world the same

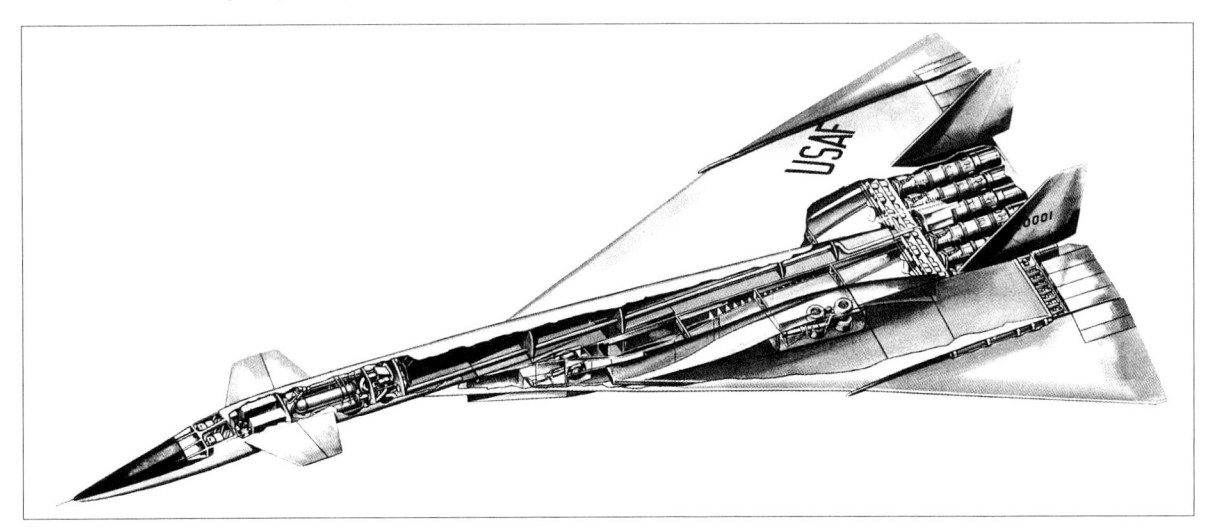

North American's XB-70 supersonic bomber had a battery of six General Electric J93-GE-5 engines between its twin fins. This engine's advanced features included the first use in the USA of air-cooled turbine blades, variable compressor stators blading and provision for the use of high-energy fuels.

FASTER, FURTHER, HIGHER

drivers applied, but the availability of increased reheat boost inherent in the extra jet-pipe airflow was also a very important attribute. Development was driven by the rapid evolution of production aircraft, supported by appropriate detailed research, rather than by specific aircraft research programmes.

The turbofan dates from the early thoughts of Sir Frank Whittle and Dr A A Griffiths in the mid-1930s. Sir Frank was a strong believer in the jet engine for high-speed propulsion, but he joined with Dr Griffiths in the search for high propulsive efficiency, and proposed several configurations of thrust augmentors, so called because of their static thrust increase as well as their cruise propulsive efficiency improvements.

Power Jets was building a Whittle turbofan (the LR.1) when it was nationalised as a design company only, and such work was terminated. Dr Griffiths joined Rolls-Royce and pursued complex contrarotating engines for use in both turboprop and turbofan configurations; and Metropolitan-Vickers bench-tested its turbofan and early propfan open-rotor engines in the mid 1940s.

In the end, considerations of requirements for the British long-range bombers (the V-bombers) resulted in the first application of the principle in production of the Rolls-Royce Conway. In 1952 funding was made available for an engine to incorporate improved thermal and propulsive efficiencies. The former was pursued with a high overall pressure and increased combustion temperature capability, and the latter by adoption of the bypass configuration. The pressure ratio was achieved by designing the engine with a two-spool configuration, as had been pioneered by Bristol, and the combustion temperature capability by cooling of the forged HP turbine blades (another production first). Unfortunately the engine was designed for a buried wing installation in the Handley Page Victor bomber and the Vickers V.1000 airliner, and its bypass ratio was therefore pitched at only 0.3. The Vickers V.1000 was never completed, but the engine was very successful in the bomber installation, and went into service as the RCo.11 for the Victor B.2 and the RCo.12 for the Boeing 707 and Douglas DC-8. Some success was achieved by the installations in the early 707 and DC-8, but the most significant effect was to spur Pratt & Whitney, in the USA, to very rapid and highly effective action in development its competing engine. The Conway went on to power the Vickers-Armstrongs VC10 in its RCo. 42 form, where its new spool of higher flow capacity increased the bypass ratio to a still low, but more respectable, 0.6. The final version for the Super VC10 was the RCo. 43, and its revised eight-stage LP compressor conferred a maximum overall pressure ratio of 15.8. Maximum take-off thrust was 22,500lb (10,212kg).

Meanwhile, in the USA, technological fallout from research on exotic projects (the chemical-fuelled and nuclear-powered bombers) enabled Pratt & Whitney to design an increased-diameter two-stage fan as a substitute for the first three stages of the JT3, and complete its conversion to a turbofan by the addition of a third LP turbine stage and a fan duct to handle the bypass flow. This produced the TF33 which, in its JT3D civil guise, had a bypass ratio of 1.4; well above that of the Rolls-Royce Conway and sufficient to eclipse any further competition from that source. Moreover, Pratt & Whitney was able to supply a kit to enable operators to make conversions of JT3 engines to JT3D standard in the field.

Another military development, driven by the need for

The Rolls-Royce RCo. 11 Conway powered the Handley Page Victor B.2 bomber. The Conway was the first bypass engine to enter service in both the bomber and, in civil guise, in the Boeing 707 and Douglas DC-8 airliners.

PROPULSION

Pratt & Whitney's impressive response to the Rolls-Royce Conway was the JT3D. A larger-diameter fan substitution, the addition of a low-pressure turbine stage and revised cowlings transformed the JT3 turbojet into a short-cowl turbofan of significant bypass ratio.

reheat-boost thrust, was also based on J57 technology. In 1956 work had been done on a J57-based duct-burning turbofan, to study short-length engine possibilities for fighter aircraft. Satisfactory stability and efficiency were not achieved, but when the USAF issued a specification for what became the TFX programme (and the swing-wing General Dynamics F-111), Pratt & Whitney was able to submit a reheated turbofan, the TF30, with a bypass ratio of 1.0. The TF30 project went ahead after it won the 1962 engine competition, and the TF30 eventually became the first reheated turbofan to enter service. Service experience was very poor for some time, but due primarily to lack of compressor surge margins (and tolerance to intake distortion) rather than deficiency in configuration.

Variations and derivations of the J57 testified to the soundness of its design. In 1954 it was scaled down to produce the J52 for the US Navy at a take-off thrust of 7,500lb (3,400kg), and it was this engine, in a manner analogous to the JT3/JT3D conversion, that was modified to produce the highly successful JT8D. Its bypass ratio was 1.0, its overall pressure ratio nearly 17 and its take-off thrust 14,000lb (6,355kg). It was to satisfy the very large demand for the Boeing 727 and Douglas DC-9, and also featured in other installations (Dassault Mercure and Sud-Aviation Caravelle).

There was innovative activity in other areas of the world. The Russians achieved a first in the short-haul market with the entry into service in 1962 of the Soloviev D-20P turbofan in the Tu-124 feeder airliner. The D-20P had a three-stage fan, an eight-stage HP compressor with automatic bleed valves to ensure adequate handling, and a bypass ratio of 1.0.

The Boeing 707 prototype departs Boeing Field with early Pratt & Whitney JT3Ds fitted. The short fan ducts are clearly visible.

FASTER, FURTHER, HIGHER

Table 2: Turbojet evolution

	Great Britain		USA		
	RB.168-25R	EJ200	TF30	F119	YF120
Shafts	2	2	2	2	2
Overall pressure ratio	20	26	17+	$\cong$35	$\cong$35
Compressor stages	17 (5+12)	8 (3+5)	16 (3+13)	9 (3+8)	?
Turbine stages	4 (2+2)	2 (1+1)	4 (1+3)	2 (1+1)	2 (1+1)
Aerofoils	3450	1800	--	--	--
Dry Weight (lb)	3,600	2,200	4,080	3,200	?
Max. Thrust (lb)	21,300	20,000	18,500	$\cong$35,000	$\cong$35,000
Thrust:weight ratio	5.9	9.1	4.5	10.9	?

Back in the USA, however, GE had attempted to enter the civil turbofan revolution by fitting an aft fan to the J79 and running it, at the end of 1957, as the CJ-805-23. A respectable improvement in fuel economy was achieved with a bypass ratio of 1.4, and a rated take-off thrust of 16,050lb (7,285kg; 71.4) ensued. Only one installation was secured, however. This was the Convair 990, which, with the poor history of the preceding Convairs, and its capacity limitations relative to the Boeing 707 and Douglas DC-8, sold only 37 examples.

In Great Britain Rolls-Royce was active in attempting to build more successfully on its Conway experience. An experiment, as with GE and its CJ-805, saw an experimental test-stand run of an aft-fan-equipped Avon. However, the turbofan engine that emerged for serious production was the RB.163 Spey, a scale-down version

The Eurojet EJ200 is the 20,000lb-thrust-class turbofan for the Eurofighter, or Typhoon. Advanced technologies are enabling thrust/weight ratios of the order of ten to be achieved with such military engines.

of the larger RB.141 Medway that had been proposed for the original Airco D.H.121 tri-jet. British European Airways specified a smaller tri-jet, which became the Trident, and the engine scale-down was the matching response from Rolls-Royce. The engine was sized for 10,000lb (4,540kg) of thrust, and became too small for the Boeing 727 (which adopted the JT8D) but it did spawn a family of both civil and military engines, the latter with reheat availability. Its bypass ratio was equal to 1.0, which reflected both the fashion of the time and the company's conclusion that this was the optimum level. The optimisation was derived by balancing the fuel consumption gain for increasing bypass ratio against the combined performance effects due to the weight and drag increases for larger ducts and cowls.

Within a few years, serious installation research, and the growing influence of noise reduction requirements, led to the adoption of much higher bypass ratios for subsonic civil engines, and radical changes in engine configuration. For military engines, supersonic speeds demanded minimum engine diameter for low-drag installation, and maximum boost from reheat for take-off, acceleration and combat. The optimum bypass ratio was therefore nearer this level, and to this day is most likely to be in the range of 0.3 to 0.6. However, the technology being incorporated is giving large increments in performance capability, and this can be illustrated, as in

Pratt & Whitney's F119 engine for the Joint Strike Fighter includes many advanced features but eschews the cycle variability solution, which alters the bypass ratio to suit operational requirements, as included by General Electric in its competing engine, the F120.

128

PROPULSION

Table 2, by comparison of early (1960s) and late (current) supersonic military turbofans.

The engine configurations are broadly the same. They are all two-shaft, low-bypass-ratio, reheated engines. But the overall pressure ratios are higher by between 30 and as much as 100%, with an order of 50% reduction in number of compressor stages. Combustion delivery temperatures cannot be confirmed, but have increased by 200-300°C. Turbine stages are reduced by 50% and, again, efficiency levels improved. The performance changes are radical: thrust:weight doubled and specific fuel consumption improved by up to 20% all in thirty years.

High-bypass turbofans.

The turbofan engines of the 1960s revolutionised the air transport and heavy-lift industries. The era of mass passenger travel had arrived, and heavy-lift capability improved dramatically. Such success raised expectations, and targets for range and fuel efficiency were revised rapidly to more ambitious levels. Radical thinking on new engine configurations demanded the removal of many long-standing inhibitions.

The USAF was instrumental in initiating the next change in transport engine technology, the subsonic, high-bypass-ratio engine. In 1962 the USAF had initiated an advanced planning project to establish priorities for the next decade. A review of contributing factors established that the US engine companies were becoming confident that large-fan engines were possible, for the required step change in thrust and fuel consumption. The result was an air force specification for the CXX large military transport with high-bypass-ratio turbofan engines. Industry recognised this event as the harbinger of things to come in the civil world, as well as in the military.

General Electric won the contract for what became the Lockheed C-5 transport for the USAF. During the pre-contract competition the company had built a half-scale demonstrator engine, which proved the dramatic specific fuel consumption improvement required by using a bypass ratio of 8.0, an impressive level no longer unthinkable.

The resultant engine for the C-5, the TF39, featured the same bypass ratio, and a fan configuration destined to stay unique in the following families of high-bypass engines. Its root performance was bolstered by a half-span stage in front of the full fan, and the associated structure generated aerodynamic noise that was unacceptable in the civil environment. The half-stage was replaced in other two-shaft high-bypass-ratio engines by boosters (on the LP shaft) in the core annulus. The core of the TF39 was traditional GE type with 16 stages, of which six had variable stators. The engine overall-pressure ratio achieved was over 26, and the thrust rating 41,100lb (18,655kg).

The competition to introduce the same technology – and benefits – into the civil transport industry became intense. The Lockheed C-5A Galaxy was a large heavy-

Lockheed's C-5A military transport is still the USAF's heavy lifter, and is still powered by the General Electric TF39, the progenitor of the high-bypass-ratio turbofans. The engine's fan configuration includes a core booster stage in front of the fan, but this had not been translated to later engines owing to the noise generation of its static supporting structure.

FASTER, FURTHER, HIGHER

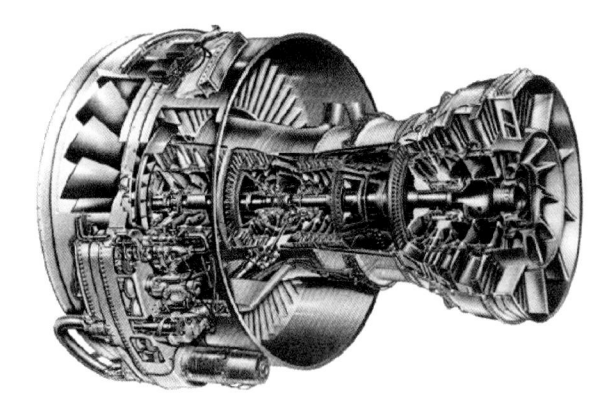

The Rolls-Royce RB211-22B, the three-shaft solution to the dictats of the high-bypass turbofan, and precursor to both the −535 and to an increasing Trent family of engines.

lifter but it was not a high-speed aircraft, and its cruise thrust, which sized the TF.39, was lower than Boeing's prospective passenger jet. Boeing chose a Pratt&Whitney engine over a GE rival to power the new 747. The JT9D engine ran on 20 December 1966. The engine contained a single-stage fan with three core booster stages on the same shaft, and an eleven-stage core HP compressor with four stages and variable stators. The turbines comprised a two-stage HP and a four-stage LP. Its bypass ratio was 5.0 and its take-off thrust 43,500lb dry (19,745kg). Development (and entry into service) demonstrated a whole raft of new problems and painful lessons of aerodynamic and mechanical origin, but the new technology was established, and it proved itself.

In Great Britain, Rolls-Royce was studying increases in bypass ratio, and by 1965 had conceived the adoption of the three-shaft configuration for the high-bypass-ratio civil engines. The basic principle of the three shafts was to allow a single-stage fan with its own turbine and a two-spool core for high pressure ratio with minimum stages and variable requirements. The core booster stages of a two-shaft engine are replaced effectively by an intermediate-pressure (IP) compressor with its own turbine. As such it can be operated at a higher rpm, and hence higher blade speeds for more work per stage, than can the equivalent LP booster, whose rpm is restricted to that allowed by fan tip speeds. The more optimum work per stage gives a reduction in the numbers of stages and a short, rigid engine and good deterioration characteristics.

Besides the three-shaft configuration, the RB.211 featured another new technology unique for its time, a wide-chord fan blade made with a carbon-fibre reinforced

The Airbus A340-500 at take-off.

PROPULSION

plastic material called Hyfil. Hyfil was of sufficient rigidity to enable omission of inter-stage supports, or clappers, and so to gain efficiency. In its initial configuration, without leading-edge strengthening, it could not pass the necessary bird-impact testing, and the company reverted to a more conventional titanium blade with inter-stage supports for certification and service.

By this time Airbus Industrie was proposing a large, twin-engined transport, and the three-shaft RB.207 was defined for this aircraft, to start at 47,500lb (21,560kg) take-off thrust. A similar specification in the USA (the American Airlines big twin) was ultimately aborted by operator concerns about the safety of big twins, and the tri-jet compromise emerged. The RB.207 was scaled to match, abandoning Airbus, and the RB.211-22 became the Rolls-Royce submission for the Lockheed and Douglas tri-jet proposals.

The subsequent large-engine story is extensive and beyond the space available, but these initial engines have established a pattern for each manufacturer to pursue the same goals; even similar engine cycles, with different configurations and the inclusion of different detailed technology. Family heritages remain visible, but bypass ratios now approach 9, overall pressure ratios approach 50 and turbine entry temperatures continue to increase. There is estimated to be still some 15% improvement to come within the family from pressure ratios up to 60:1, possible increase in turbine entry temperatures of 300 degrees, and some component efficiency gains, despite the ever-reducing inefficiencies remaining to be expunged.

The advent of the high-bypass-ratio produced an unprecedented and immediate 25 percent step downward in engine fuel consumption. Such a radical step change is visible now only in the in the form of the propfan. This type of engine can be regarded as a turboprop driving a very large supersonic propeller, open rotor or ducted. Modern design techniques can now allow such propellers to retain efficiency to supersonic levels of airflow relative to their aerofoils, and therefore to higher aircraft speeds. Propfan engines have been run and flight tested and one is in use in a military transport from Ukraine, the An-70; but, in the absence of a solution to high noise and vibration and owing to certification problems with blade-off compliances, their civil use is restricted.

Future technology
In the early days of the gas-turbine era, most technological advances and innovations were driven by military

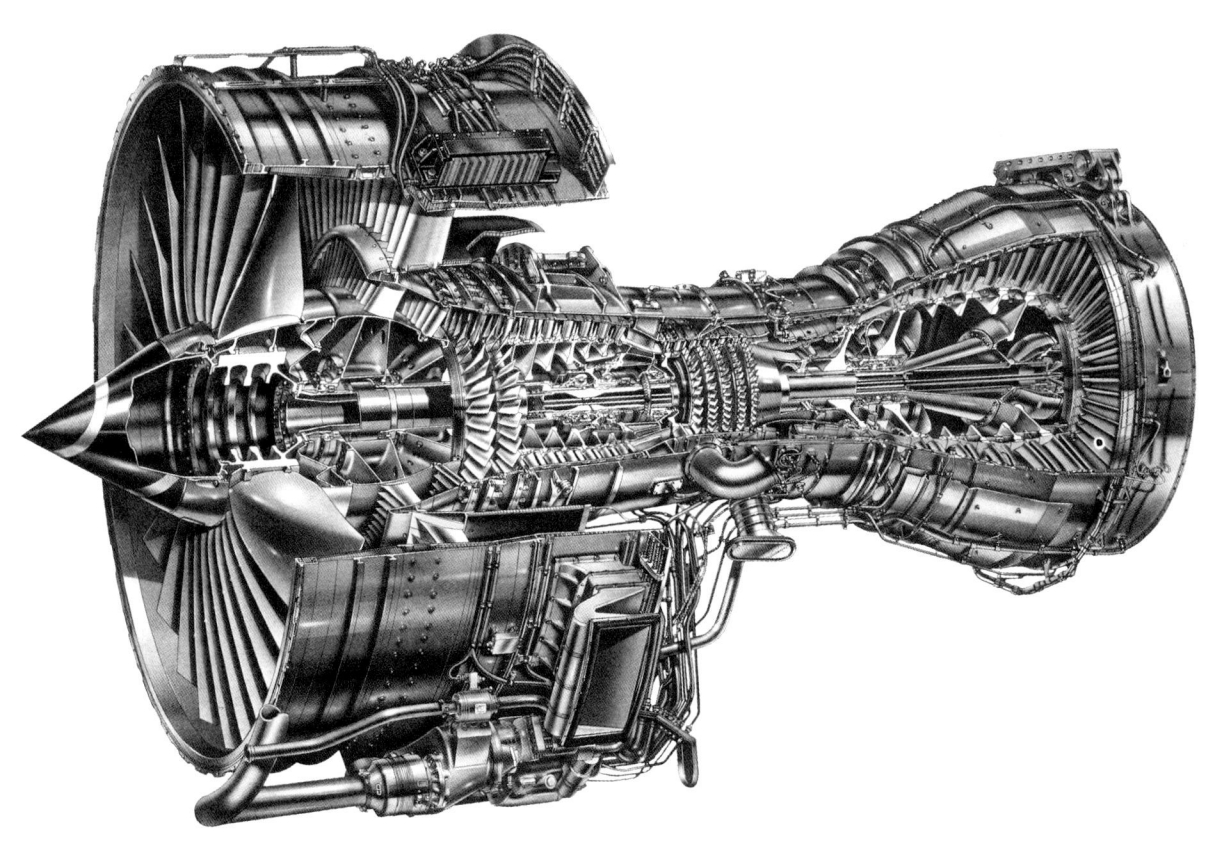

A cutaway drawing of the Trent 500, the engine for the Airbus A340-500 and –600, which retains the highly successful and unique Rolls-Royce three-shaft configuration.

requirements. However, the military (primarily supersonic), and civil (primarily subsonic), turbofan configurations have diverged markedly with the large difference in prevalent bypass ratios. Commercial pressures have also accelerated and driven civil development times to levels below those available to much of the military industry. Technical solutions, therefore, are as likely to appear from civil as from military sources. Nevertheless, their applicability will remain relevant to both branches of the industry.

Material properties have been improved dramatically, and the process is continuing. Turbine blade temperature capability has been improved by the order of 250° Centigrade over the past thirty years by progression through wrought, cast, directionally solidified and single-crystal materials. The current materials are complex nickel-base alloys with numerous additives such as tungsten, rhenium, tantalum and aluminium. These additives variously contribute to base properties and oxidation prevention, but nickel-based alloys are regarded generally as nearing the limit of their development. Extension of their capability to the higher temperatures being demanded will depend on the satisfactory development of thermal barrier coatings to insulate the metal from the hot gas stream. This will require coatings with strain tolerance, to avoid spalling and exposure of the base metal. It is expected that the use of a fibrous grain structure and an effective bond-coat will enable considerable increase in temperature capability. The next step will involve blades made of ceramic matrix composite (CMC) materials. Today's versions of these materials are, for example, silicon and carbon fibres in a silicon-carbon matrix, but other, more capable, options are under development. Engine cold section alloys are not being neglected, and improved titanium alloys are becoming available.

Component configuration changes are conferring weight and cost savings, typified by the incorporation of blisks (single-piece disks and blades), which eliminate the traditional blade-fixing structure. Titanium metal matrices (with silicon/carbon fibres) will enable the extension of this technology to the production of blings (diskless blade rings) for more weight saving.

Aerodynamics is the primary discipline in engine design and, with the very large growth continuing in computer capacity, computation fluid dynamics (CFD) has emerged as an extremely powerful design tool. Modelling of boundary layers, leakage and bleed effects has allowed considerable improvement in blade design and this has resulted in the use of reduced stages for given work, and the ability to optimise for efficiency. Associated configuration changes include the swept fan and the ability to design better contrarotating turbines without inter-stage vanes. The gas turbine engine is not yet at its limit of development.

Alternative fuels

Considerations of alternative fuels have intensified since 1945. The stimulus for the search and its directions has changed with time, from early performance considerations to the more recent dominance of environmental concerns. One of the early post-war performance programmes that could loosely qualify as driven by alternative fuel considerations was the investigation into the potential of nuclear power. After 1945 the nuclear reactor was perceived very quickly to be a source of virtually unlimited power. It was therefore inevitable that thought should be given to whether it could be used in aviation,

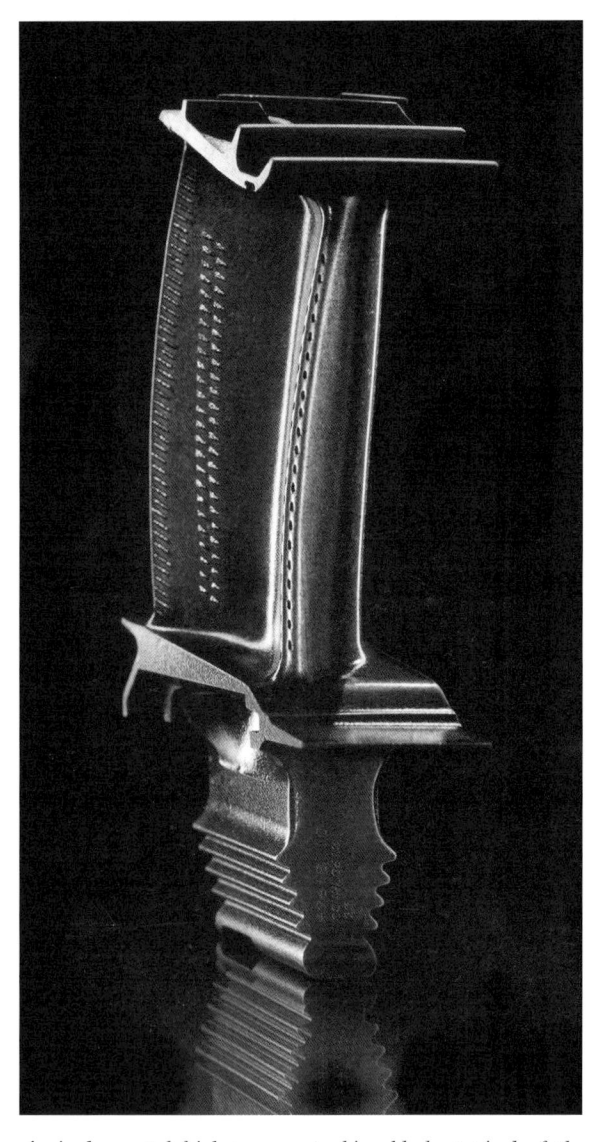

A single-crystal high-pressure-turbine blade, typical of the Rolls-Royce blades for Trent civil engines. It features extensive cooling air usage, plus a tip shroud for optimum tip clearance control and maximum efficiency.

PROPULSION

The Convair NB-36H flew extensively with a live reactor as part of the programme to study radioactive shielding requirements and the problems inherent in the aspects of maintaining reactors in aircraft. Its companions in the programme, the X-6s, which would have used the reactor power for propulsion, were not completed and therefore never flew.

and by early 1951 the USAF and the Atomic Energy Commission (AEC) concluded that a research programme was warranted to investigate the feasibility of a nuclear-powered propulsion system for military aircraft.

General Electric was awarded a contract to develop a nuclear-powered turbojet, AEC a contract to develop a compatible reactor, and Convair a contract for modification of three B-36H bombers. Two of the aircraft were to carry the full nuclear powerplant. The third, with its retained conventional engines, was to carry a live reactor for in-flight shielding tests. This test was regarded as vital for investigating radiation effects on propulsion systems, electrical and components systems and crew (flight and maintenance) protection. As this research programme was sponsored by the armed forces, NACA and industry, the aircraft were identified as X-planes, and given the designation X-6. The Joint Chiefs of Staff endorsed the military necessity for such a programme and launched other development initiatives in the industry, so that Pratt & Whitney began an investigation of alternative propulsion systems to those of GE. In the end, the preferred GE system was for large twin turbojets with their compressor

Table 3: Alternative fuels

	Specific energy relative to kerosene		Performance index
	By volume	By weight	
Kerosene	1.0	1.0	1.0
Alternate solids			
Berylium	3.34	1.54	5.2
Boron	3.97	1.37	5.4
Carbon	2.02	0.76	1.5
Magnesium	1.17	0.58	0.7
Lithium	0.63	1.00	0.7
Alternate liquids			
Decaborane ($B_{10}H_{14}$)	1.71	1.55	2.7
Pentaborate (B_5H_9)	1.16	1.58	1.8
Diborate (B_2H_6)	0.85	1.69	1.5
Lithium Bora-hydride ($LiBH_4$)			
	1.02	1.31	1.3
Methane	0.6	1.16	0.7
Liquid Natural Gas	0.5	1.12	0.6
Liquid hydrogen	0.23	2.79	0.6

delivery air-diverted through the reactor (when live) instead of through the engine combustor. A very large demonstrator representing half of the proposed unit was built and run, but on conventional fuels. The eventual Pratt & Whitney system was to use reactor heating into J.91 gas turbines by means of molten salt heat transfer. Again, two demonstrators were built and run on kerosene.

As for the aircraft, despite the commitment of the USAF to an ultra-long-range bomber implied by the issue of a nuclear powered bomber specification, WS 125A, doubts about both viability and necessity continued. The shielding test aircraft flew in July 1955, but not with the reactor critical until September. Its flight programme was regarded as successful in identifying the hazards associated with the use of nuclear power in aircraft. In the event this only served to emphasise the difficulties that would arise in working with nuclear fuel in service conditions. The nuclear-powered aircraft were never built, and by 1958 the project was being starved of funds. In 1961 the whole programme was finally terminated by the Kennedy administration. Nuclear energy, despite some work in Russia, disappeared as an alternative aviation fuel source.

The general definition of an alternative fuel for aviation is of one that would replace oil-based fuels such as kerosene. Research in the 1950s for such replacements was primarily performance-driven, i.e. searching for enhanced long-range capability. In parallel with the nuclear-powered bomber specification (WS-125A), the USAF issued specification WS-110A, for a chemically powered bomber. This specification reflected the recognition of the range and load carrying limitations of the Convair B-58, the first supersonic bomber, and identified a range requirement of a Boeing B-52 at B-58 speeds. It did not need much studying to conclude that, along with utilisation of all available technology, some form of fuel-energy enhancement would be required for the resultant B-70. Serious research programmes followed. What was needed was a fuel with a higher oxidation energy release for its weight than that of kerosene. Berylum (Be) and boron (B) looked the most promising initially. Berylium has the greatest heat release, but is toxic when combined with hydrogen and, therefore, boron was ultimately favoured as more practical. Initial thoughts were for use in elemental form (powder) or as a kerosene slurry, but pumping and abrasion appeared insurmountable, and the effort switched to the liquid boron hydrites. Table 3 (page 133) lists the specific energies of the primary alternatives, by weight and volume, relative to that of kerosene. Also listed is the product of the two which has been seen as a initial measure of merit. It is crude, in that it allocates equal weight to each component, whereas in reality the applicable weighting will vary with use and installation.

In the end, the manufacture of the hydrites was shown to be difficult and expensive. They were also variously toxic, they yielded malodorous exhaust fumes and they deposited glassy products of combustion on to engine parts: and their performance gains were, in the end, marginal. The whole programme was abandoned in 1959, after considerable expenditure on research and on production facilities. Low volatility and heat-tolerant versions of conventional jet fuel were developed for the exotic aircraft programmes, such as the XB-70 and SR-71, and proved adequate in service with the latter.

The emphasis has now switched to environmental considerations. The current industry jet fuel is a fossil fuel and its supply, therefore, is subject to eventual exhaustion, albeit now thought to be at a time considerably in the future. Also, the burning of fossil fuels increases atmospheric CO_2 levels, with potential impacts on global warming and climate changes. The only fuel that will burn without some atmospheric pollution is liquid hydrogen, which also has a high energy-release for its weight. It is also easily produced by electrolysis of water, to which combustion returns it to complete a closed-cycle system. Gas turbines are easily adapted to burn it. However, it requires maintenance at a very low temperature (- 252° C) and its density is low. In Europe, trials of accommodating the necessarily large insulated fuel tanks in hydrogen-powered aircraft are continuing, and ground and flight tests of running engines on hydrogen are also in progress. The adoption of hydrogen world-wide, however, would require a huge investment in production facilities at airports, and this alone would appear to be a major inhibition to selection of hydrogen as a universal fuel in the near future.

Summary

In 1952 a NACA committee decided that a research programme should be undertaken to explore flight characteristics at speeds of up to a Mach Number of 10, and altitudes of up to 250,000ft (76,200m). North American won the design contest, and three aircraft were built under the designation X-15. Power was by rocket, definitively the Reaction Motors XLR99 of 50,000lb (22,695kg) thrust at sea level. It used ethyl-alcohol and liquid oxygen as its fuels, and was capable of being throttle-controlled between 30 and 100% power. It was also capable of relighting in flight. The X-15 was air-launched from a B-52 and landed on a simple nose-wheel/skid undercarriage. During its powered flight programme between 17th September 1959 and 24th October 1968 it demonstrated a maximum Mach Number of 6.7 and attained an altitude of 354,200ft (107,960m).

The X-15 was a true research aircraft, and carried instrumentation for many experiments up to hitherto unreachable conditions. The experimentation included some initial attempts to test supersonic combustion ramjet (scramjet) models. The Space Shuttle, launched by rocket boosters and its own rockets but reuseable on return to Earth in an X-15-like manner, has extended the envelope into space orbit.

However, sustained endurance flights by piloted aircraft within the Earth's atmosphere have been demonstrated only to speeds of Mach 3.5 and altitude of 85,000ft (25,900m) (with the SR-71), and this, at the moment, is the effective limit of the gas turbine. To extend the sustained endurance mission to hypersonic conditions (Mach 12 plus), a modified air-breathing engine will have to be developed. Considerable research effort in both the USA and Europe has already gone into investigating the problem, as short-duration, ultra-long-range travel has both commercial and military attractions. The ramjet can take speeds to the order of Mach 6.0, where deceleration of its intake air to subsonic conditions creates enormous duct pressure and gaseous dissociation to inhibit combustion. The prime solutions for study and research initiatives have included hydrogen-fuelled rockets that would use oxygen from the atmosphere as their oxidant, or scramjets. The latter demand the capability to burn fuel in a supersonic airflow, which will demand hydrogen as the only fuel with a reaction time sufficiently quick to sustain burning. It is doubtful whether the success of such schemes has yet been demonstrated, but it is hoped that any review such as this will, in the future, be able to include intelligence of their contribution to a dramatic extension of the current flight envelope.

Bibliography

Baker, Dr A 'NASP – Waveriders in the Sky', *Air International*, Vol 44/1 and 44/2, Jan & Feb 1993. A review of the National Aero-Space Plane programme and the X-30, which represented it at the time.

Baxter, A, *Olympus: the First Forty Years* (Rolls-Royce Heritage Trust, Derby, 1990). A comprehensive history of the two-spool Olympus engine and its installations.

Braybrook, R, 'The Cyroplane Project', *Air International*, Vol 53/5, Nov 1997). An article reviewing progress in Russia and Germany with LNG and liquid hydrogen options.

Golley, J, *Whittle, the True Story* (Airlife, Shrewsbury, 1987). A full history of Sir Frank and his efforts to get the gas turbine accepted as a viable engine.

Goodger, E M, & Vere, R, *Aviation Fuels Technologies* (Macmillan, London, 1985). A good explanation of alternative fuels, with tabulations of properties to support the explanations.

Goodger, E M, *Transport Fuels Technology*, (Portland Press, London, 2000). An update of the previous volume, with much very specialist detail.

Gunston, B, *The Development of Jet and Turbine Aero Engines* (Patrick Stephens, Sparkford, 1997). Contains both some readable technical background and a concise history.

—— *Faster than Sound* (Patrick Stephens, Sparkford, 1992). A concise but very extensive review of supersonic research programmes and of ensuing production aircraft.

—— *World Encyclopaedia of Aero Engines* (Patrick Stephens, Sparkford, 1986). A listing of companies that have been involved in the engine industry, piston and gas turbine, and notes on each of their products.

Jane's All the World's Aircraft and Aircraft Engines (Editions as required). The essential reference yearbook.

Miller, J, *The X-Planes* (Midland Publishing, Hinckley, 2001). Third edition of an encyclopaedic review of the US Armed Forces, NASA and Industry research programmes, X-1 to X-45.

Pohl, H.W, *Hydrogen and other Alternative Fuels for Air and Ground Transportation* (John Wiley, Chichester, 1995). A series of articles covering the reasons why the search for alternatives is, or will be, necessary, and how the demand may be satisfied.

St Peter, J, *The History of Aircraft Gas Turbine Development in the United States* (ASME, Atlanta, 1999). Commissioned by an impressive list of sponsors (the US Armed Forces, NASA and the ASME International Gas Turbine Institute), this is an excellent and comprehensive review of the American industry. It gives due deference to the early European initiatives by starting with chapters on the UK and German pioneers, and noting other inputs appropriately as they occurred.

Smith, G, *Gas Turbines and Jet Propulsion* (Iliffe, London, 1955). The sixth edition of the early classic on the subject. Contains some very good basic technical information and some interesting early history.

Smith, M, *Aviation Fuels* (G T Foulis & Co, Henley-on-Thames, 1970). Good for chemical and thermal details of fuel alternatives to kerosene.

7
New-age Systems
Ian Moir

Introduction

New-age systems have resulted from the continuous evolution of new technologies, rather than a revolutionary process. Many of the basic techniques used in avionics systems today were conceived during the tremendous acceleration of technology spawned by the Second World War. Further impetus came in the early 1960s, following the introduction of jet aircraft for long-range civil transportation. The speed and navigation demands of these long-range aircraft placed additional demands on navigation systems, and the adoption of militarily developed inertial navigation systems became commonplace in the early 1970s

The microelectronics revolution and the introduction of digital computers led to further improvements in the late 1970s and early 1980s. The digital computer allowed more accurate calculations to be performed. The development of the accompanying digital data buses, such as ARINC 429 (the standard digital data highway for civil aircraft), permitted computers to conduct the timely exchange of important system data. At the same time the introduction of data buses – the channels by which information is exchanged between digital devices – greatly reduced the amount of wiring required to interconnect the systems, thereby reducing aircraft manufacturing costs.

The widespread availability of suitable microelectronic devices, and the introduction of standardised methods and system design implementations using the ARINC (from Aeronautical Radio, Inc., a private company based in the USA) and military specifications, brought further advantages. It enabled industry to concentrate upon the application of the new technologies within an overall framework of standardisation, minimising wasted and nugatory development effort.

The new applications have benefited from the adoption of military technology in the areas of radar and navigation sensors, head-up display (HUD) and so on. Also, microelectronic technologies developed for the telecommunications and computer industries have enabled commercial off-the-shelf (COTS) technologies to be adopted in many areas, not always without difficulty. These applications and technologies are enabling the development of more advanced and more capable systems to be brought to fruition.

The industry has also learned hard lessons regarding the development of high-integrity software and the integration of highly complex systems. The civil industry and the regulators have worked hard to develop specifications and guidance to ameliorate the risk inherent in developing such systems as the level of systems integration moves ahead at a rapid pace.

Rapidly increasing air traffic is forcing new solutions in the way aircraft fly through the airspace. Increased navigational accuracy and reporting methods are being evolved to enable more aircraft to be fitted into the existing airspace while preserving levels of safety. The integration and fusion of technology advances such as the global positioning system (GPS), together with other advances, as well as some of the long-established techniques, are leading to the development of a future air navigation system (FANS) that will supply these solutions.

This chapter addresses the following main subject areas:

- Sensors
- Communication and navigation aids
- Navigation
- Displays
- Safety aids
- Flight control
- CNS/ATM (FANS) developments

In each subject area an attempt has been made to portray some a measure of historical perspective. In addition, as the chapter unfolds, the importance and degree of interaction and integration of all of the technologies that comprise modern civil avionics systems should become apparent

Sensors

The primary autonomous sensing methods available to the modern civil airliner are:

- Air data
- Magnetic
- Inertial
- Radar sensors

Each of these sensor types has its own attributes and strengths and shortcomings. A modern system will take account of these attributes, matching the benefits of one particular sensor type against the deficiencies of another. On-board autonomous sensors are also used in conjunction with external navigation aids and systems to achieve the optimum performance for the navigation system. As

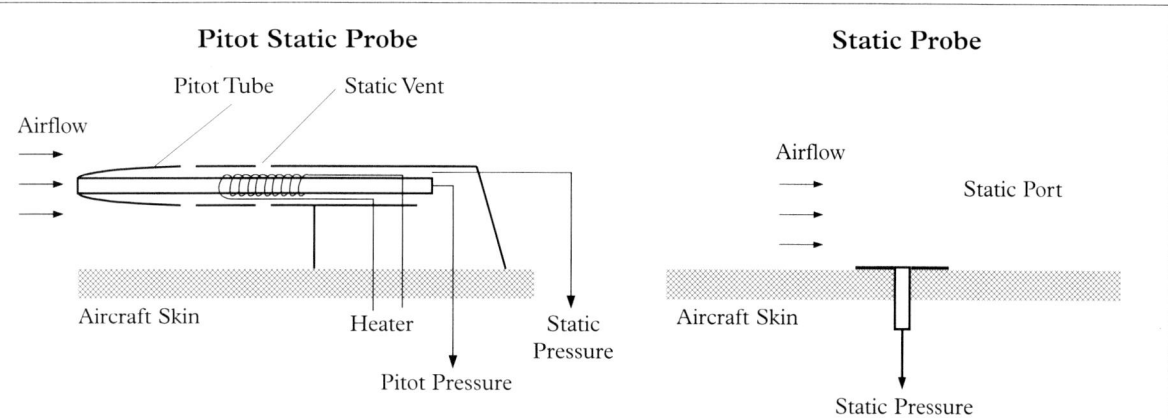

Figure 1. *Air data probes*
Pitot pressure is the dynamic head of pressure created by the forward movement of the aircraft during flight. The dynamic pressure varies according to the square of the forward velocity of the aircraft. Static pressure is the local pressure surrounding the aircraft at a given altitude, and may be used to determine the aircraft's altitude. Static pressure may be measured by means of ports in the side of the pitot static probe or by means of static ports in the side of the aircraft skin. The combination of sensed pitot and static pressure may be used to derive aircraft data, as shown in Figure 2.

will be seen, the capabilities of modern integrated navigation systems blend the inputs of sensor types to attain high levels of accuracy that enable new navigation and approach procedures to be used. Indeed, in today's crowded skies such highly integrated systems are becoming essential to assure safe and smooth traffic flow.

Air Data
Air data, as the name suggests, involves the sensing of the medium through which the aircraft is flying. Typical parameters are dynamic pressure, static pressure and the rate of change of pressure. Air data is sensed by means of pitot static probes and static sensors, as shown in Figure 1.

Originally, the portrayal of aircraft airspeed, altitude and rate of change of altitude was accomplished using discrete instruments: the airspeed indicator, altimeter and vertical speed indicator (VSI). As aircraft systems became more sophisticated, and control laws were adopted, the number of systems requiring air data increased. Therefore the provision of air data in various navigation, flight control and other subsystems required a more integrated approach. This led to the introduction of one or more air data computers (ADCs), which centrally measured air data and provided corrected data to the recipient subsystems. Consequently, while the pilot still had the necessary air data presented to him, more accurate and more relevant forms of the data could be provided to the aircraft systems. Initially this was achieved by analogue signalling, but with the evolution of digital data buses in the late 1970s, data was provided to the aircraft subsystems by this means: notably by the use of the standard ARINC 429 data buses. The function of an air data computer is shown in Figure 3.

Other air data producers include the airspeed direction detector (ADD), which measures the direction of the airflow relative to the aircraft and permits the angle of incidence to be detected. Typically, this information is of use in the flight control system to modify the pitch control laws or in a stall warning system, to warn the pilot of an impending stall. A typical airstream detection sensor is shown in Figure 4.

The small-bore pneumatic sensing lines associated with routing the sensed pitot or static pressure throughout the aircraft posed significant engineering and maintenance penalties. As the air data system is critical to the safe operation of the aircraft, it was typical for three, four, or more alternative systems to be provided. The narrow bore of the sensing lines necessitated the positioning of water drain traps at low points in the system where condensation could be drained off, avoiding blockage of the lines due to moisture accumulation. Finally, following the replacement of an instrument or disturbance of any section of tubing, pitot-static leak checks were essential to ensure that the sensing lines were intact and leak-free and that no corresponding instrumentation sensing errors were likely to occur.

The advent of digital computing and digital data buses such as ARINC 429 meant that computation of the various air data parameters could be accomplished in air data modules (ADMs) closer to the pitot-static sensing points. Widespread use of the ARINC 429 data buses enabled this data to be rapidly transmitted throughout all the necessary aircraft systems. Virtually all civil transport aircraft designed within the last fifteen years or so have adopted the air data module implementation. Figure 5. shows a typical air data system for an Airbus aircraft.

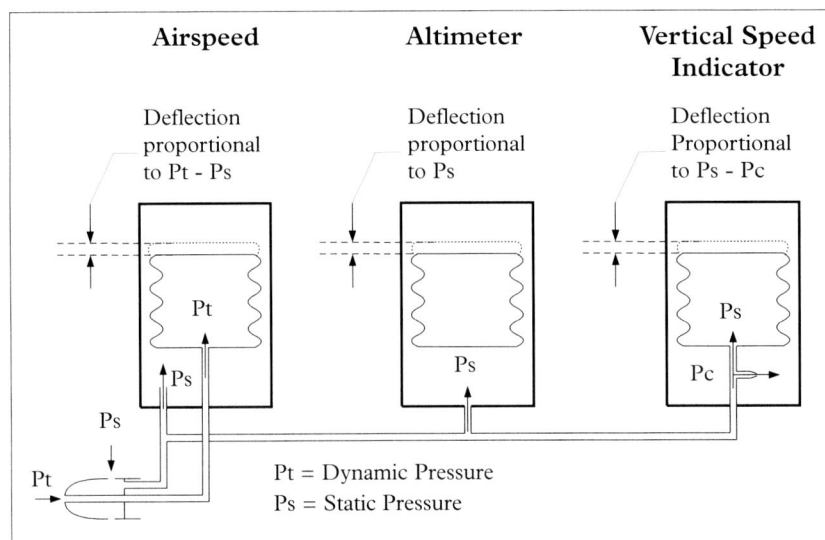

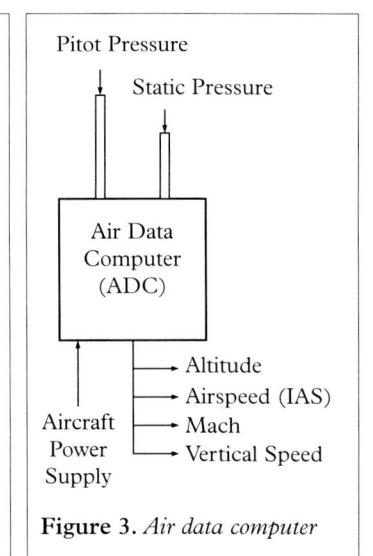

Figure 2. *Use of pitot and static pressure*
By using the capsule arrangement shown on the left, dynamic pressure is fed into the capsule while static pressure is fed into the case surrounding the capsule. The difference between these two parameters, represented by the deflection of the capsule, represents the aircraft's airspeed. This permits airspeed to be measured.

In the centre capsule configuration, static pressure is fed into the case of the instrument while the capsule itself is sealed. In this case capsule deflection is proportional to changes in static pressure and therefore aircraft altitude. This allows aircraft barometric altitude to be measured

In the arrangement shown on the right, static pressure is fed into the capsule. It is also fed via a calibrated orifice into the sealed case surrounding the capsule. In this situation the capsule defection is proportional to the rate of change of altitude. This permits the aircraft rate of ascent or descent to be measured

Figure 3. *Air data computer*

Air data is generally regarded as accurate, but in the longer term rather than the short term. The fact that air data sensing involves the use of relatively narrow-bore tubing and pneumatic capsules means that there are inherent delays in the measurement of air data as opposed to some other forms of air data

Magnetic sensing
The use of the Earth's magnetic field to sense direction and use north-seeking devices to establish the direction of magnetic north for the purposes of navigation is one of the oldest forms of sensor. A magnetic sensing device, called a flux valve, is usually located in the outer section of one of the aircraft's wings, well clear of any aircraft induced sources of spurious magnetism. The relationship of magnetic north to true geodetic north varies considerably over the Earth's surface, the difference in some areas being ~ 40°. However, the variation occurs slowly with time and is constantly surveyed, such that suitable corrections may applied be locally depending upon one's location on the globe. This does not apply to areas in the vicinity of the Earth's magnetic poles, where readings are erratic due to the proximity to the poles and the high angles of declination that apply. In general magnetic compasses can be a useful navigation tool

Magnetic sensing therefore provides a very simple heading reference system which is used by virtually all aircraft of modest performance today. In light aircraft it is likely to be the only heading reference

Inertial sensing
Inertial sensors are associated with the detection of motion in a universal (non-earth) referenced set. Inertial sensors comprise:

- Position gyroscopes
- Rate gyroscopes
- Accelerometers

Gyroscopes are most commonly implemented as a spinning mass or wheel which remains in place in a specific attitude. Position gyroscopes or gyros use this property to provide a positional or attitude reference; typically pitch position, roll position or yaw position. Position gyros are used in heading and reference systems to provide vital information regarding the aircraft's attitude.

Gyroscopes can also be used to provide information relating to the aircraft's pitch rate, roll rate and yaw rate. These rate gyros use the property of gyroscopic precession. When a gyro is rotating and the frame of the gyro is moved, the gyro moves or 'precesses'. By balancing and measuring this precession force, the angular rate of

NEW-AGE SYSTEMS

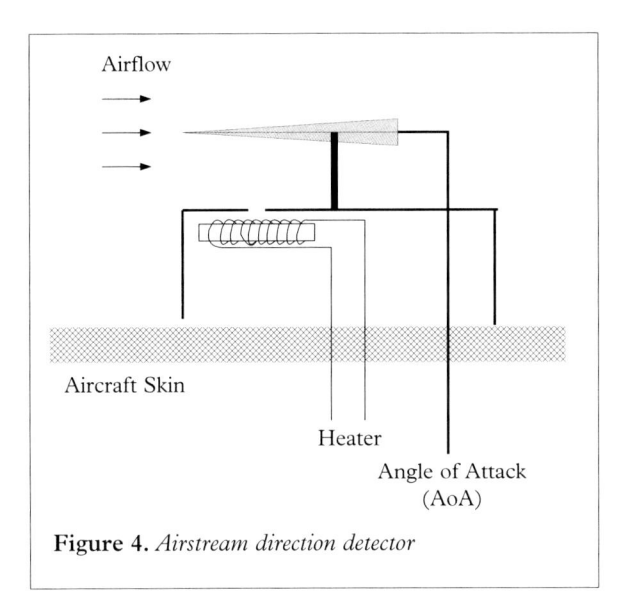

Figure 4. *Airstream direction detector*

(Diagram labels: Airflow, Aircraft Skin, Heater, Angle of Attack (AoA))

movement applied to the gyroscope frame may be measured. Therefore, rate gyros enable an aircraft's pitch, roll and yaw rates to be measured. This information, together with the attitude data provided by the position gyros, is crucial in the performance of modern flight control or fly-by-wire (FBW) systems.

In early systems, gyroscopes were air driven; then electrically driven gyroscopes became the norm. In modern systems gyroscopes are likely to be laser-ring devices.

Accelerometers are devices which measure acceleration along a particular axis. The measurement of acceleration can be integrated using computers used to derive aircraft velocity and position.

Magnetic heading reference system
The magnetic sensor or flux valve described earlier can be combined with a directional gyro (DG) to provide a magnetic heading and reference system, as shown in Figure 6.

The flux valve and DG provide magnetic and inertial heading respectively to the magnetic heading and reference system. According to the type of navigation being used, either magnetic or true (inertial) heading may be selected and displayed on the aircraft's heading displays. This type of system was commonly used for navigation until the late 1960s/early 1970s, when inertial platforms became common.

Inertial navigation
The inertial platform uses a combination of gyros and accelerometers to provide a platform with a fixed reference in space. By using the combined attributes of position and rate gyros and accelerometers, a stabilised platform can be constructed with a fixed attitude reference in space, which when fitted in an aircraft can provide information about aircraft body rates and acceleration in all three axes. Suitable computation can also provide useful information relating to velocity and distance travelled in all three axes. See Figure 7.

This is a significant achievement in establishing the

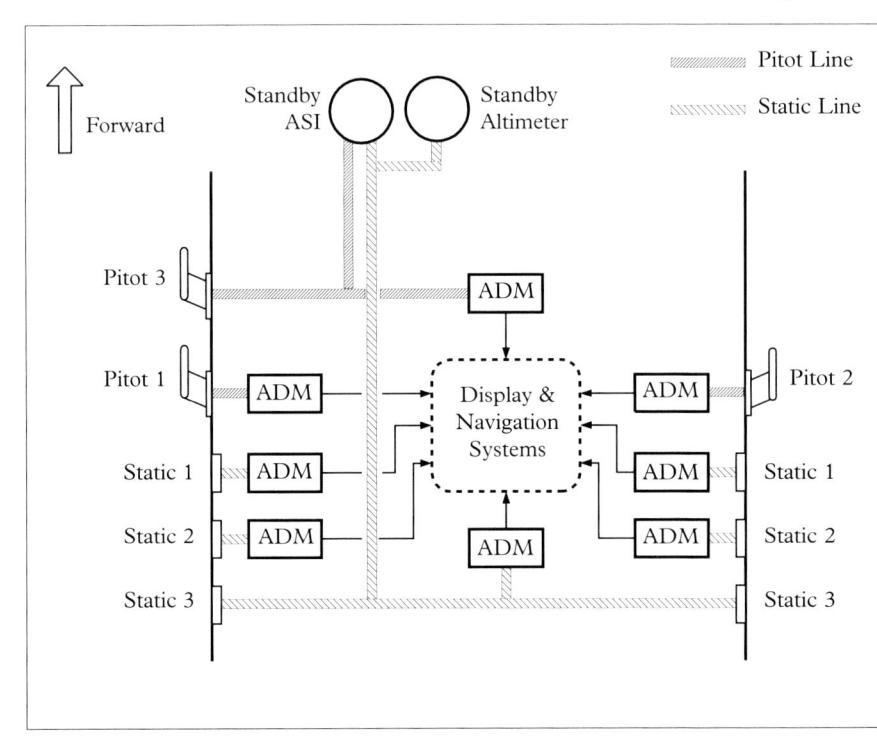

Figure 5.
Airbus air data system
This system comprises three pitot and six static sensors. Pitot sensors 1 to 3 and static sensors 1 and 2 on each side of the aircraft are connected to their own air data module. Each of these seven air data modules provides pitot or static derived air data to the display, flight control and navigation systems, among others. Static sensors, three on each side, are connected to a common line and a further air data module. Pitot 3 and the combination of the static sensors are also used to provide pitot and static pressure to the aircraft standby airspeed indicator and standby altimeter.

FASTER, FURTHER, HIGHER

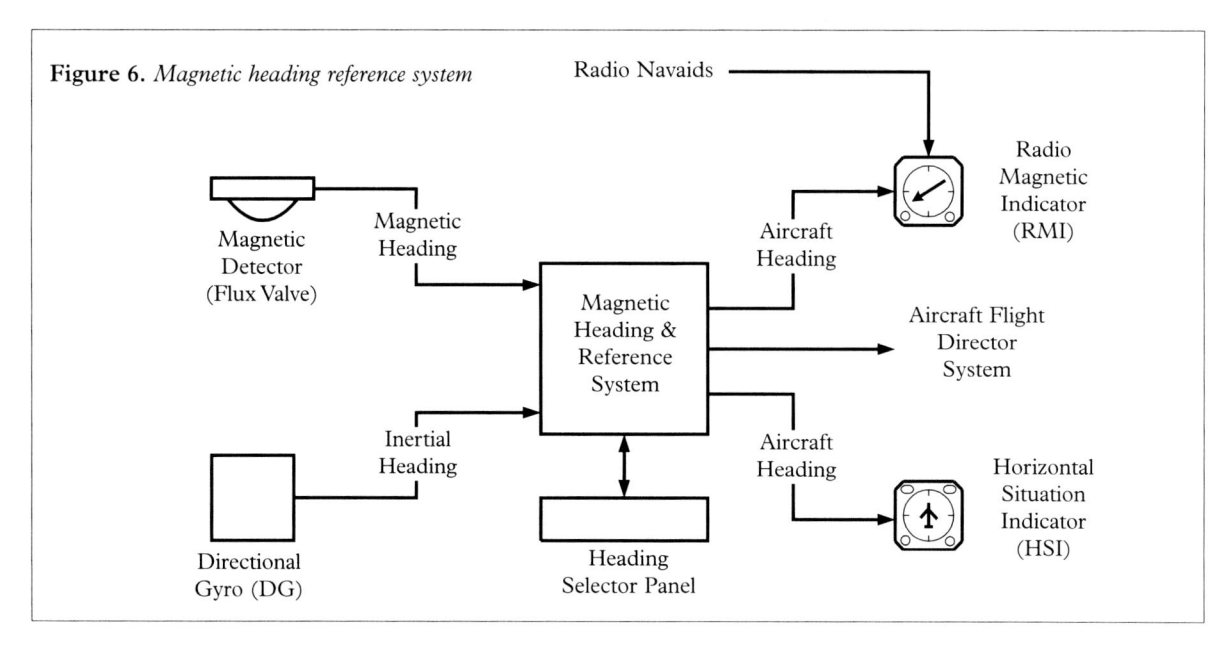

Figure 6. *Magnetic heading reference system*

movement of the aircraft, but it suffers a major disadvantage. The reference for the aircraft body data is set in space, whereas, to be useful in navigating on Earth, the system needs to use a global reference set. The development of the inertial navigation system (INS) provided the additional computation to provide this essential capability. The INS performs a series of transformations on a meaningful Earth reference set for the inertial platform. This is how it works:

- Aligns the vertical axis (Z axis) with the local Earth vertical
- Aligns the horizontal (Y axis) with north: the X axis will now point east

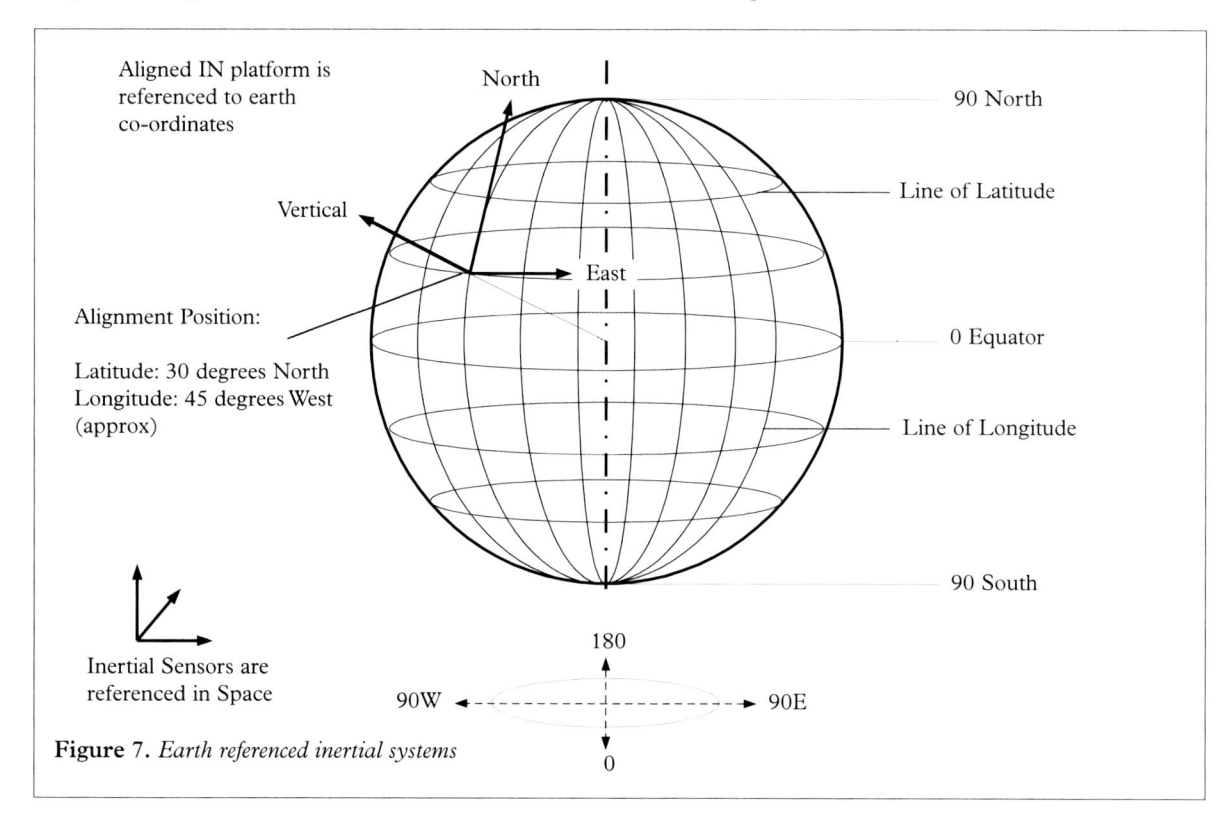

Figure 7. *Earth referenced inertial systems*

NEW-AGE SYSTEMS

- Calculates movement across the surface of the Earth
- Allows for the Earth's rotation variation with longitude. To do this it needs an Earth starting point in terms of latitude and longitude.

The alignment of the INS occurs following start-up, and usually takes several minutes. During this process the orthogonal axis set is aligned with north, local vertical and, by implication, east. The platform is initialised by the flight crew inserting the latitude and longitude co-ordinates provided at the aircraft departure gate. This process is completed before aircraft start-up, and the INS is therefore able to provide accurate information regarding aircraft position, velocities and accelerations throughout the flight. Typical information would include:

- Three-axis accelerations
- Three-axis velocities
- Present position – latitude and longitude
- Distance along and across track
- Angle of drift
- Time to next waypoint and subsequent waypoints
- Calculated wind speed and direction

On an aircraft such as a Boeing 747 three INSs would be provided, and additional calculations performed to establish the best estimate of aircraft position according to information provided by all three systems.

Inertial systems have the advantage that they are very accurate in the short term. No settling time is required as for air data. Conversely, inertial sensors have a tendency to drift with time and so become progressively less accurate as the flight continues. Often the flight crew will use other navigation sensors to update the INS during flight and minimise these effects.

With the advent of ADMs, a combined unit called air data and inertial reference System (ADIRS) performs all the necessary calculation on air data and inertial data to provide the navigational information that the aircraft requires

Radar sensors
Civil aircraft carry a number of radar sensors which permit the aircraft to derive data concerning the flight of the aircraft. The principle radar sensors used on civil aircraft are:

- Radar altimeter
- Doppler radar
- Weather radar

Radar altimeter
The radar altimeter reflects radar transmissions off the ground immediately below the aircraft, thereby providing an absolute reading of altitude. This contrasts with the

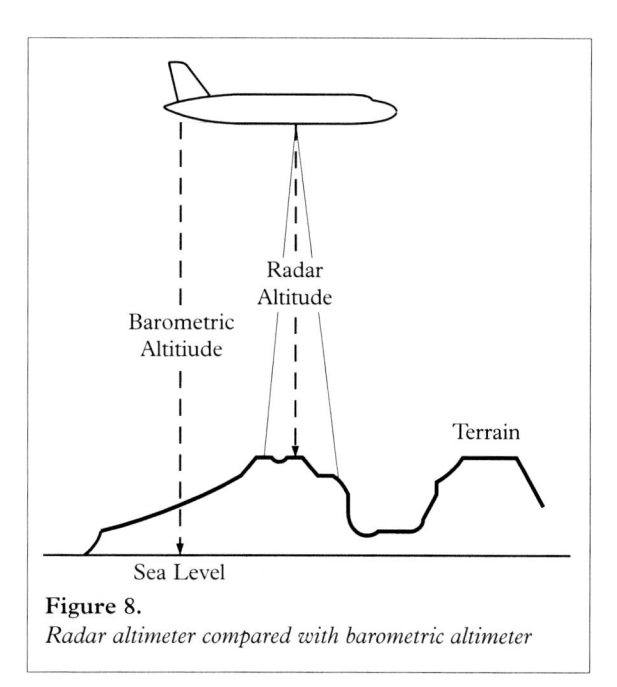

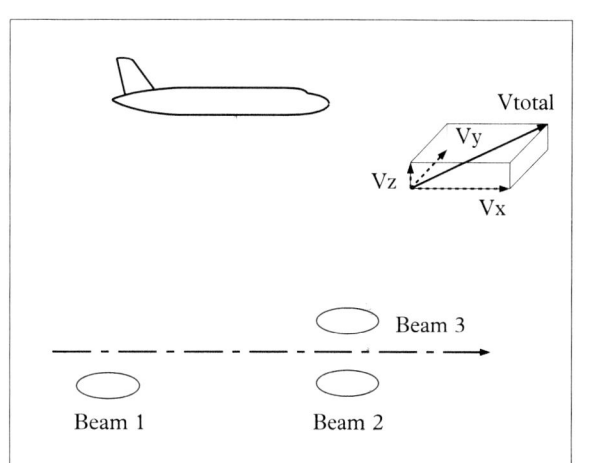

Figure 8.
Radar altimeter compared with barometric altimeter

Figure 9. *Principles of Doppler radar*
Beams 2 and 3 will return higher frequencies where the frequency increase is proportional to the aircraft ground speed, while beam 1 will detect a lower frequency where the frequency decrease is also proportional to ground speed. This enables the aircraft groundspeed to be derived. If the aircraft is drifting across-track due a crosswind, then the beams will also detect the lateral frequency difference component, and cross-track velocity may be measured. Finally, by using computation, the aircraft V_x, V_y and V_z velocity components may be determined, as may the overall aircraft velocity vector. Doppler radar velocity outputs may be compared with those from the INS, thereby making possible a more accurate estimate of aircraft velocities and position.

FASTER, FURTHER, HIGHER

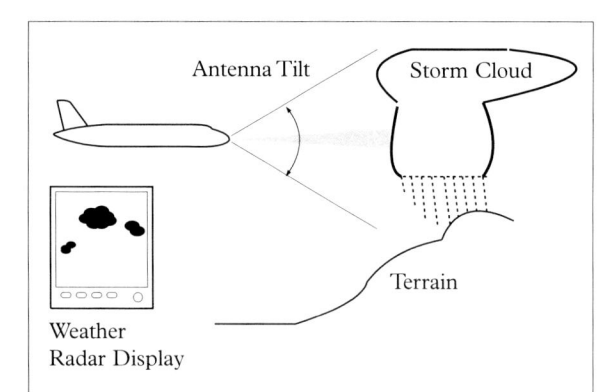

Figure 10. *Operation of weather radar*
The radar beam (shown in grey) is pointing horizontally ahead of the aircraft and will detect the storm cloud through which the aircraft is about to fly. By referring to his weather radar display, the pilot will be able to see if the storm cells can be avoided by altering course left or right. The use of the antenna tilt function is crucial. In the example given, if the antenna is fully raised the crew will not gain any information relating to storm cloud, precipitation or rising terrain. If the antenna is fully depressed, the radar will detect the rising terrain but not the storm cloud or precipitation ahead. Many weather radars therefore incorporate an automatic tilt feature, so that the radar returns are optimised for the flight crew.

barometric or air data altimeter, where the altitude may be referenced to sea level or some other datum. The radar altimeter is therefore of particular value in warning the pilot that he or she is close to the terrain and might need to take corrective action. Alternatively, it may provide the flight crew with accurate altitude with respect to terrain during the final stages of a precision approach. The principle is shown in Figure 8.

Most radar altimeters use a triangular modulated frequency technique on the transmitted energy. Comparison of the frequency of the reflected energy with the transmitted energy yields a frequency difference that is proportional to the time taken for the radiated energy to return; hence radar altitude may be calculated.

Doppler radar
Doppler radar transmits energy in three beams skewed to the front and rear of the aircraft, as shown in Figure 9. The beams are also skewed laterally to the sides of the aircraft track. As for the radar altimeter, it depends upon the radiated energy being reflected from the terrain within the Doppler beams, and, as before, a frequency difference between the radiated and reflected energy carries vital information. Owing to the effects of the Doppler principle, energy reflected from beams facing forward will be returned with a higher frequency than the radiated energy. Conversely, energy from a rearward facing beam will have a lower frequency than that radiated.

Doppler radar suffers from one significant drawback. In certain circumstances the terrain over which the aircraft is flying may not reflect enough energy for the aircraft velocities to be determined. Such conditions may occur while flying over an expanse of water where the surface is very smooth, giving a 'mill pond' effect, or flying over snow-covered or glacial terrain may cause the radar to 'loose lock' and produce unreliable readings.

Doppler radar was commonly used in the 1960s, but the advent of inertial systems and, more recently, GPS means that this technique is little used in systems produced today.

Weather radar
The weather radar has been in use for over forty years to alert the flight crew to the presence of adverse weather or terrain in the aircraft's flightpath. The weather radar radiates energy in a narrow beam with a width of ~3° which may be reflected from clouds or terrain ahead of the aircraft. The radio waves are beamed either side of the aircraft centreline to give a radar picture of objects ahead. The antenna may also be tilted in elevation by around ±15° from the horizontal to scan areas above and below. The principle of operation is shown in Figure 10, which shows a storm cloud directly ahead of the aircraft, with some precipitation below and also steadily rising terrain. Precipitation can be indicative of severe vertical windshear, which can cause a hazard.

Most modern weather radars can use Doppler processing to detect turbulence ahead of the aircraft. This is a very useful feature, as maximum windshear does not necessarily occur coincidentally with the heaviest precipitation. In fact, some of the most dangerous windshear can occur in clear air with the aircraft flying nowhere near any clouds or precipitation

The radar picture may be displayed on a dedicated radar display or overlaid on the pilot or first officer's navigation display. Displays are typically in colour, which helps the flight crew to interpret the radar data, have various selectable range markers and are usually referenced to the aircraft heading. Separate displays may be provided for weather or turbulence modes.

The usefulness of weather radar depends greatly upon interpretation by the flight crew. As in other areas, the flight crew are unlikely to depend upon the information provided by the weather radar alone, but are also likely to confer with air traffic controllers and take account of status reports from aircraft that have already flown through the area.

Communications & Navigation Aids

The sensors described so far are those that are on board or are autonomous to the aircraft. However, the aircraft

NEW-AGE SYSTEMS

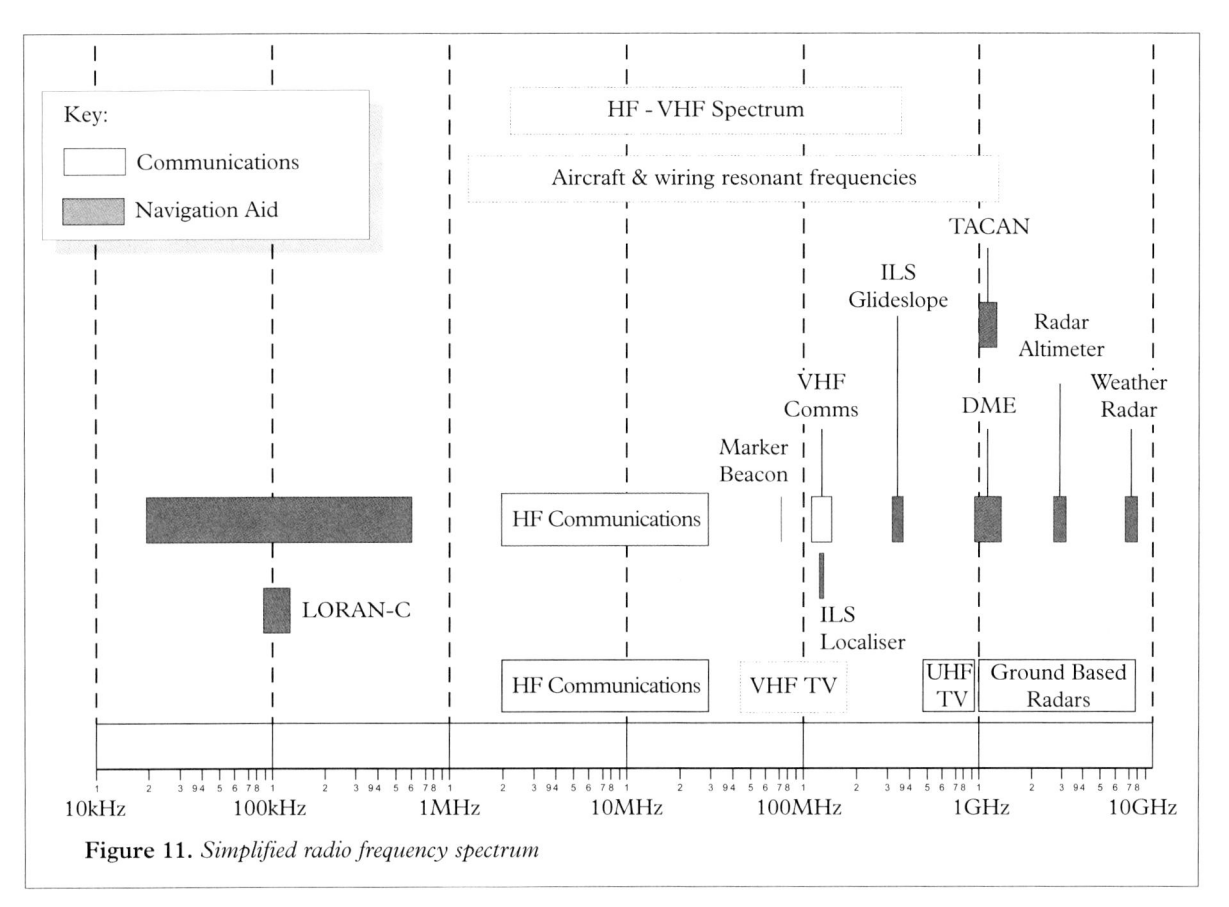

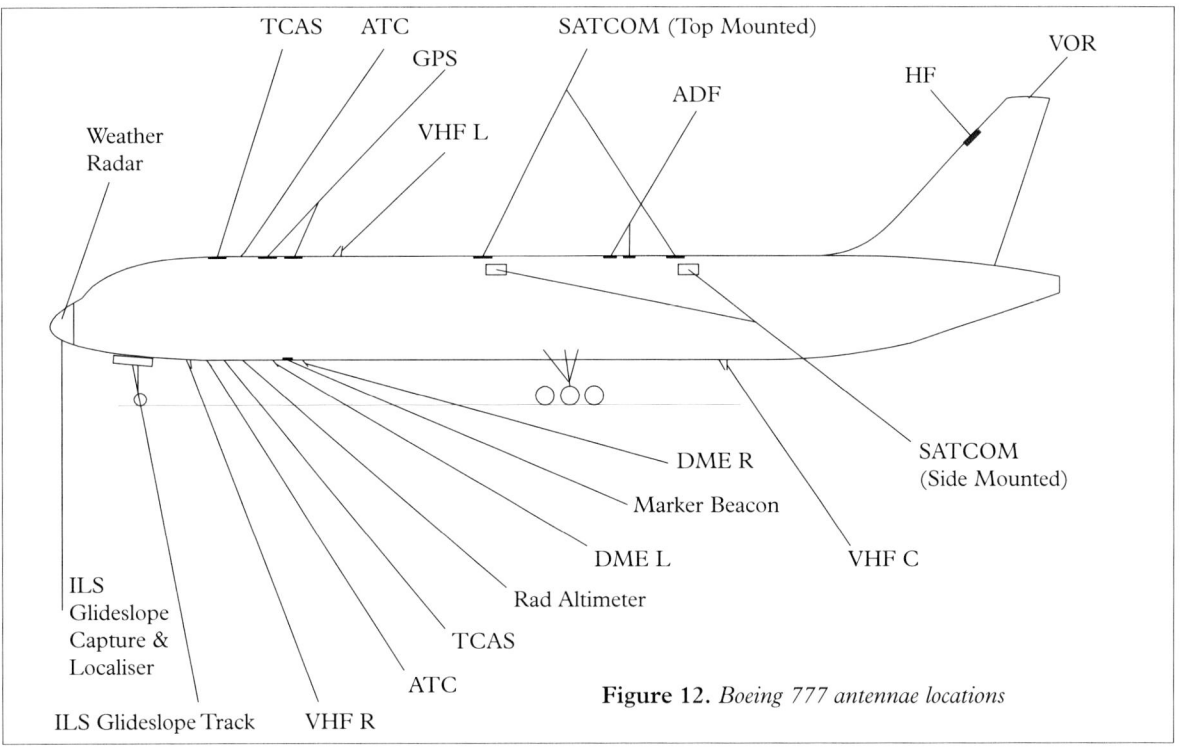

Figure 11. *Simplified radio frequency spectrum*

Figure 12. *Boeing 777 antennae locations*

FASTER, FURTHER, HIGHER

also uses a number of other systems, either for communications or for navigational assistance, that depend upon external agencies.

Communications systems comprise the following:

- High frequency (HF)
- Very high frequency (VHF) and aircraft communications and reporting system (ACARS)
- Satellite communications (SATCOM)
- Air traffic control (ATC) Mode S
- Traffic collision and avoidance system (TCAS)
- Communications control

Common navigation aids are:

- Very-high-frequency omni-range (VOR)
- Distance measuring equipment (DME)
- Automatic direction finding (ADF)
- TACAN
- VOR/TACAN (VOR/TAC)
- Instrument landing system (ILS)
- Microwave landing system (MLS)
- Hyperbolic navigation systems

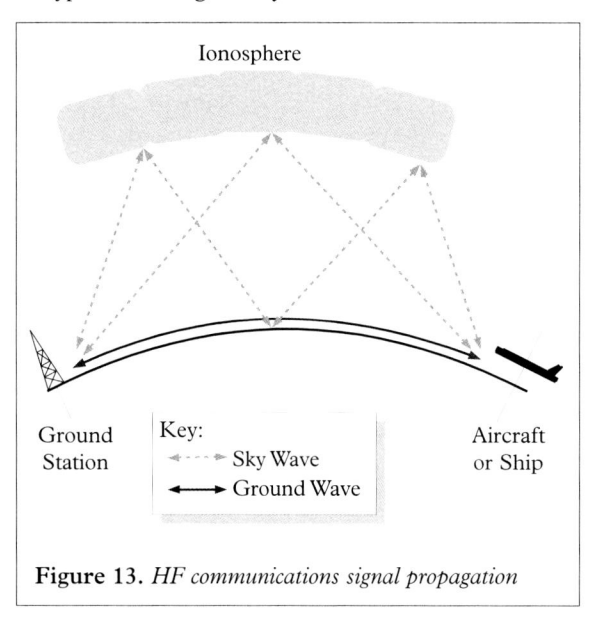

Figure 13. *HF communications signal propagation*

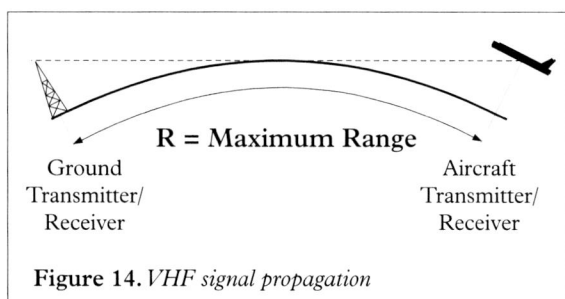

Figure 14. *VHF signal propagation*

Radio frequency spectrum

The radio frequency (RF) spectrum from 10kHz (1 x 10⁴ Hertz) up to 10 GigaHertz (1 x 10¹⁰ Hertz) is shown in a simplified form in Figure 11. This spectrum, stretching over six decades, covers the range in which most of the civil aircraft equipment operates. For military aircraft the spectrum will be wider. The wide frequency coverage of this spectrum, and the nature of radio wave propagation, mean that the performance of different equipment varies according to the conditions of operation. The figure distinguishes between communications and navigation aids. It can be seen that the part of the spectrum at which aircraft equipment and wiring systems are most susceptible to emissions covers a wide band. Therefore care has to be taken when designing and operating the aircraft that any mutual interference effects are kept to a minimum. The figure also shows that various ground based domestic equipments, namely HF communications, VHF TV and radio transmissions, UHF TV transmissions and ground based radars, can have an adverse effect.

A considerable number of antennae are required on board an aircraft to handle all the sensors, communications and navigation aids. This is compounded by the fact that many of the key equipments may be replicated in duplicated or triplicated form. This is especially true of VHF, HF, VOR and DME equipments. Figure 12 shows typical antennae locations on a Boeing 777, but is indicative of the installation on most modern civil aircraft.

Communications Systems – High Frequency

High Frequency (HF) covers the communications band between 3 and 30MHz, and is a very common communications means for land, sea and air. Such communications have an idiosyncrasy regarding the means of signal propagation.

Figure 13. shows that there are two main means of propagation; known as the sky wave and the ground wave. The sky wave method relies upon single or multiple path bounces between Earth and the ionosphere until the signal reaches its intended location. The behaviour of the ionosphere is itself greatly affected by radiation falling upon Earth, notably solar radiation. High sunspot activity adversely affects the ionosphere's ability as a reflector. It may also be affected by the time of day and other atmospheric conditions. The sky wave as a means of propagation may therefore be severely degraded by a variety of conditions; occasionally to the point of being unusable.

The ground wave method of propagation relies upon the ability of the wave to follow the Earth's curvature until it reaches its intended destination. Like the sky wave, the ground wave may on occasions be adversely affected by atmospheric conditions. On occasions, therefore, HF voice communications may be unreliable, though HF data links are more resistant to these propagation upsets.

HF communications are one of the main methods of communicating over long ranges between air and ground during oceanic and wilderness crossings, when there is no line-of-sight between the aircraft and ground communications stations. Most long-range civil aircraft are equipped with two HF sets.

Communications Systems – Very High Frequency
Very-high-frequency communication is probably the most heavily used method of communication used by civil aircraft. The VHF band operates in the frequency range 30 to 300MHz.

The VHF band also experiences limitations in the method of propagation, illustrated in figure 14. VHF signals will normally only propagate over line-of-sight, the signal only being detected by the receiver when it has line of sight or can 'see' the transmitter. As VHF transmissions possess neither of the qualities of HF transmission, neither sky-wave nor ground-wave properties apply. This line-of-sight property is affected by the relative heights of the radio tower and aircraft. For an aircraft flying at 35,000ft (10,600m), transmissions will generally be received by a radio tower 100ft (30m) high if the aircraft is within a range of around 235 nautical miles (330km). Additionally, VHF transmissions may be masked by terrain, such as range of mountains. These limitations also apply to equipments operating in higher frequency bands and mean that VHF communications, and similar equipment such as the navigation aids VOR and DME, may not be used except over large land masses, and then only when there is adequate transmitter coverage. Most long-range aircraft have three VHF equipments, one usually being used for ACARS though not dedicated to that purpose.

The ARINC communications and reporting system (ACARS) is a specific variant of VHF communications operating on 131.55MHz and using a data link rather than voice transmission. As will be seen during the discussion on future air navigation systems, data link rather than voice transmission will be increasingly used for air-to-ground and air-to-air communications. ACARS is dedicated to down-linking operational data to the airline's operational control centre. Originally only four basic parameters were transmitted, but now data such as fuel state, aircraft serviceability, arrival and departure times, weather, crew status and so on are also included in the data messages. ACARS is dedicated to assisting the operational effectiveness of an airline. Future data link applications will allow the transfer of more complex data relating to air traffic control routing and flight planning. On board the aircraft, ACARS introduces a dedicated management unit, control panel and printer to provide the interface with the flight crew for formatting, dispatching, receiving and printing messages.

Communications systems – satellite
Satellite communications provide a more reliable method of communications using the INMARSAT satellite constellation, originally developed for maritime use. The system forms a useful component of aerospace communications.

The principles of operation of SATCOM are shown in Figure 15. The aircraft communicates via the INMARSAT constellation and remote ground Earth station by means of C-Band and L-Band uplinks and downlinks. In this way communications are routed from the aircraft via the satellite to the ground station, and on to the destination. Conversely, communications to the aircraft travel in the reverse fashion. Provided the aircraft is within the area of coverage or footprint of a satellite, communication may be established.

The coverage offered by the INMARSAT constellation comprised four satellites in 2001. Further satellites are due to be launched in the near future. Their geostationary nature does confer disadvantages. Owing to low grazing angles, coverage begins to degrade beyond 80° north and 80° south, and fades completely beyond about 84°. Therefore no coverage exists in the extreme polar regions, a fact that assumes more prominence as airlines seek to expand northern polar routes. A second limitation is posed by the performance of the on-board aircraft system in terms of antenna installation. Nevertheless, SATCOM is proving to be a very useful addition to the airborne communications suite, and promises to be an important component as FANS procedures are developed.

Air traffic control Mode S
As a means to aid the identification of aircraft and facili-

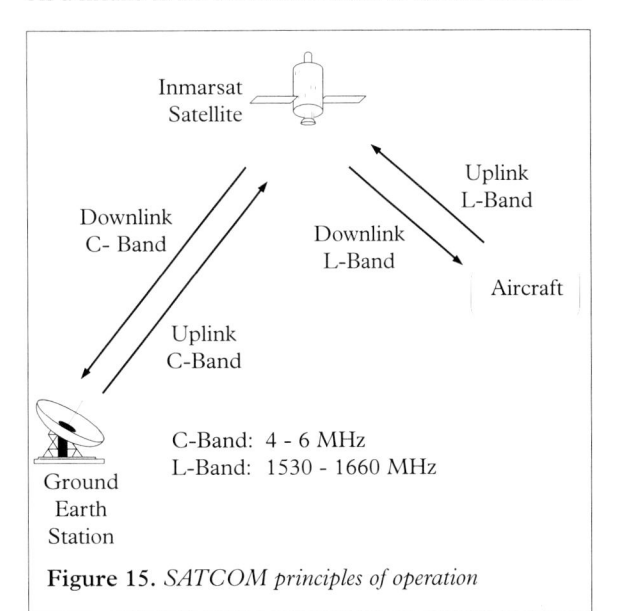

Figure 15. *SATCOM principles of operation*

tate their safe passage through controlled airspace, an ATC transponder allows ground surveillance radars to interrogate aircraft and formulate data enabling a radar track to be correlated with a specific aircraft. In its present form the ATC transponder allows aircraft identification (usually the airline call sign) together with height and track information. ATC Mode S is intended to expand this capability to provide:

- Air-to-air as well as air-to-ground communication
- The ability of aircraft to determine, autonomously, the precise whereabouts of other aircraft in their vicinity
- When used together with TCAS, AC Mode S will provide an important feature for FANS; that of automatic dependent surveillance – A (ADS-A), which will assist the safe passage of aircraft when operating in a 'free-flight' mode.

Traffic collision and avoidance system
The TCAS is intended to help aircraft avoid collisions. It has been mandatory for a number of years in the USA, and is now expanding into Europe. The principles and use of TCAS are explained under 'Safety Aids'

Communications control system
The control of an aircraft's suite of communications systems, including internal communications, has become increasingly complex. This task has expanded as speeds have increased and the breadth of communications types has expanded. The communications control function is increasingly being absorbed into the flight management function as the management of communications type, frequency selection and intended aircraft flight path become more interwoven. Now the flight management system can automatically select and tune the communications and navigation aids required for a particular flight leg, reducing crew workload and allowing the crew to concentrate more on managing the on-board systems.

Navigation aids – very high frequency Omni-Range
This comprises a widely used set of radio beacons operating in the VHF frequency band over the range 108 to 118MHz, with 100kHz spacing. Each beacon emits a morse code modulated tone which may be provided to the flight crew for the purposes of beacon identification. The beacon provides a phase modulated output such that the aircraft's VOR receiver may determine the bearing of the beacon from the aircraft. Overland in the North American continent and Europe, VOR beacons are widely situated to provide an overall coverage. Usually they are arranged to coincide with major airway waypoints and intersections in conjunction with DME stations (see below) such that the aircraft may navigate for the entire flight using the route/beacon structure. By virtue of the transmissions within the VHF band, these beacons are subject to the line-of-sight and terrain-masking limitations of VHF communications. Typical VOR accuracy is around 2.5 to 3 degrees.

Navigation aids – distance measuring equipment
Distance measuring equipment is a method of pulse-ranging used in the 960 to 1215MHz band. The aircraft equipment interrogates a ground-based beacon and upon the receipt of retransmitted pulses – unique to the on-board equipment – determines the range to the DME beacon. The beacons can service simultaneous requests from a large number of aircraft, but are generally understood to be able to handle approximately 110 aircraft at once. Specified DME accuracy is ±3% or ±0.5 nautical miles, whichever is the greater.

Throughout the airway route structure DME beacons are paired with VOR beacons, so that aircraft can navigate the airways by a having a combination of VOR bearing to, and DME distance to run to the next beacon in the airway structure. A more recent development, scanning DME, allows the airborne equipment to scan a number of DME beacons rapidly, thereby achieving greater accuracy by taking the best estimate of a number of distance readings. This combination of VOR/DME has served the flying community well in those countries where the beacon structure has been established and maintained. The progress of the global positioning system (GPS), described in the 'Navigation' section, is more likely to prove attractive to developing countries that have not been able to invest heavily in VOR/DME beacons.

Navigation aids – automatic direction finding
This involves the use of a loop direction-finding technique to establish the bearing to a radiating source. This might be to a VHF beacon or a non-directional beacon (NDB) operating in the 200 to 1600kHZ band. Non-directional beacons in particular are the most prolific and widely spread beacons.

Navigation aids – TACAN
Tactical air navigation (TACAN) is a military omni-bearing and distance measuring equipment using techniques similar to DME for distance measurement. The bearing information is accomplished by amplitude modulation achieved within the beacon, which imposes 15Hz and 135Hz modulated patterns and transmits this data together with 15Hz and 135Hz reference pulses. The airborne equipment can therefore measure distance using DME interrogation techniques while using the modulated data to establish bearing.

Because TACAN beacons operate in the frequency band 960 to 1215MHz, as opposed to the 108 to 118MHz used by DME, they are smaller and thus suitable for shipborne and mobile tactical use. Some airborne equipments have the ability to offset to a point

NEW-AGE SYSTEMS

ILS Characteristics:

Localiser transmitting between 108-112MHz

Glideslope transmitting between 329-335MHz

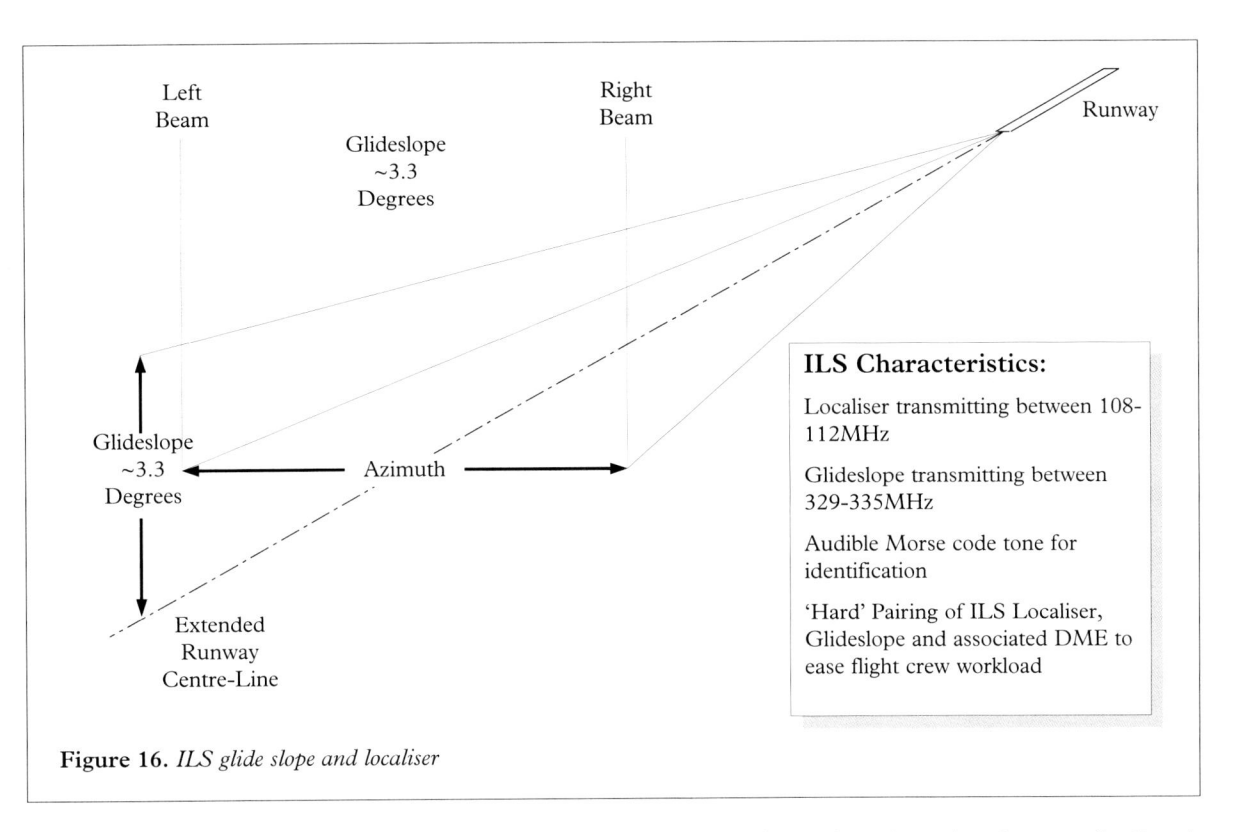

Audible Morse code tone for identification

'Hard' Pairing of ILS Localiser, Glideslope and associated DME to ease flight crew workload

Figure 16. *ILS glide slope and localiser*

remote from the beacon, which facilitates recovery to an aerodrome when the TACAN beacon is not co-located. The system is reportedly accurate to within ±1 per cent in azimuth and ±0.1 nautical miles in range, so it offers accuracy improvements over VOR/DME.

Navigation aids – VORTAC

As most military aircraft are equipped with TACAN, some countries provide VORTAC beacons, which combine VOR and TACAN beacons. This allows interoperability of military and civil air traffic. Military users use the TACAN beacon while civil users use the VOR bearing and TACAN (DME) distance measuring facilities.

Navigation aids – instrument landing system

The ILS is an approach aid that has been in widespread use since the 1960s and 1970s. The main elements of the system include:

- A localiser antenna centred on the runway to provide lateral guidance; 40 operating channels are available within the band 108 to 112MHz
- A glide slope antenna located beside the runway threshold to provide lateral guidance. Forty operating channels are available within the frequency band 329 to 335MHz.

The ILS localiser, glide slope and DME channels are paired in such a way that only the localiser channel needs to be tuned for all three channels to be correctly aligned. [figure 16]

Marker beacons located at various points down the approach tell the pilot what stage on the approach has been reached. These are outer, middle and inner markers. The marker beacons all fan beams radiating on 75MHz, and provide different morse code modulation tones which can be heard through the pilot's headset [figure 17].

A significant disadvantage of the ILS system is its susceptibility to beam distortion and multi-path effects. This distortion can be caused by local terrain effects or large artificial structures, and even taxying aircraft can cause unacceptable beam distortion, the glide slope being the most sensitive. At times on busy airports and during periods of limited visibility, this may preclude the movement of aircraft in sensitive areas, leading, in turn, to a reduction in airport capacity. More recently interference by high-power local FM radio stations has presented an additional problem.

Navigation aids – microwave landing system (MLS)

The MLS approach aid was conceived to redress some of the shortcomings of the ILS. The specification of a time-reference scanning beam MLS was developed through the late 1970s/early 1980s, and a transition to MLS was envisaged to begin in 1998. However, with the emergence of satellite systems such as GPS there was also a realisation that both ILS and MLS could be rendered obsolete

FASTER, FURTHER, HIGHER

ILS Markers:

All markers transmit on 75MHz

Outer marker: 400Hz - 2 tones/sec

Middle marker: 13000Hz - dash-dot/sec

Inner marker: 3000Hz - six dots/sec

Figure 17. *ILS approach markers*

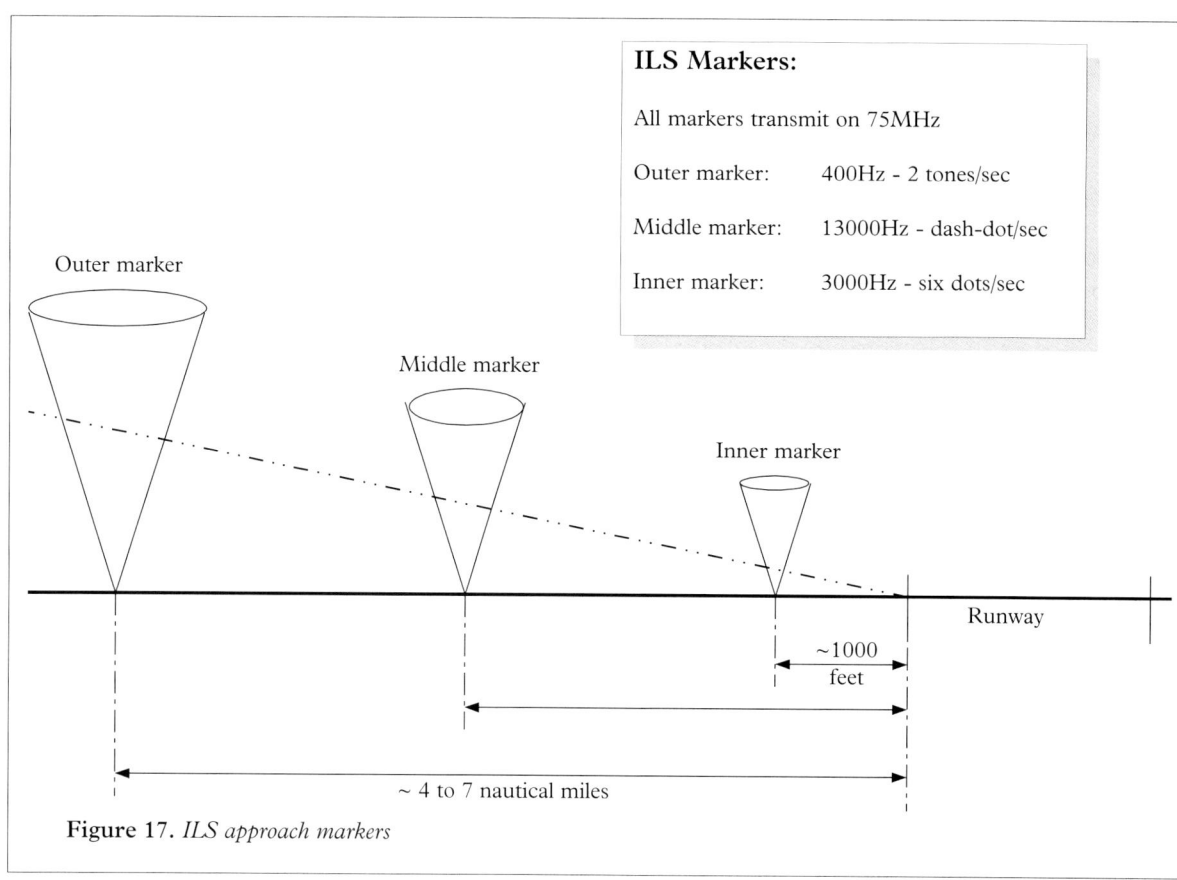

Figure 18. *Principle of operation of a hyperbolic navigation system*

The hyperbolic solid lines represent points equidistant from the blue-master and blue-slave stations. These points will all have the same time difference between the arrival of signals from the blue-master and blue-slave stations. This in itself will not yield position, but if a second pair of stations is used – angled approximately 45° to the first – shown as dashed lines, then position can be obtained. The relative positioning of the lines in this dual chain example shows that three outcomes are possible:

• *At point A the lines cross at almost 90° and this represents the most accurate fix*

• *At point B the lines cross at a much more acute angle and the result is a larger error ellipse*

• *At point C there are two possible solutions and an ambiguity exists which can only be resolved by using a further station.*

NEW-AGE SYSTEMS

when such systems matured. In the event, the US civilian community is embarking upon higher-accuracy developments of the basic GPS system; the wide area augmentation system (WAAS) and local area augmentation system (LAAS), which will be described later. Europe; especially the UK, The Netherlands and Denmark have embarked upon a modest programme of MLS installations at key airports.

The MLS operates in the 5031.0 to 5190.7MHz frequency band and offers some 200 channels of operation. It has wider field of view than ILS, covering ±40° in azimuth and up to 15° in elevation. The addition of a DME beacon permits 3D positioning with regard to the runway, and the combination of higher data rates mean that curved arc approaches may be made, as opposed to the somewhat pedestrian linear approach offered by ILS. This offers advantages when operating into aerodromes with confined approach geometry and tactical approaches favoured by the military.

Navigation aids – hyperbolic navigation systems
Hyperbolic navigation systems operate upon hyperbolic lines of position rather than circles or radial lines. Figure 18 illustrates the principle of operation of a hyperbolic system in a very elementary manner.

The system in use today is LORAN-C, conceived in principle around the beginning of the Second World War. Worldwide coverage existed in 1996, and new facilities were being planned in the late 1990s. Operating in the frequency band 90 to 110kHZ, LORAN is a pulsed system which enables the ground wave to be separated from the sky wave, the ground wave being preferred. A LORAN chain will comprise at least three stations, one being nominated as the master. The time difference of arrival between the master and slaves allows position to be determined. Each of the stations in a chain transmits unique identifiers which allow the chain to be identified. Within the defined area of coverage of the chain, LORAN-C will provide a user with a predictable accuracy of 0.25 nautical miles. A typical chain will have an operating range greater than 870 miles (1,400km). Also capable of relaying GPS positional error within the transmissions, LORAN-C is expected to remain in commission until at least 2008.

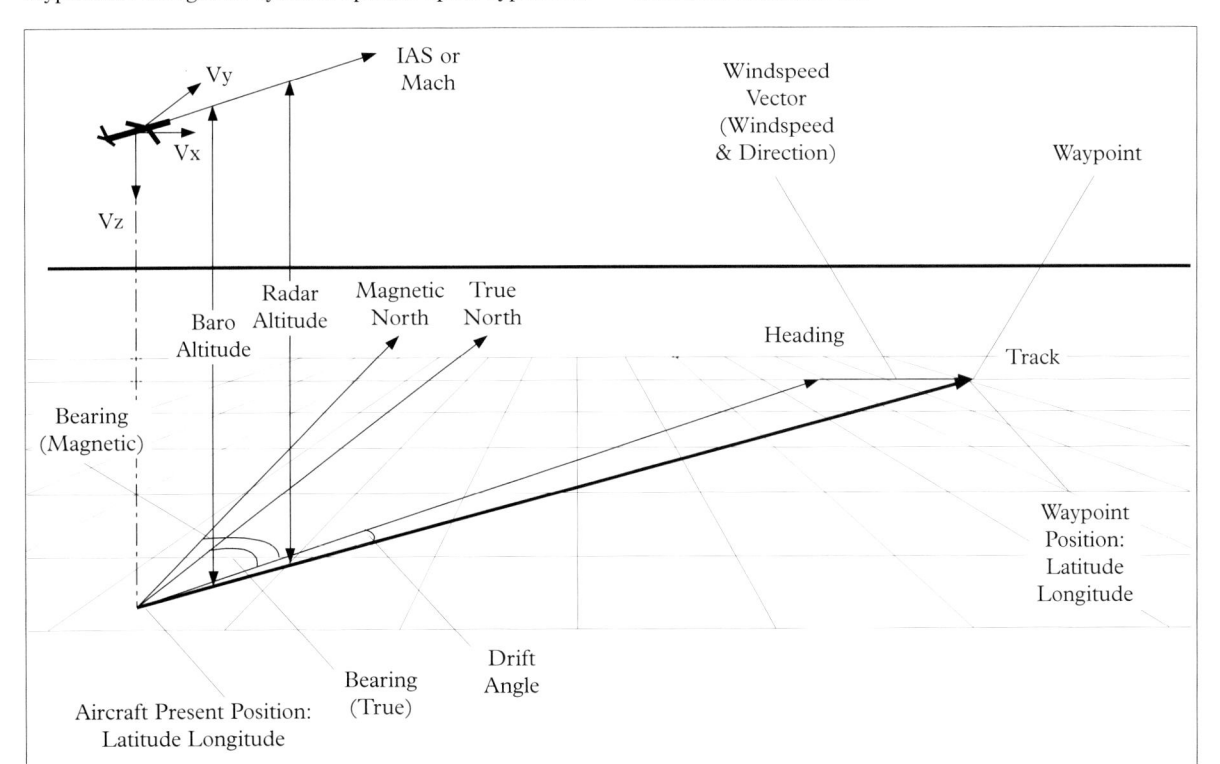

Figure 19. *Basic navigation parameters*
The classic method of navigation, in use for many years, is to use a combination of the magnetic and inertial directional gyros shown in Figure 6, together with airspeed information derived from the air data sensors. This is subject to errors in both the heading system and the effects of prevailing winds, which can cause along-track and across-track errors. In the 1930s it was recognised that the use of radio beacons and navigation aids could significantly reduce these errors by providing the flight crew with navigation assistance related to precise points on the ground.

Figure 20. *VOR/DME route navigation*

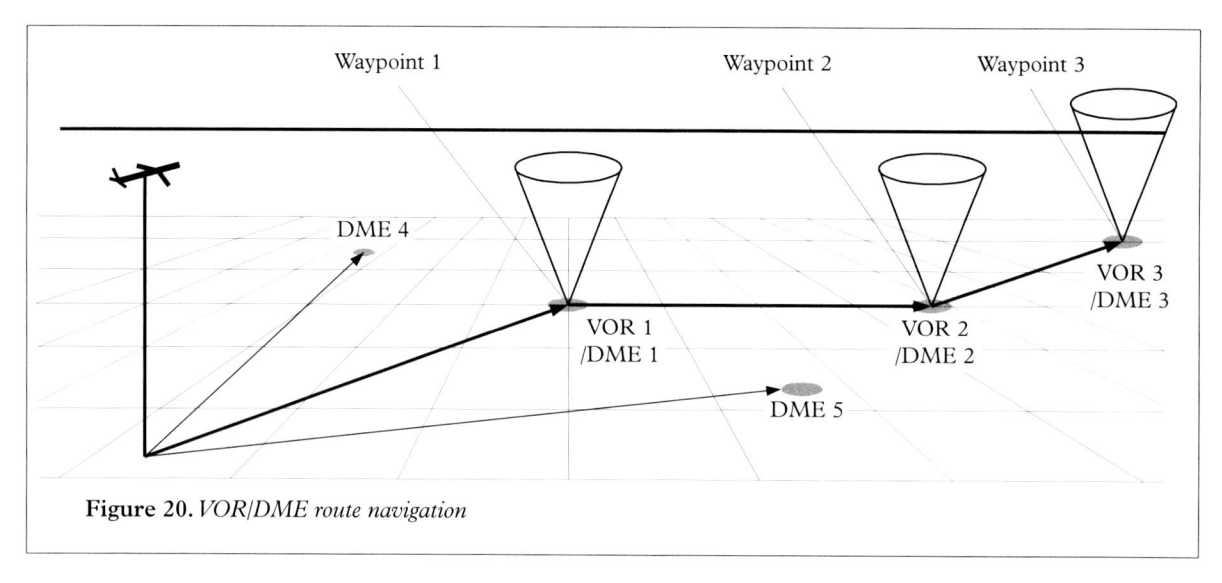

NAVIGATION

While the principles of navigation have not changed since the early days of sail, the increased speed of flight, particularly with the advent of the jet age, has placed greater emphasis upon accurate navigation. The increasingly busy skies and rapid technological developments have both brought the need for higher accuracy navigation and the means to accomplish it. This section addresses some of the modern methods of navigation, and leads to a later section on the FANS.

The main methods of navigation as practised today may be summarised and simplified as follows:

- Classic magnetic-inertial navigation using magnetic and inertial (gyro) sensors
- Radio navigation using navigation aids
- Barometric-inertial navigation using a combination of air data and INS
- Satellite navigation using a global navigation satellite system (GNSS), more usually GPS
- Integrated navigation using a combination of all of the above.

Classic navigation

The basic navigation parameters are shown in Figure 19, and may be briefly summarised:

- An aircraft will be flying at a certain height or altitude relative to a barometric datum (barometric altitude) or terrain (radar altitude)
- The aircraft may be moving with velocity components in the X (Vx), Y (Vy) and Z (Vz) axes. Its speed through the air may be characterised as indicated airspeed (IAS) or Mach number
- The aircraft will be flying on a certain heading, though the prevailing wind speed and direction will modify this to the aircraft track. The aircraft track will lead to the aircraft's destination or next waypoint
- The aircraft heading will be defined by a bearing to magnetic (compass) north or to true north relating to Earth related geographic co-ordinates
- The aircraft will be flying from its present position, defined by latitude and longitude, to a waypoint also characterised by latitude and longitude
- A series of flight legs, defined by waypoints, will determine the aircraft's designated flight path from the departure airport to the destination airport

As has already briefly been described, a combination of sensors and navigation techniques may be used solely or in combination to navigate the aircraft throughout the mission.

Radio navigation

For many years the primary means of navigation over land, at least in Europe and on the North American continent, was by means of radio navigation routes defined by VOR/DME beacons, as shown in Figure 20. By arranging the location of these beacons at major navigation or crossing points, and in some cases aerodromes, it was possible to construct an entire airway network which could be used by the flight crew from take-off to touchdown.

In addition to using navigation information from the 'paired' VOR/DME beacons that define the main navigation route, bearing or cross-fix information could also be derived from DME or NDB beacons in the vicinity of the planned route by using automatic direction finding techniques. As has already been described, VOR and DME operate at frequencies requiring a line-of-sight contact between transmitter and receiver: terrain or distance can block the radio beacons' signals. Their reliability and

Figure 21. *Principles of baro-inertial navigation*

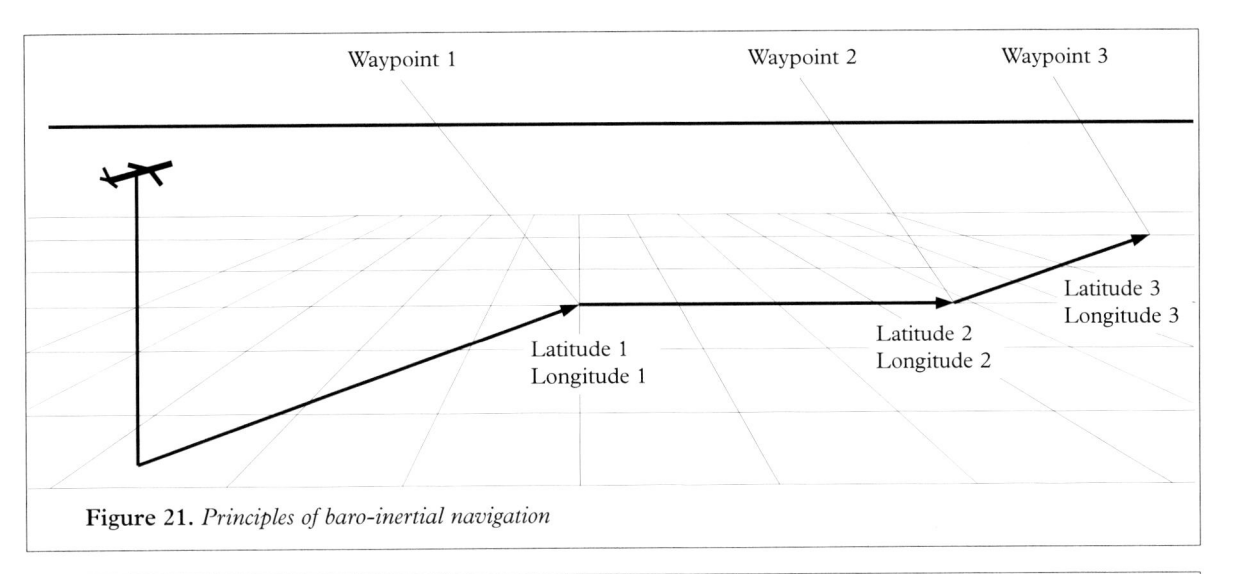

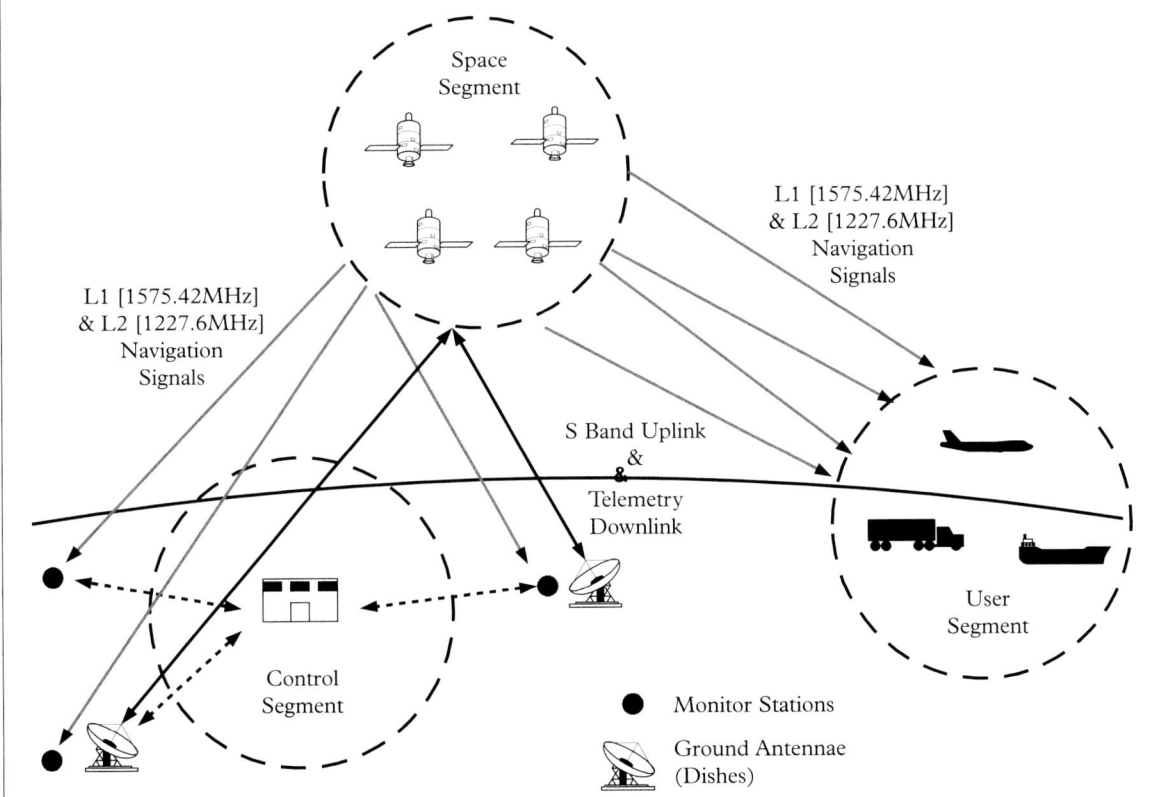

Figure 22. *Principles of GPS satellite navigation*
GPS comprises three major components:
- The control segment, which embraces the infrastructure of ground control stations, monitor stations and ground-based satellite dishes that exercise control over the system
- The space segment, which includes the satellite constellation, presently around 25 satellites, which form the basis of the network
- The user segment, which includes all the users: ships, trucks, automobiles, aircraft and hand-held sets. In fact anyone in possession of a GPS receiver is part of the user segment.

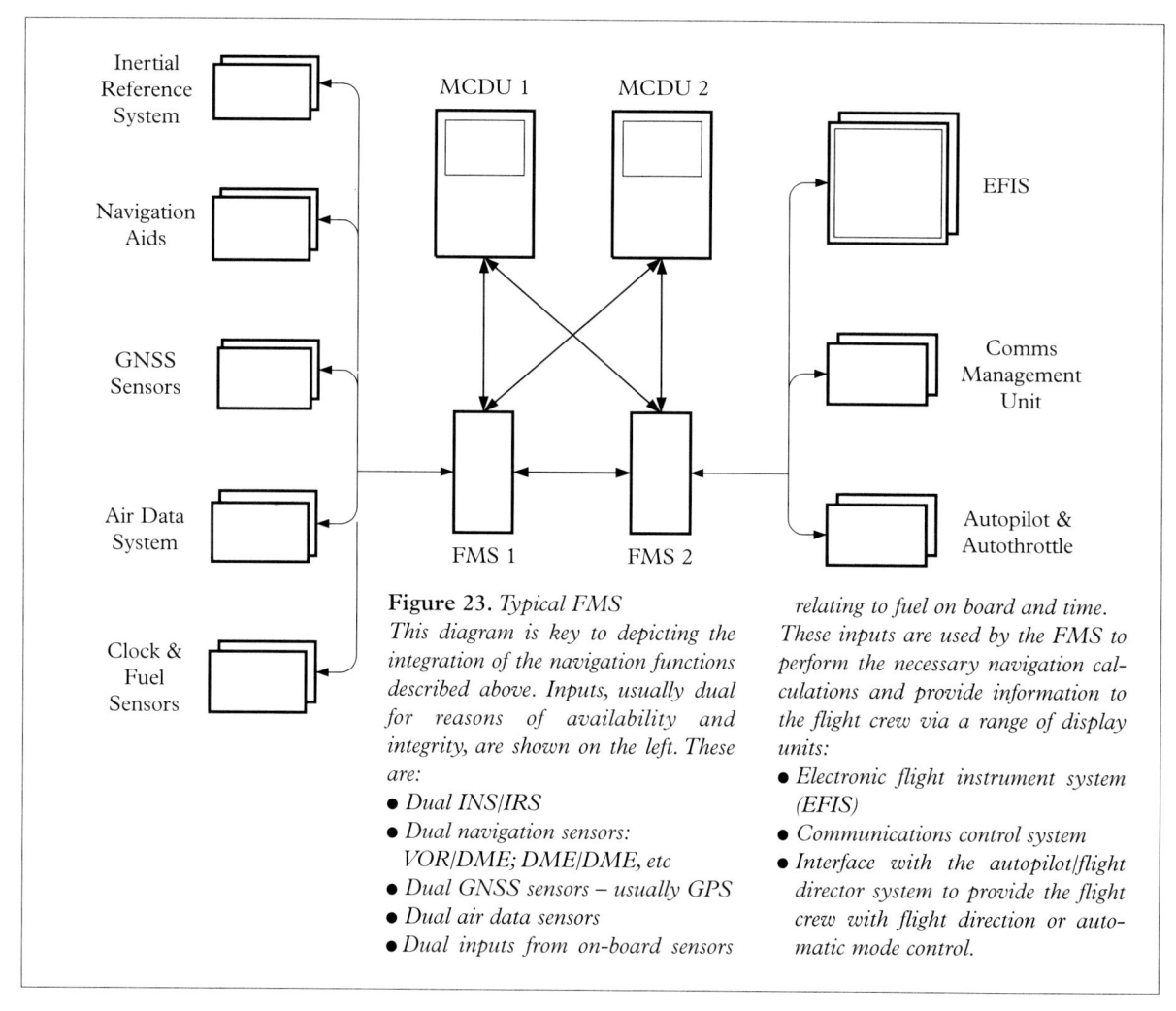

Figure 23. *Typical FMS*
This diagram is key to depicting the integration of the navigation functions described above. Inputs, usually dual for reasons of availability and integrity, are shown on the left. These are:
- *Dual INS/IRS*
- *Dual navigation sensors: VOR/DME; DME/DME, etc*
- *Dual GNSS sensors – usually GPS*
- *Dual air data sensors*
- *Dual inputs from on-board sensors relating to fuel on board and time.*

These inputs are used by the FMS to perform the necessary navigation calculations and provide information to the flight crew via a range of display units:
- *Electronic flight instrument system (EFIS)*
- *Communications control system*
- *Interface with the autopilot/flight director system to provide the flight crew with flight direction or automatic mode control.*

accuracy can also be severely affected by electric storms. Over longer ranges LORAN-C could be used.

Barometric-inertial navigation

The availability of INS to the civil aviation community during the late 1960s added another dimension to the navigation equation. Flight crew could now navigate by autonomous means using an on-board INS employing inertial sensors. By aligning the platform to Earth-referenced co-ordinates and present position during initialisation, it was now possible to fly for long distances without relying upon VOR/DME beacons overland or hyperbolic navigation systems elsewhere. Waypoints could be specified in terms of latitude and longitude as arbitrary points on the globe, more suited to the aircraft's intended flight path rather than a specific geographic feature or point in a radio beacon network.

By combining the air data information with the inertially derived flight information, the best features of barometric and inertial systems can be combined. For reasons of availability and accuracy, systems were developed with dual and triple INS installations. Means of taking external fixes were evolved such that longer-term inaccuracies could be corrected by updating the INS position during long flights. Some systems, such as Tornado, also added a Doppler radar so that Doppler-derived data could be included in the navigation process. The availability of on-board digital computers enabled statistical techniques to be used to calculate the best estimated position using all of the sensors available.

Satellite navigation

These techniques were prevalent from the 1960s until the 1990s, when satellite navigation became commonly available. The use of GNSS offers a cheap and accurate navigational means to anyone possessing a suitable receiver. Although the former Soviet Union developed a system called GLONASS, the US ground positioning system (GPS) is the most widely used. Its principles are illustrated in Figure 22.

NEW-AGE SYSTEMS

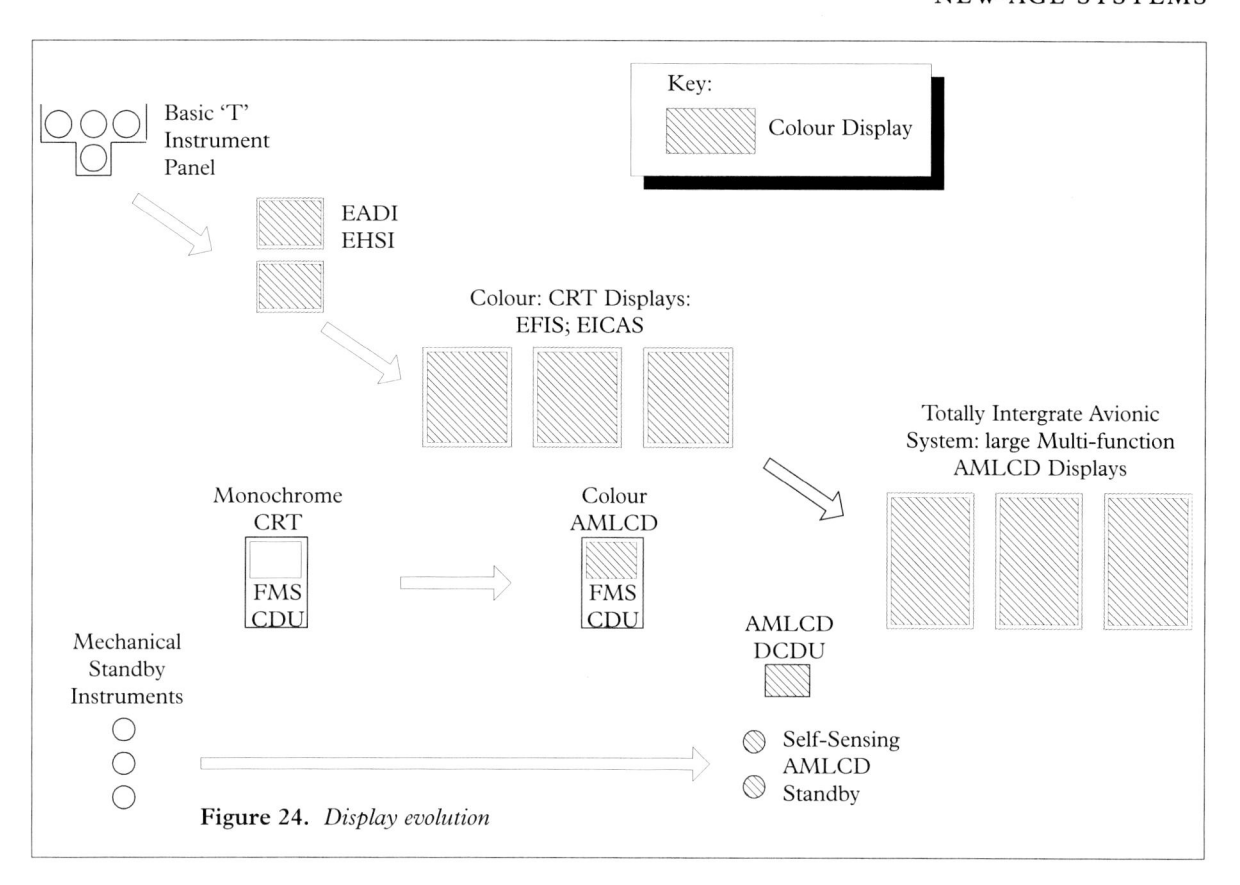

Figure 24. *Display evolution*

The baseline satellite constellation downlinks data in two bands: L1 on 1575.42 MHz and L2 on 1227.60 MHz. A GPS modernisation programme will provide a second civil signal in the L2 band for satellites launching in 2003 and onwards. In addition a third civil signal, L5, will be provided on 1176.45 MHz on satellites launched in 2005 and beyond. Finally, extra signals for military users (Lm) will be included in the L1 and L2 bands for satellites launched in 2005 and beyond.

Users are able to detect the data from a number of GPS satellites in the constellation and are able to determine their precise position in latitude and longitude. There is discrimination between military users (for whom the system was originally conceived) and civilian users. Until recently, a selective availability feature that provided increased accuracy to military users was denied to civil users. This feature is presently available to all users, pending the development of differential or augmented GPS systems that will provide high accuracy to civilian users while preserving the accuracy and security demanded by military users.

The basic accuracy without selective availability is about ±100m (30ft) as opposed to ±1m (3ft) when the full system is available. Developments are under way in the USA to improve the accuracy available to civilian users. These are:

- A wide area augmentation system (WAAS) to improve accuracy en route. The WAAS improves the GPS accuracy to around ±7m (23ft), a considerable improvement over the 'raw' signal. This level of accuracy is sufficient for Cat I approach guidance. Some problems were experienced during initial system tests during 2000. Commissioning of the WAAS now appears unlikely before 2003 at the earliest in some parts of the USA.
- A local area augmentation system (LAAS) to improve terminal guidance. The LAAS improves the GPS accuracy to about ±1m (3ft), close to the higher GPS level of accuracy and sufficient to permit Cat II and Cat III approaches. Implementation is expected to begin in 2003, with final deployment in 2006, though these timescales are apt to slip, as has the implementation of WAAS.

Integrated navigation

As the name suggests, integrated navigation employs all of the features and systems described so far. It is a precursor to the introduction of the advanced navigation capabilities that will comprise the FANS, which is designed to make more efficient use of the existing airspace so that the expected increases in air traffic may be accommodated. Some elements of the FANS have already been implemented; others will take several years to attain maturity. A key prerequisite is the installation of a high-grade flight management system (FMS) to integrate all the necessary functions and provide a suitable interface with the flight crew. A typical FMS will embrace dual computers and dual multi-function control and display units (MCDUs), as shown in Figure 23 (page 152).

DISPLAYS

The aircraft flight deck displays are the key pilot-to-system interface. They provide the flight crew with the information needed to fly and navigate, details of system serviceability and status, and warnings and situational awareness. The evolution and application of civil aircraft flight deck display technology has moved swiftly over the past twenty years, and is briefly summarised in Figure 24 (page 153). In the early 1980s the pilot's basic instruments still comprised the 'basic six' flight instruments that had been the standard for the previous twenty-five years or more. The first change to this format came when colour, electronic versions of the primary attitude and horizontal situation displays were adopted. The electronic attitude director indicator (EADI) and electronic horizontal situation indicator (EHSI) represented the first step in the introduction of flightdeck colour displays.

The evolution of the FMS to assist with the increasingly demanding navigation tasks led to the introduction of monochrome cathode-ray-tube (CRT) displays for the FMS control and display unit. The availability of colour active matrix liquid crystal display (AMLCD) technology in the mid-to-late 1990s enabled the control and display unit to use colour.

The advent of large colour CRT display technology allowed the major aircraft manufacturers to adopt colour increasingly for the main suite of six aircraft displays. This technology has been the norm for the past decade, being used on virtually all but the smallest civil aircraft. Further developments in AMLCD technology, some spawned from the commercial PC laptop market, have led to increasingly large, and fewer, displays to provide the flight crew interface. The latest business-jet developments by Honeywell, in conjunction with Gulfstream and Dassault, are applying commercial laptop technology using 10in x 13in displays.

Other applications of AMLCD technology have occurred in some of the specialist display areas. Airbus, together with Smiths, has developed a datalink control and display unit (DCDU) to act as a dedicated display in the Airbus FANS 1 upgrade package to enable air-traffic-control-related messages to be displayed. Standby instruments, which have been dedicated electromechanical 3in displays for many years, are using AMLCD technology to offer colour standby displays with greatly increased functionality and reliability.

Finally, though it has taken a long time to reach fruition, civil aircraft flight decks are increasingly employing HUDs that have been used in military fighter aircraft for many years. The civil HUD increases pilot awareness during the later stages of low-visibility approaches and can reduce decision height. The use of military forward-looking infrared (FLIR) technology is also being developed as an enhanced vision system, due be certificated by Gulfstream in the near future.

Basic six instruments

Figure 25 shows the standard instrument layout until advanced display technology challenged this orthodoxy and led to more-integrated display systems. One of the first steps in this process was the use of electronic versions of the ADI and HSI. Where EADIs and EHSIs were introduced, these displays were still included in the 'T' format, replacing the conventional ADI and HSI electro-

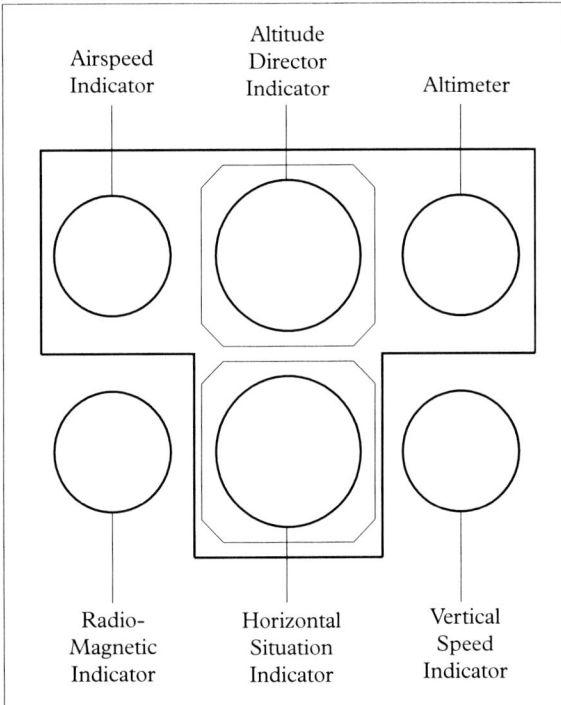

Figure 25. *Basic six flight instruments*
The basic six flight instruments, with the four most important in the classic 'T' layout. These four are:
 • *The ASI, using pitot and static air data*
 • *The attitude director indicator (ADI), using pitch and roll gyro derived attitude information*
 • *The altimeter, using static air data*
 • *The horizontal situation indicator (HSI), using azimuth heading reference data.*
Also included, outside the 'T', are:
 • *The radio magnetic indicator (RMI), using VOR derived bearing data*
 • *The VSI, using static air data.*

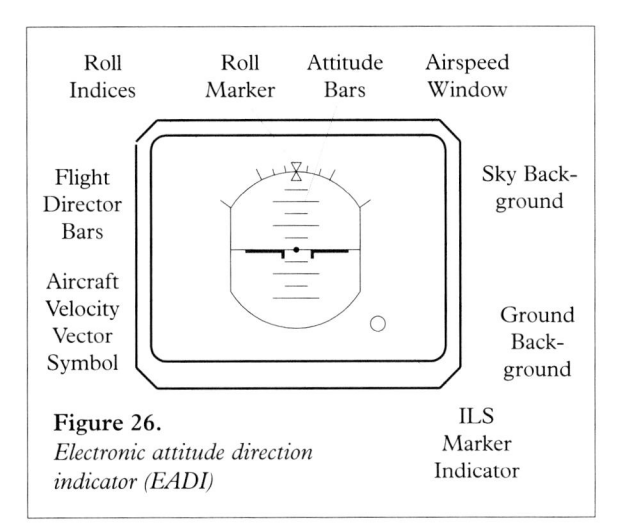

NEW-AGE SYSTEMS

Figure 26. *Electronic attitude direction indicator (EADI)*

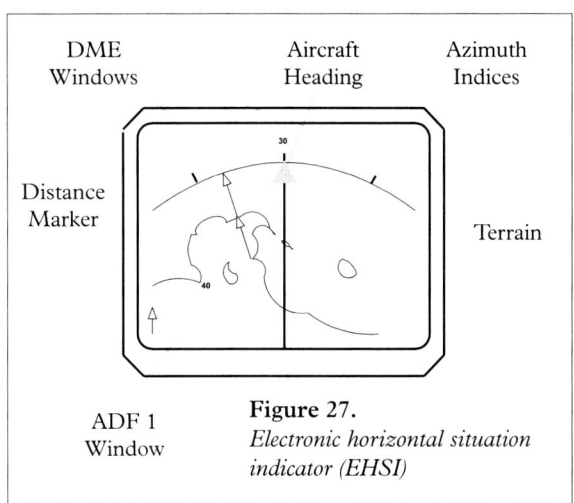

Labels: Roll Indices, Roll Marker, Attitude Bars, Airspeed Window, Flight Director Bars, Aircraft Velocity Vector Symbol, Sky Background, Ground Background, ILS Marker Indicator

Figure 27. *Electronic horizontal situation indicator (EHSI)*

Labels: DME Windows, Aircraft Heading, Azimuth Indices, Distance Marker, Terrain, ADF 1 Window

mechanical instruments. A typical EADI layout is shown in Figure 26. This layout is also very similar to that used for the primary flight display colour formats used in most integrated display systems.

Key features of the EAHI are:

- Sky (blue) and ground (brown) backgrounds shown above and below the horizon line respectively
- Pitch attitude bars
- Roll marker and roll indices
- Flight director bars
- Aircraft velocity vector symbol.

In addition, other information such as airspeed windows and ILS marker windows can be used to enhance the usefulness of the display. The EHSI layout is typified by the picture in Figure 27. Key display parameters are:

- Aircraft heading, in some cases with an additional heading window
- Azimuth indices
- Distance or range markers
- VOR/DME beacons information, shown in a cruise configuration above. However, the display may also show the beacons to be followed in an approach to a runway
- Terrain data may also be displayed.

A variety of formats may be specified but all effectively present navigational data to the flight crew in a bird's-eye format, assisting the flight crew's awareness of the horizontal positioning of the aircraft.

Typical flight deck layout

Following introduction of the two-crew concept and the consequent removal of the flight engineer's position, there was a need to modify the flight deck to facilitate the captain and first officer's interface with the aircraft systems. The development of integrated display suites meant that much of the aircraft systems could be displayed on multifunction colour displays. These systems displays could take the form of synoptic displays, showing a basic system layout with appropriate captions, or status pages listing aircraft system parameters. The typical civil aircraft flight deck began to resemble the layout shown in Figure 28.

The key elements of the flight deck displays and controls as follows:

- An overhead panel
- A glareshield, used for mounting key control panels and annunciators, permitting the display of information within the pilot's field of view when looking out of the aircraft
- An integrated display suite comprising captain's displays, first officer's displays and centrally located aircraft systems displays

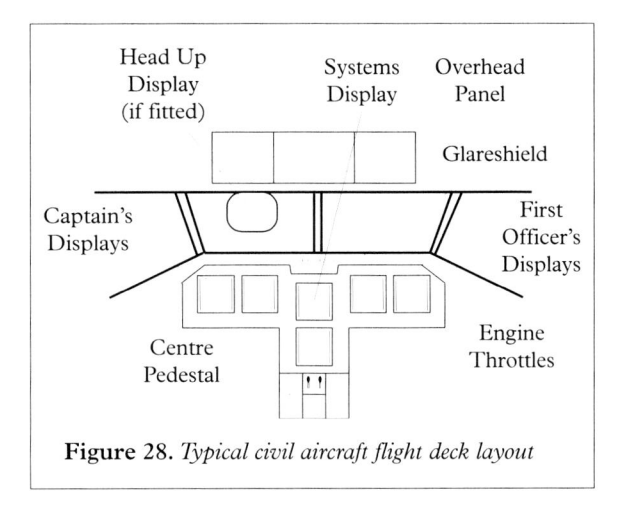

Figure 28. *Typical civil aircraft flight deck layout*

Labels: Head Up Display (if fitted), Systems Display, Overhead Panel, Glareshield, Captain's Displays, First Officer's Displays, Centre Pedestal, Engine Throttles

FASTER, FURTHER, HIGHER

- A centre console housing the aircraft throttles and other control panels. It is also usual to locate the FMS CDUs on this console
- Head-up display (HUD) if fitted.

Overhead panel

A typical overhead panel is shown in Figure 29. This is the main control panel via which the pilots may make system selections for the basic aircraft system. The various switches and push buttons required to select aircraft systems functions are grouped on the overhead panel, on a system-by-system basis. While a limited amount of display data may be provided on the overhead panel by way of a few dials, gauges or digital read-outs, most of the display information – indicator lights, illuminated push buttons, etc. – relates to system selections. Much of the aircraft system data appears in systems synoptic or status pages presented on the aircraft systems displays on the centre multifunction colour displays

Centre pedestal

The design and layout of the centre pedestal varies among aircraft manufacturers. However, some of the key controls and displays are shown in Figure 30. Typical examples are:

- Two or three FMS control and display units may be provided
- Engine controls: throttles, reverse thrust and engine start
- Speed brakes
- Flaps
- VHF communications and aircraft trim control (ATC) panels

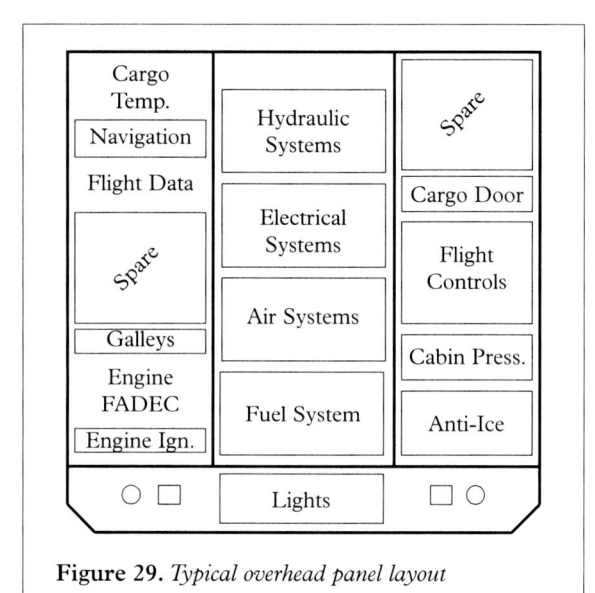

Figure 29. *Typical overhead panel layout*

- Communications control – an increasingly greater proportion of communications control is being given to the FMS
- Weather radar control
- EFIS (Electronic Flight Instrument System) control
- ATCs

Display suite

The integrated display suite format is fairly standard across all manufacturers. It comprises a six-display system:

- The two right-hand displays comprise the captain's primary flight display (PFD) and navigation display (ND)
- The two left-hand displays are the first officer's PFD and ND
- The two centre displays are related to system displays, alerts and status

The nomenclature for the centre displays differs among manufacturers. In Airbus the displays are termed electronic checkout and maintenance (ECAM); in Boeing terminology the system is called engine indication and crew alerting system (EICAS). The Airbus implementa-

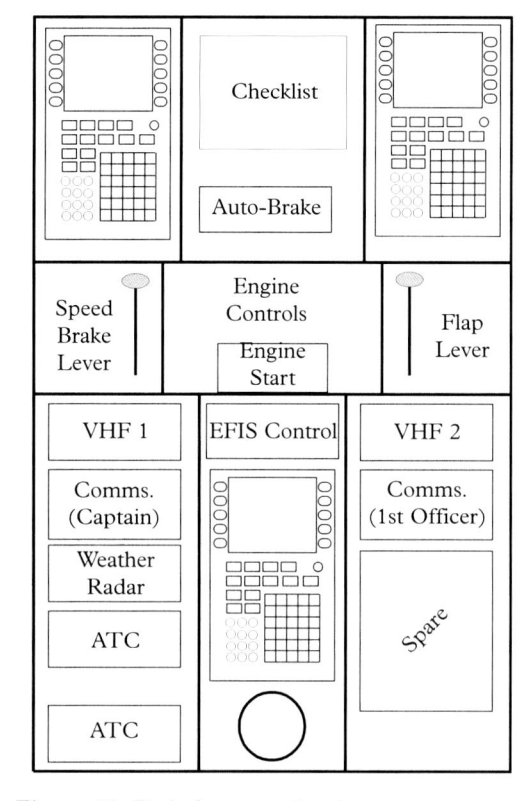

Figure 30. *Typical centre pedestal*

NEW-AGE SYSTEMS

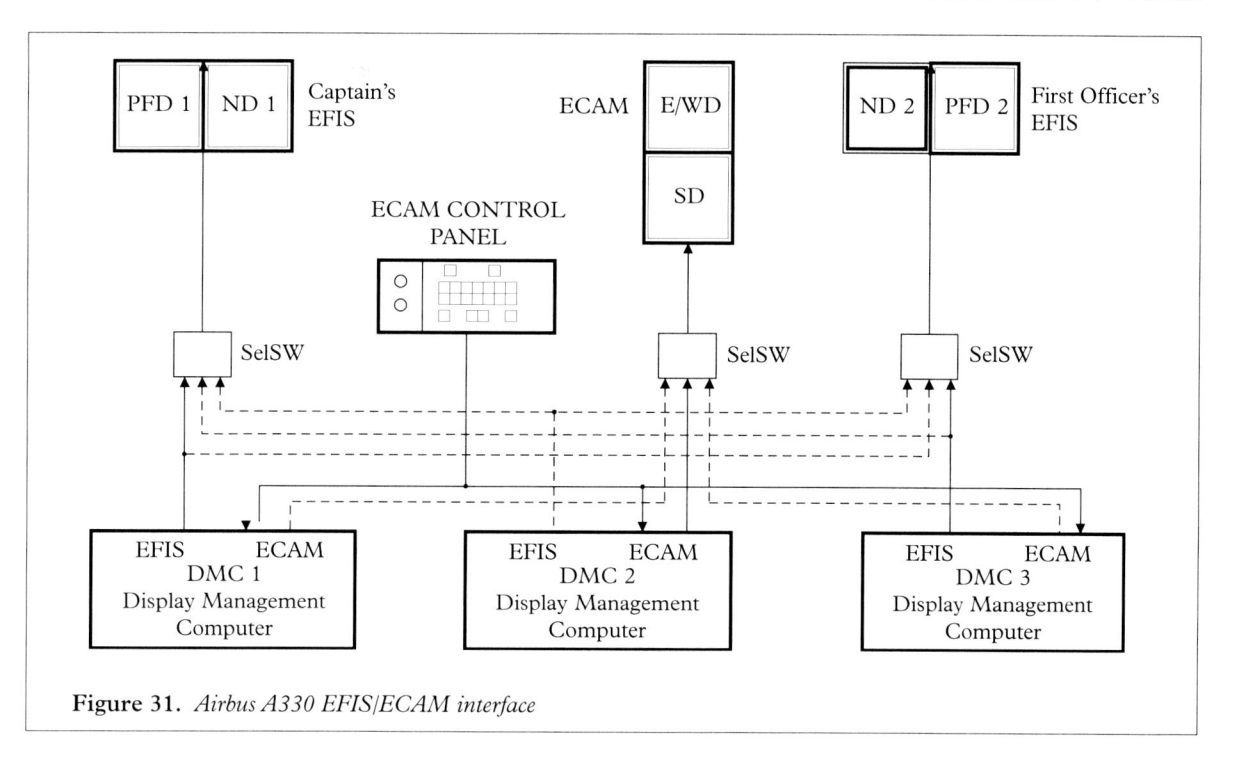

Figure 31. *Airbus A330 EFIS/ECAM interface*

tion for the A330 EFIS and ECAM systems is shown in Figure 31.

The displays are driven by three display management computers (DMCs), each of which may provide all of the EFIS and ECAM functions. The selections for display information is made via the ECAM control panel. The system routes display information via ARINC 429 data links to the EFIS and ECAM display heads respectively. Selector switches provide alternate sources of display information should one of the DMC processing sections fail.

Airbus is also introducing additional displays as part of its FANS A upgrade package. This introduces two datalink control and display units (DCDUs) just above the FMS control and display units as shown in Figure 32. This will permit the display of controller-to-pilot datalink messages as some of the FANS VHF datalink communications features are introduced. The Airbus view is that the separation of communications functions on the DCDUs from navigation functions on the FMS displays will enhance flight crew awareness and performance. This will particularly apply in high-workload air traffic environments.

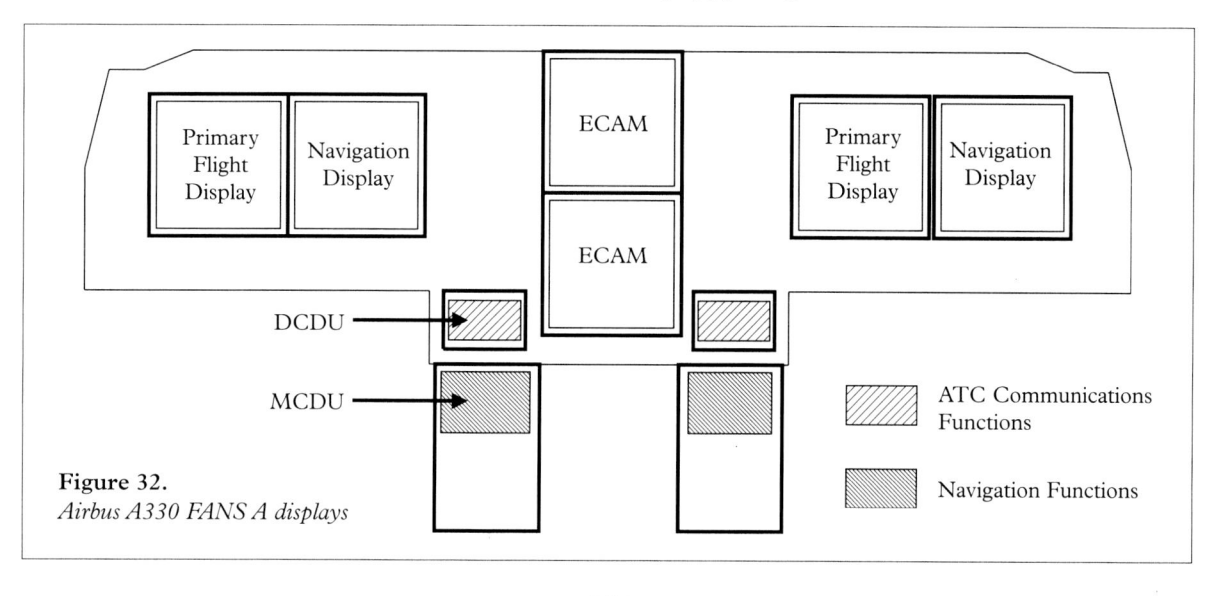

Figure 32. *Airbus A330 FANS A displays*

157

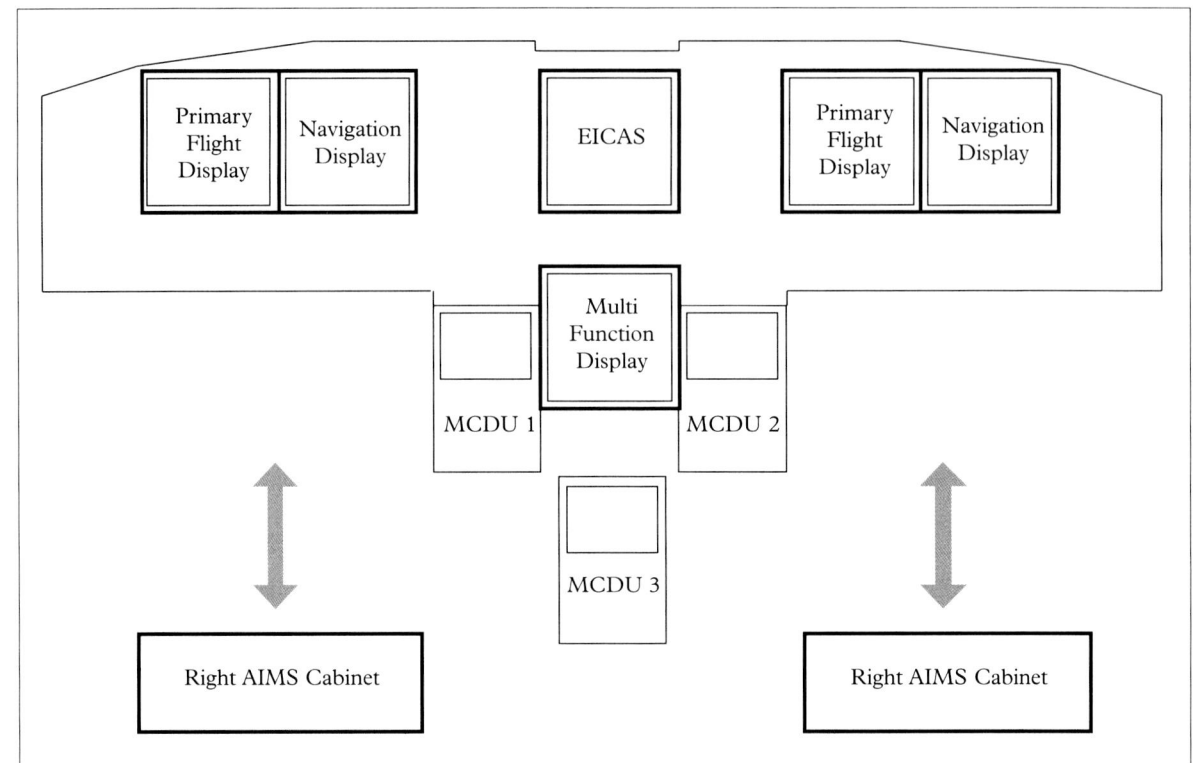

Figure 33. *Boeing 777 AIMS display interface*
A pair of right and left AIMS cabinets provide all of the avionics functions in a dual-cabinet packaging arrangement. The AIMS cabinets also provide the feeds to the six-display integrated display suite and the FMS control and display units.

Boeing

The Boeing display system implementation is shown in Figure 33. This models the Boeing 777 architecture. The Boeing 777 avionics functions are contained within a system called the airplane information management system (AIMS). Where the Airbus approach is to have functional areas interconnected by ARINC 429 digital data buses, Boeing took a more radical step, using a more integrated system interconnected by ARINC 629 data buses. ARINC 429 connects one transmitter with up to 31 receivers, usually operating at data rates around 110 kbits/second. By contrast, ARINC 629 operates at much higher data rate of 2 Mbits/second, almost twenty times the data rate of ARINC 429. Furthermore, ARINC 629 is capable of handling up to 128 terminals. The result is that ARINC 629 is a more capable vehicle for providing the data buses interfaces on a highly integrated modern avionics system.

The result shown is not atypical for any of the modern civil jet aircraft. However, the Boeing FANS 1 approach differs from the Airbus implementation in that no dedicated data link display units are provided. The Boeing approach is to integrated communications messages on the lower multi-function display (MFD) and FMS control and display units.

Civil HUD and EVS

The application of HUD techniques to civil aircraft, shown in figure 34, has taken a long time. The benefits claimed for the use of a HUD in a civil environment are:

- Improved situational awareness
- Precise flight path guidance
- Reduced tail strikes and heavy landings
- Display of warning alerts
- Lower take-off minima
- Improved energy management
- Take-off and roll-out guidance
- Unusual attitude recovery.

New developments include the capability of also overlaying FLIR imagery upon the combiner so that the pilot receives an enhanced IR image. This further enhances his situational awareness in low-visibility weather conditions and assists the probability of concluding a successful approach.

Standby instrument displays

Aircraft have to carry standby flight instruments in case the main display suite should fail. Hitherto, these standby instruments have been small 3 ATI (3in) mechanical

NEW-AGE SYSTEMS

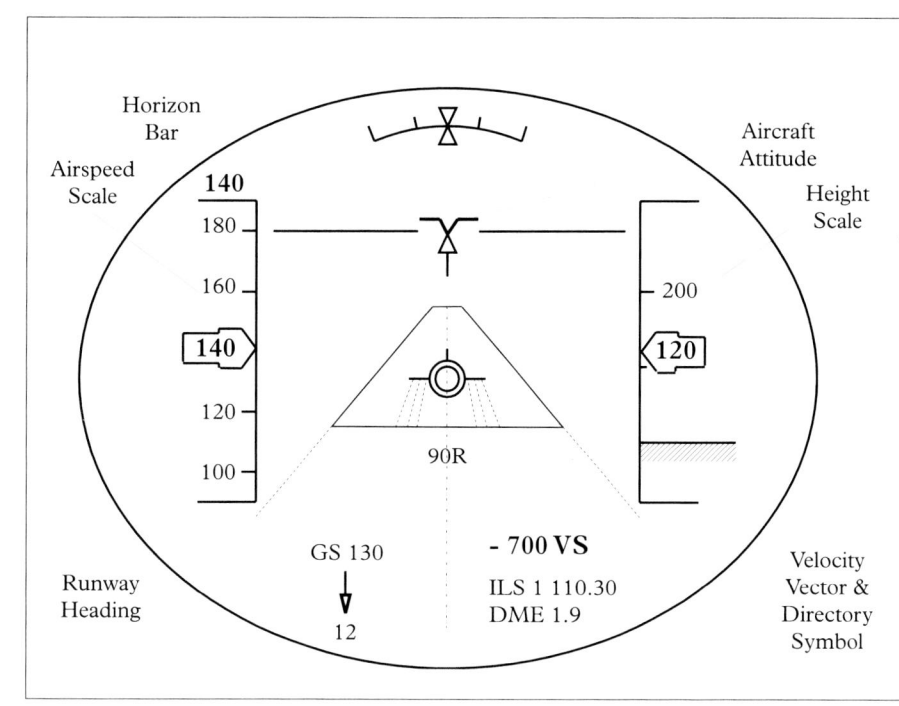

Figure 34. *Typical civil HUD presentation*
A notional HUD display format is shown in Figure 35, with typical parameters identified. ,This image is projected, focused at infinity, on the combiner glass, so the effect to the pilot is for the symbology to be overlaid upon the runway and terrain ahead. In the example given, supplementary data regarding wind direction and groundspeed, ILS and DME data, is shown as well as the usual airspeed, height and attitude data. Of particular value is the velocity vector and coincident director symbol, which indicates the point of touchdown on the runway ahead.

instruments with separate instruments provided for attitude, airspeed and altimeter functions. These instruments are small and contain all the complex linkages necessary to display the flight data, so reliability is not high and the mean time between failures is probably around 3,000hr. New solid-state AMLCD displays are now certificated which also contain miniature accelerometers as well as accepting pitot static data in the conventional sense. Consequently, the new standby instruments can combine all three functions in one 3 ATI display with a reliability several times greater. Refer to Figure 35 (page 160).

SAFETY AIDS

Traffic alert and collision warning system (TCAS)

The TCAS comprises two elements; a surveillance system and collision avoidance system, and also exists in two forms; TCAS I and TCAS II. The TCAS is based upon a beacon-interrogator principle similar to that employed by ground-based surveillance radars. It detects the range, bearing and altitude of aircraft in the near proximity for display to the pilots

TCAS I indicates the range and bearing of aircraft within selected parameters, usually 15-40nm forward, 5-15nm aft and 10-20nm on each side. The system also warns of aircraft within ± 8,700ft (2,650m) of the aircraft's own altitude.

The collision avoidance system predicts the time to, and separation at, the intruder's closest point of approach. These calculations are undertaken using range, closure rate, altitude and vertical speed. Should the TCAS ascertain that certain safety boundaries will be violated it will issue a Traffic Advisory (TA) to alert the crew that closing traffic is in the vicinity by displaying certain coloured symbols. Upon receiving a TA the flight crew must visually identify the intruding aircraft and may alter their altitude by up to 300ft (90m). A TA will normally be advised between 20 and 48 seconds before the point of closest approach with the simple warning: 'TRAFFIC, TRAFFIC'. TCAS I does not offer any de-confliction solutions, but does provide the crew with vital data in order that they may determine the best course of action.

TCAS II offers a more comprehensive capability with the provision of Resolution Advisories (RAs). It determines the relative motion of the two aircraft and determines an appropriate course of action. The system issues an RA via Mode S, advising the pilots to execute the necessary manoeuvre to avoid the other aircraft. Usually, an RA will be issued when the point of closest approach is within 15 and 35 seconds, and the de-confliction symbology is displayed coincident with the appropriate warning. A total of ten warnings may be issued. Examples are:

- 'CLIMB, CLIMB, CLIMB'
- 'DESCEND, DESCEND, DESCEND'
- 'REDUCE CLIMB, REDUCE CLIMB'
- Finally, when the situation is resolved: 'CLEAR OF CONFLICT'

TCAS II clearly requires a high level of integration between the active equipment. Figure 36 (page 161) shows the inter-relationships between them:

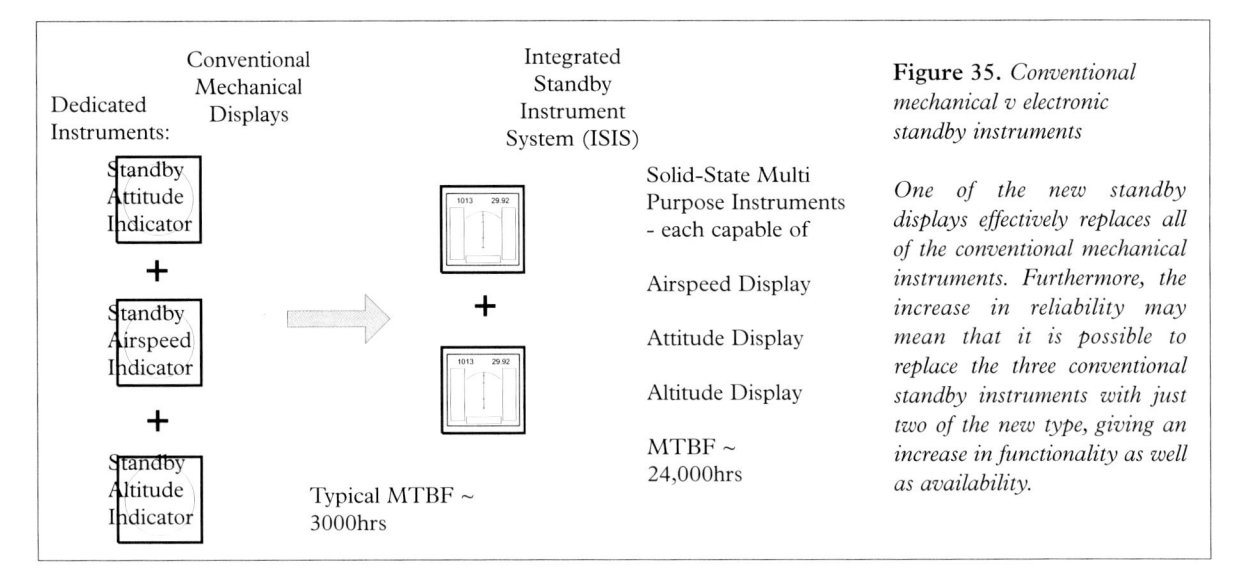

Figure 35. *Conventional mechanical v electronic standby instruments*

One of the new standby displays effectively replaces all of the conventional mechanical instruments. Furthermore, the increase in reliability may mean that it is possible to replace the three conventional standby instruments with just two of the new type, giving an increase in functionality as well as availability.

Dedicated Instruments: Standby Attitude Indicator + Standby Airspeed Indicator + Standby Altitude Indicator (Conventional Mechanical Displays, Typical MTBF ~ 3000hrs)

Integrated Standby Instrument System (ISIS): Solid-State Multi Purpose Instruments - each capable of Airspeed Display, Attitude Display, Altitude Display. MTBF ~ 24,000hrs

- TCAS interrogator
- ATC Mode S transponders
- VSI display showing vertical guidance for TAs and RAs
- Optional horizontal situational indicator for RAs that could be the navigation display
- Audio system and annunciators
- Antennae for ATC Mode S and TCAS

This is indicative of the level of integration required between ATC Mode S transponders, TCAS, displays and annunciators.

TAWS

The terrain awareness and warning system (TAWS) embraces the overall concept of providing the flight crew with prediction of a potential controlled flight into terrain (CFIT). The new term is a generic one, since enhanced ground proximity warning system (GPWS) became associated mainly with an Allied Signal (now Honeywell) product. The latest implementation is designed to provide the crew with an improved prediction compared with previous systems. The US Federal Aviation Administration (FAA) specifies that turbine equipped aircraft with six seats or more will have to be equipped with TAWS by 2003.

GPWS and EGPWS

While TCAS is designed to prevent air-to-air collisions, the ground proximity warning system (GPWS) is intended to prevent unintentional flight into the ground, or CFIT, which is the cause of many accidents. The term describes conditions where the crew are in control of the aircraft but, perhaps being unfamiliar with the locality, are unaware that they are about to crash into the terrain. The GPWS takes data from various sources and generates a series of audio warnings when a hazardous situation is developing. In this sense it is very similar to TCAS, but specific to terrain avoidance. The GPWS uses radar altimeter information together with other information relating to the aircraft flight path. Warnings are generated when the following scenarios are unfolding:

- Flight below the specified descent angle during an instrument approach
- Excessive bank angle at low altitude
- Excessive descent rate
- Insufficient terrain clearance
- Inadvertent descent after take-off
- Excessive closure rate to terrain – the aircraft is descending too quickly or approaching higher terrain.

The installation of GPWS in airliners flying in US airspace was mandated by the FAA in 1974, since when the number of CFIT accidents has dramatically decreased. More recently, enhanced versions (EGPWS) have become available. The EGPWS offers a much greater situational awareness to the flight crew, as it provides more quantitative information, together with earlier warning of the situation arising. It uses a world-wide terrain data base which is compared with the aircraft's present position and altitude. Within the terrain data base the Earth's surface is divided into a grid matrix with a maximum altitude assigned to each square within the grid.

The aircraft's intended flight path and manoeuvre envelope for the prevailing flight conditions are compared with the terrain matrix, and the result graded according to the proximity of the terrain as shown in Figure 37.

Terrain responses are graded as follows:

- No display for terrain more than 2,000ft (600m) below the aircraft
- Light green dot pattern for terrain between 1,000 and

NEW-AGE SYSTEMS

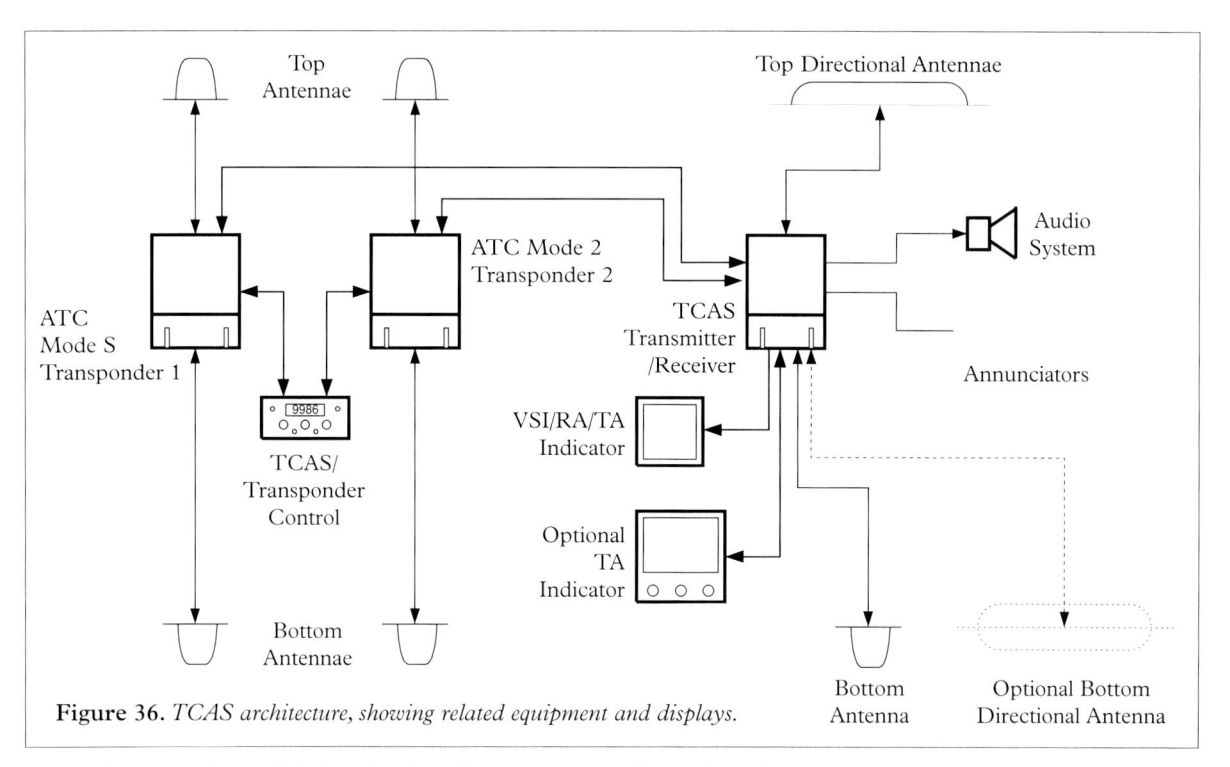

Figure 36. *TCAS architecture, showing related equipment and displays.*

- 2,000ft (300 and 600m) below the aircraft
- Medium green dot pattern for terrain between 500 and 1,000ft (150 and 300m) below the aircraft
- Medium yellow dot pattern for terrain between 1,000ft above and 500ft below the aircraft
- Heavy yellow display for terrain between 1,000 and 2,000ft above the aircraft
- Heavy red display for terrain more than 2,000ft above the aircraft.

This portrayal, very similar to that for the weather radar, is usually shown on the navigation display. It is far more informative than the audio warnings given by GPWS.

The EGPWS also gives audio warnings, but much earlier than those given by the GPWS. The earlier warning, together with the quantitative display, gives the flight crew much better overall situational awareness and more time to react positively to their predicament.

FLIGHT GUIDANCE
Inter-relationship of flight control functions

There is sometimes some confusion regarding the inter-relationship of FBW, autopilot and FMS functions, which Figure 38 seeks to correct. These separate but intertwined functions may be described as three nested control loops, each with its own distinct tasks.

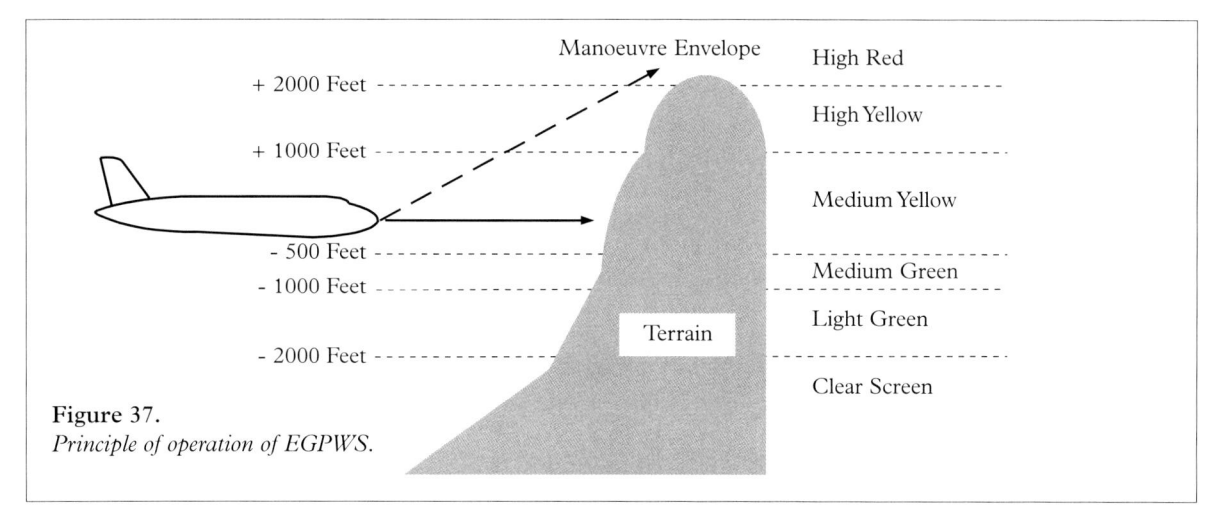

Figure 37. *Principle of operation of EGPWS.*

161

Figure 38. *Inter-relationship of flight control functions*

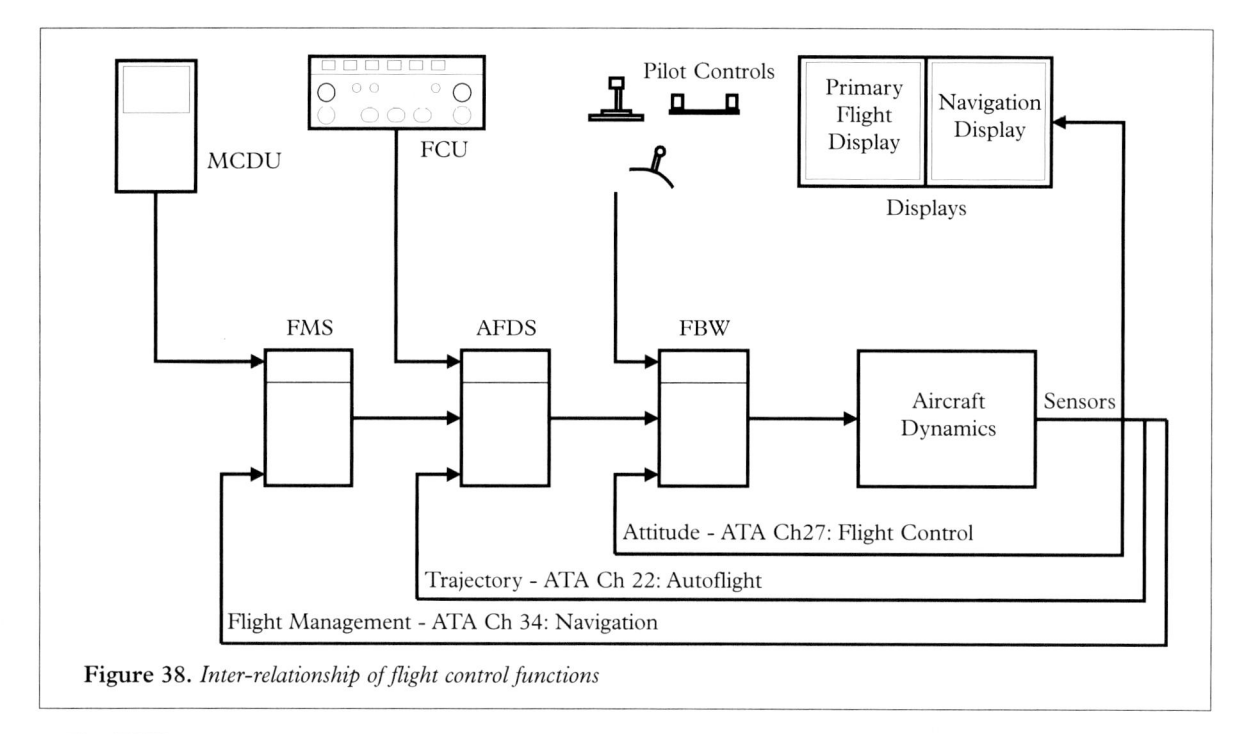

The FBW system, comprising the inner loop, is concerned with controlling the attitude of the aircraft. Inputs from the pilot's control column (or sidestick), rudders and throttles determine, via the aircraft dynamics, how the aircraft will respond at various speeds and altitudes. Inertial and air data sensors determine the aircraft response and close the pitch, roll and yaw control loop to give the aircraft well harmonised control characteristics throughout the flight. In some aircraft relaxed stability modes of operation may be invoked by using the fuel system to modify the aircraft's centre of gravity, reducing trim drag and reducing aerodynamic loads on the tailplane. The aircraft pitch, roll and yaw (azimuth) attitude are presented on the primary flight display and navigation display.

The autopilot flight director system (AFDS) performs additional control loop closure to control the aircraft trajectory. The AFDS controls the speed, height and heading at which the aircraft flies. Navigation functions such as holding and acquiring headings are also included. Approach and landing guidance is provided by coupling the autopilot to the ILS or MLS systems. Control and indication associated with these multiple autopilot modes is provided by a flight control unit (FCU) which enables the selection of the principal modes and also provides information confirming that the various modes are correctly engaged and functioning properly.

The final loop closure is that undertaken by the FMS, which performs the navigation function, ensuring that the FBW and AFDS systems position aircraft correctly in the sky to coincide with the waypoints along the aircraft route. The pilot uses a control and display unit to start and monitor the aircraft's progress.

Fly-by-wire systems

A generic FBW system is shown in Figure 39. Although the precise number of control surfaces and means of flight control computer implementation may vary from aircraft to aircraft, all modern FBW systems accord to this generic form. The computers use a number of parameters to achieve the necessary handling characteristics.

Both the Airbus and Boeing implementations have the ability to fly the aircraft in a manual backup mode. In Airbus aircraft the mechanical trim wheel can alter the position of the tailplane surface for pitch control, and inputs from the rudder pedals can alter the inputs to the three rudder actuators. A combination of pitch trim and rudder pedals, therefore, allows the aircraft to be controlled by manual means in an emergency. In the Boeing 777 system, alternate pitch trim levers are mechanically connected directly to the horizontal stabilator. A direct mechanical link from the rudder to one pair of spoilers allows roll control to be maintained in a standby mode.

A top-level comparison of the Boeing and Airbus approaches to FBW is shown in Figure 40. The Boeing system employs three primary flight computers (PFCs), each of which contains three independent computer sections. The PFC demands are fed to one of four actuator control electronics (ACE) units via triple-redundant ARINC 629 flight control data buses. The ACE units provide the digital-to-analogue interface for all except the mechanical link actuators, effectively closing the actuator control loop for elevator (4), flaperon (4), aileron (4), and spoiler (10) flight control actuators. So far, the Boeing 777 is the only Boeing aircraft to have adopted a FBW system. The Airbus approach is to provide a number of command/monitor flight control computers which control the actuators in a number of ways. There are differences of implementation between the A320 family and A330/340, but the overall philosophy is the same.

Airbus will carry forward its existing FBW philosophy to the A380. It is believed that, in place of mechanically signalled systems of the A320 and A330/340, electrically signalled and powered backup systems are being planned.

Autopilot flight director system (AFDS)

The workload on modern flightdecks, together with the precision with which the aircraft needs to be flown, make an autopilot an essential part of the avionics system. On board aircraft with conventional cable-and-pulley-operated flight control systems, the autopilot takes the form shown in Figure 42 (page 165).

The two digital computers are controlled by means of a flight guidance control panel which enables autopilot modes to be engaged and various speed, altitude and heading datums to be adjusted and displayed.

Both computers output two lanes of autopilot demand in the pitch, roll and yaw channels according to the autopilot mode(s) selected. These demands are fed to an appropriate dual servo motor which enables the autopilot to be coupled into the appropriate flight control run in pitch, roll and yaw. Feedback from the servo motor enables the autopilot to ensure that its demands are satisfied. An example of the dual servomotor is shown in more detail in Figure 43 (page 166).

As well as having their pitch, roll and yaw motion con-

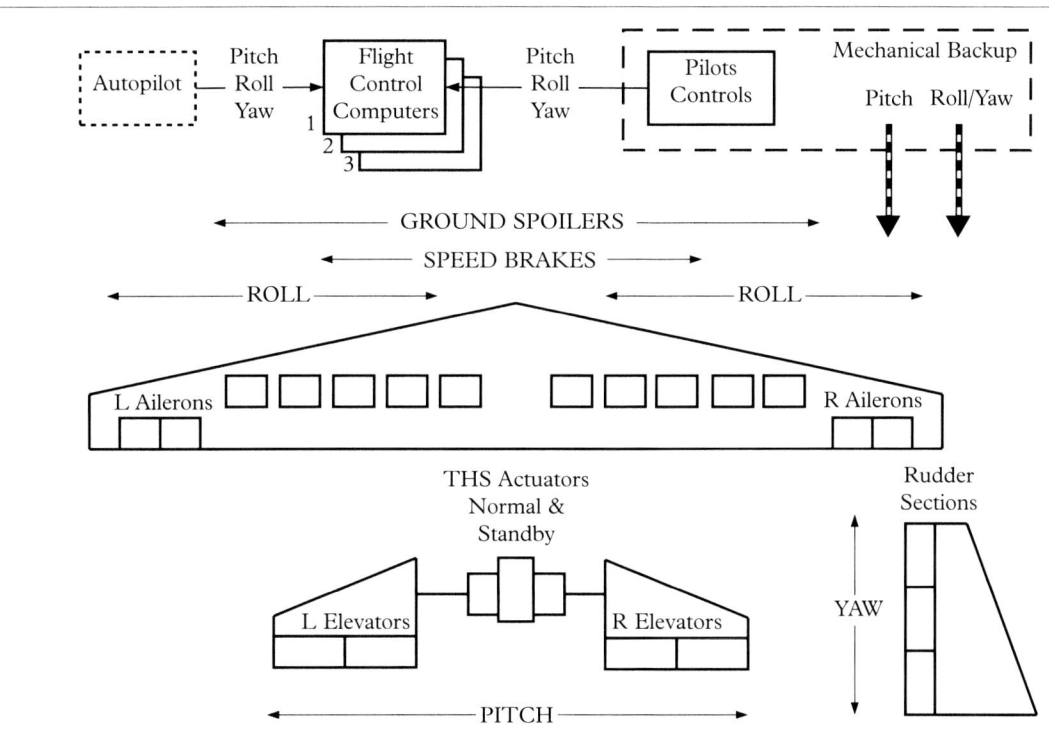

Figure 39. *Generic FBW system.*
The key control surfaces are:
- Pitch control is effected by four powered flight control actuators powering four elevator sections
- Pitch trim is undertaken by means of two tailplane horizontal stabiliser (THS) actuators (operating as normal and standby) which move the entire horizontal tailplane surface
- Roll control is invoked using the left and right ailerons; augmented as required by the extension of spoilers on the inboard wing
- Yaw control by means of two or three rudder sections
- The spoilers also perform a speed brake function in flight and a ground spoiler or lift-dump function during the landing roll
- In addition, secondary flight control or high-lift augmentation is provided by leading-edge slats and trailing-edge flaps for take-off and landing as appropriate. On the Boeing 777 the aileron and flap functions are combined by the use of two inboard flaperons, whereas conventional ailerons are used outboard.

FASTER, FURTHER, HIGHER

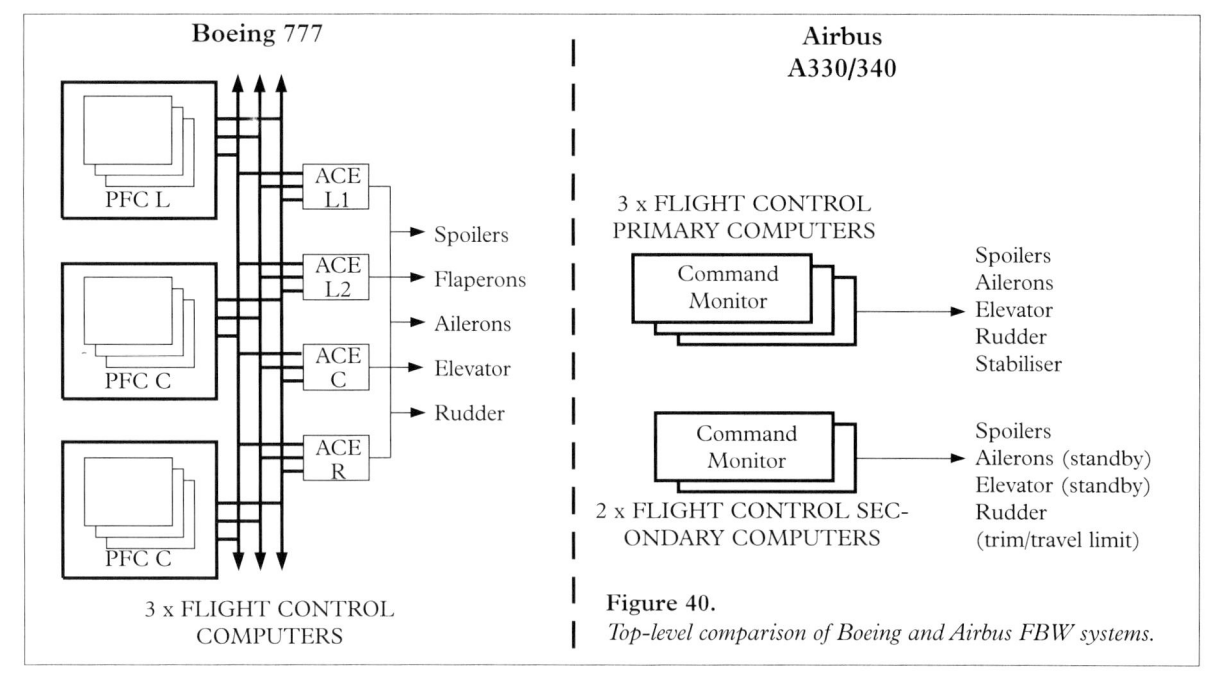

Figure 40.
Top-level comparison of Boeing and Airbus FBW systems.

Boeing 777: PFC L, PFC C, PFC C connected to ACE L1, ACE L2, ACE C, ACE R driving Spoilers, Flaperons, Ailerons, Elevator, Rudder. 3 x FLIGHT CONTROL COMPUTERS.

Airbus A330/340: 3 x FLIGHT CONTROL PRIMARY COMPUTERS (Command/Monitor) driving Spoilers, Ailerons, Elevator, Rudder, Stabiliser. 2 x FLIGHT CONTROL SECONDARY COMPUTERS (Command/Monitor) driving Spoilers, Ailerons (standby), Elevator (standby), Rudder (trim/travel limit).

trolled by coupling into the aircraft flight control, most aircraft have an auto-throttle capability such that the aircraft speed may also be maintained at a fixed datum as required. Figure 44 shows all those sensors whose inputs are necessary for the operation of a speed control mode.

The calculations associated with the auto-throttle control laws are peformed within lanes A and B of each digital flight guidance computer. In this case, however, the outputs are fed to an auto-throttle actuator able to alter the position of the engine throttle linkage. The resultant changes in engine power enable the autopilot to control the aircraft speed.

Autopilot modes

There are many autopilot modes that can be used singly or in combination; a modern autopilot will possess some twenty to thirty modes of operation. Autopilot modes are usually selected by means of a mode selector panel located centrally on the glare-shield. There is also an instinctive cut-out button which the pilot can use to disengage the autopilot in an emergency. The different modes of autopilot operation each require a suite of sensors providing the necessary data so that the relevant autopilot control laws may be performed. In several modes of operation dual sensors may be required. Typical autopilot modes are:

- Roll modes: Heading hold, heading select, VOR, localiser
- Pitch modes: Take-off, go-around, glideslope, Mach hold, IAS hold, altitude pre-select, attitude hold, vertical speed
- Land modes: Localiser, glideslope

Specific sensors will be required for each autopilot mode. In the event that any of these sensors are unavailable, the autopilot engagement logic will prevent the selection of an associated mode. In the cruise, while flying airway routes using VOR beacons, it is likely that the flight crew will have VOR selected, together with altitude hold. This combination of modes will ensure that the aircraft flies directly to the next VOR (waypoint) at the altitude for which clearance has been obtained.

Flight management system

The final element in the overall flight guidance equation is the FMS. The interfaces with the FMS were briefly described in Figure 23 in the section of this chapter covering navigation. A first-generation FMS produced by Smiths is shown in Figure 41.

This system comprises dual redundant FMS computers and CDUs. Whereas the FMS computers are shown

Figure 41. *First-generation FMS (Smiths Group).*

NEW-AGE SYSTEMS

as stand-alone 4 Modular Concept Unit (MCUs), most modern applications could now be hosted on a single card, such has been the degree of integration of microelectronics. The CDUs in this case are the older monochrome CRT units, whereas colour AMLCD displays would be common today.

The FMS receives inputs from:

- INS
- Navigation aids
- GPS
- Air data system
- Fuel sensors and clock

In a modern high-performance system most of these sensors will be dual redundant

The FMS outputs data to:

- EFIS displays
- Communications management unit
- Autopilot and auto-throttle

The FMS controls the navigation of the aircraft by integrating the functions of all of these units. This does not merely include steering the aircraft from waypoint to waypoint; it also controls the tuning of all of the appropriate aircraft receivers to navigation beacons and communications frequencies via the communications control units. The flight plan that resides within the FMS memory will be programmed for the entire route profile, including aircraft diversion and all other eventualities. More advanced capabilities include 3-D navigation and the ability to adjust the aircraft speed reach a waypoint within as a short a time as ±6 seconds.

The key crew-to-FMS interface is achieved through the CDU. This provides a very powerful tool using mode and function keys, soft keys and alpha-numeric keys which enable the crew to use a powerful menu-driven

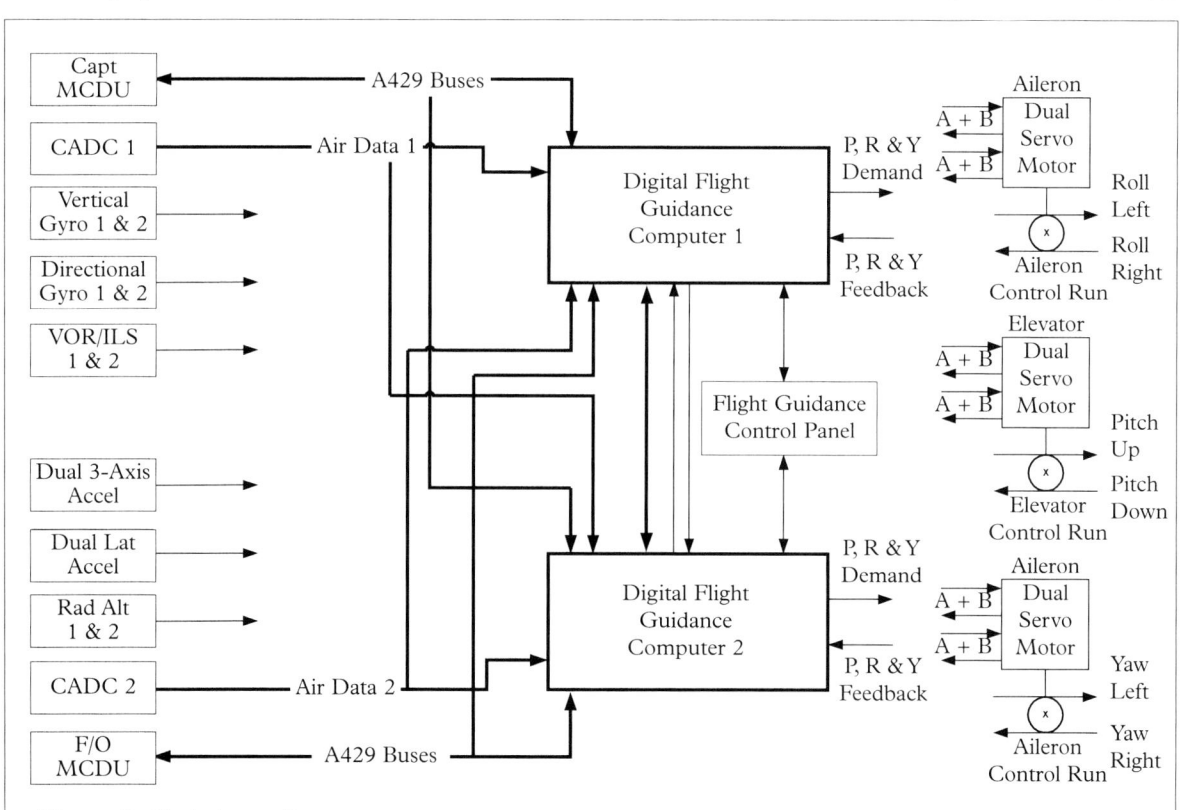

Figure 42. *Typical autopilot system.*
This depicts a system with two digital flight control computers, each of which has dual lanes of operation. Relevant sensor and input data is fed from the units shown on the left side of the diagram. These include:

- Captain and first officer's FMS control and display units
- Dual central air data computers
- Dual vertical gyros providing a pitch attitude reference
- Dual directional gyros furnishing a directional attitude reference
- Dual VOR/ILS units to provide radio navigational and approach guidance
- Dual radar altimeters
- Two accelerometers units to provide aircraft acceleration data for advanced control loop closure.

165

FASTER, FURTHER, HIGHER

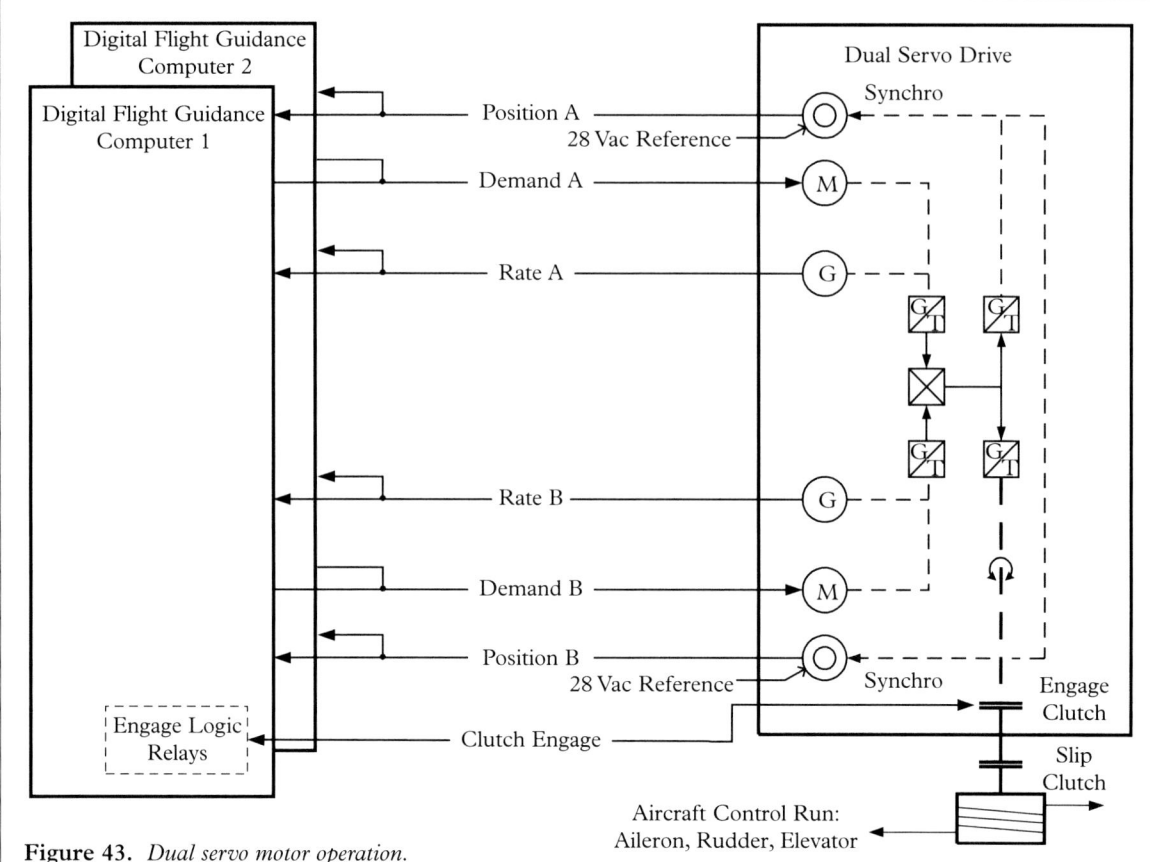

Figure 43. *Dual servo motor operation.*
This figure shows how autopilot demands and position and rate feedback are duplicated for lanes A and B. A series of gearboxes converts these to a single rotary demand shown at the bottom of the servomotor. The output to the appropriate flight control run drives through two clutch mechanisms. The first is the engage clutch, which closes and engages the autopilot when the necessary control logic conditions have been satisfied. The second is a slip clutch, which enables the control run to be moved if the servomotor jams, once an established break-out force has been exceeded.

system. The CDU also displays key information relating to navigation, radio navigation and communications frequencies, time to waypoints and destination, and navigation accuracy. As the functional capabilities of the FMS increase, so the display of the relevant data becomes more crucial.

Future Air Navigation System
The rapidly increasing density of air traffic is leading to a pressing need to move on from the techniques and technologies which have served the air transport management (ATM) system for the last forty years. This evolution will embrace the use of new technologies mixed with existing capabilities to offer improved air traffic management. The aims of ATM may be summarised as follows:

- To maintain or increase levels of safety
- To allow dynamic accommodation of user-preferred 3-D and 4-D flight trajectories
- To improve the provision of information to users in terms of weather, traffic situation and services
- Increase user involvement in ATM decision-making including air-ground computer dialogues
- To organise airspace in accordance with ATM procedures
- To increase system capacity to meet traffic demand
- To accommodate a full range of aircraft types and capabilities
- To improve navigation and landing capabilities to support advanced approach and departure procedures
- To create, to the maximum extent possible, a seamless continuum of airspace where boundaries are transparent to the user

To this end the air traffic control authorities, airline industry, regulatory authorities and airframe and equipment manufacturers are working to create the future air navigation system (FANS) to develop the necessary equipments and procedures. The areas where improvements may be made relate to communications, navigation and surveillance (commonly referred to as CNS). The key attributes of the these improvements may be briefly summarised as:

- Communication: The use of data links to increase data flow and permit the delivery of complex air traffic control clearances
- Navigation: The use of GPS in conjunction with other navigational means to improve accuracy and allow closer spacing of aircraft
- Surveillance: The use of data-links to signal aircraft position and intent to the ground and other users

These headings form a useful framework to examine the CNS/ATM improvements already made and those planned for the future.

Communications

The main elements of improvement in communications are:

- Air-ground data link for domestic communications. To enable ATC flight service and surveillance, VHF Data Link (VDL) communications will be used for domestic communications. As has already been explained, VHF communications are line-of-sight limited. There are a number of options:
  VDL Mode 1: Compatible with existing ACARS transmitting at 2.4kbps
  VDL Mode 2: Data only transmitted at 31.5 kpbs
  VDL Mode 3: Simultaneous data and voice
  VDL Mode 4: Used with the 1090 MHz signal of ATC Mode S
- Air-ground communications for oceanic communications. This may comprise SATCOM, which, as already explained, is limited at very high latitudes, or the use of HF data links. Modern technology enables HF data link transmissions to be more robust than HF voice, and therefore less susceptible to the effects of the sunspot cycle. HF data link provides primary coverage out to 2,700nm and secondary coverage beyond that given favourable propagation conditions. There is extensive cover by ground stations located in the northern hemisphere such that HF data link is a viable alternative to SATCOM for north polar transitions
- 8.33 kHz VHF voice communications. Conventional VHF voice channels are spaced at intervals of 25 kHZ throughout the spectrum. A denser communications environment has resulted in the introduction of digital radios which permit spacing at 8.33 kHz, allowing three channels to be fitted in the spectrum where only one could be used previously. These radios have already been introduced in Europe and will follow in the USA within a number of years.

Navigation

A number of navigational improvements are envisaged:

- Introduction of required navigation performance (RNP) and actual navigation performance (ANP) navigation performance criteria. This defines absolute navigational performance requirements for various flight conditions and compares this with the actual

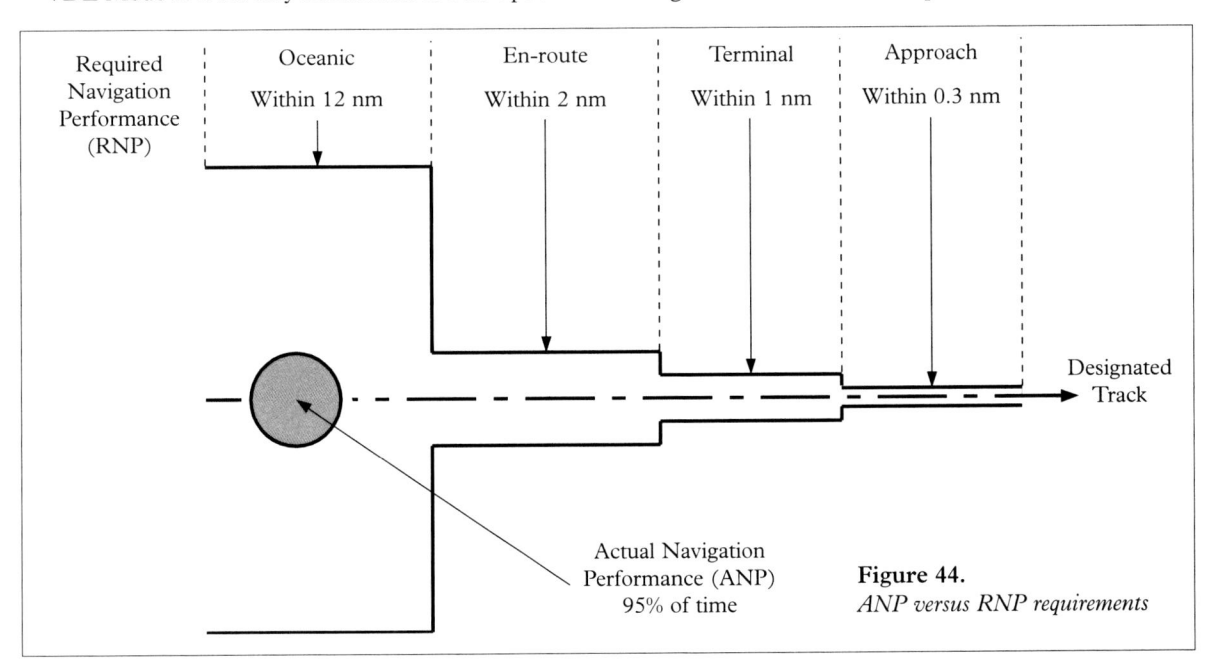

Figure 44.
ANP versus RNP requirements

performance the aircraft system is capable of providing.

- The ANP is represented by a circle which embraces the accuracy of the aircraft navigation system for 95% of the time. The RNP defines the lateral track limits within which the ANP circle should be constrained for various phases of flight. The general requirements are:

  For oceanic crossings the RNP is ±12nm, also be referred to as RNP-12
  For en-route navigation the RNP is ±2nm (RNP-2)
  For terminal operations the RNP is ±1nm (RNP-1)
  For approach operations the RNP is ±0.3nm (RNP-0.3)

It is clear that this represents a more definitive way of specifying aircraft navigational performance, versus the type of leg being flown, than has previously been the case. Other more specific criteria exist:RNP-5 (also known as BRNav or area navigation) has already been introduced in parts of the European airspace, with the prospect that RNP-1 (also known as PRNAV or precision navigation) will be introduced in a few years. There are precision approaches in being (noticeably those in Juneau, Alaska) where RNP-0.15 is required for new precision approaches developed for mountainous terrain.

Introduction of reduced vertical separation minima (RVSM). Aircraft flight levels have operated with a 2,000ft vertical separation at flight levels between FL290 and FL410 for many years. As traffic density has increased this has proved to be a disadvantage for the busiest sections of airspace. The RVSM was introduced to increase the available number of flight levels in this band and effectively permit greater traffic density. The principle is to introduce additional useable flight levels such that the flight level separation is 1,000ft throughout the band, as shown in Figure 45.

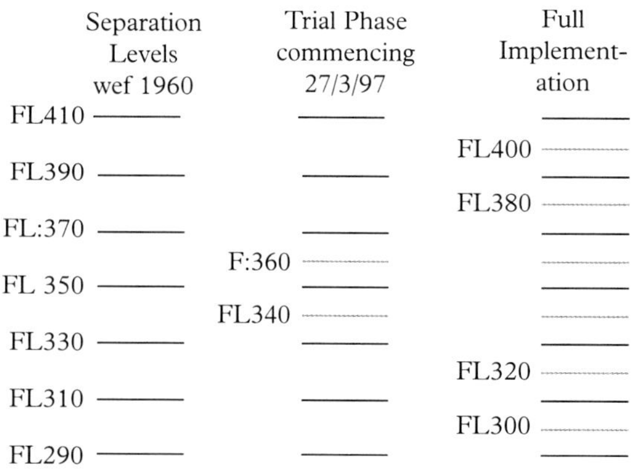

Figure 45. *Reduced separation vertical minima*

Originally a trial was mounted in 1997 to test the viability of the concept on specific flight levels – FL 340 and FL360 as shown in Figure 45. RVSM is now implemented throughout most of Europe from FL290 to FL410, introducing six new flight levels. All the specified flight levels were implemented on the North Atlantic by the end of 2001. Other regions in the globe will have RVSM selectively implemented to increase air traffic density.

For RVSM operation an aircraft must possess two independent means of measuring altitude, and an autopilot with an accurate height hold capability. The operators of RVSM equipped aircraft are not taken on trust: independent height monitoring stations survey aircraft passing overhead, measuring actual height compared with flight plan details and the individual aircraft fin number and route plan. RVSM implementation therefore embraces a watchdog function that ensures that all users are conforming to the RVSM provisions.

GPS enhancements for en-route and precision landings in the USA. The GPS enhancements, WAAS and LAAS, have already been described.

WAAS is anticipated to yield an accuracy of ~7m, which will be sufficient for Cat I approaches

LAAS is expected to provide enhanced accuracies ~1m, which will be sufficient for precision approaches

Protected ILS introduction in Europe. Within Europe some ILS installations suffer interference from high-power FM local radio stations. Modifications will be introduced to protect the ILS systems from this interference.

Surveillance
Surveillance enhancements include the following:

TCAS. The operation of TCAS has already been described. When operating together with a Mode S transponder and a stand-alone display or EFIS presentation, TCAS is able to monitor other aircraft in the vicinity by means of airborne interrogation and assessment of collision risk. TCAS II provides vertical avoidance manoeuvre advice by the use of RAs. TCAS II will soon be made mandatory for civil airliners – aircraft with a weight exceeding 15,000kg or 30 or more seats - operating in Europe. This will be extended to aircraft exceeding 5,700kg or more than 10 seats, probably by 2005.

ATC Mode S transponder. The use of the ATC Mode S transponder in providing digital air-ground and air-air data link (VDL Mode 4) has already been described. This provides a basic ADS-B capability

Figure 46. *Principle of 'free flight'*

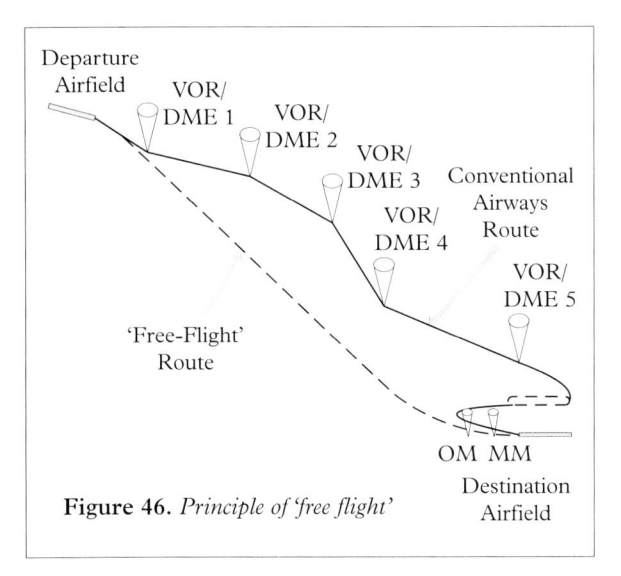

Automatic dependent surveillance – address mode (ADS-A). ADS-A will be used to transmit aircraft 4-D position and flight plan intent based upon GPS position during oceanic crossings. The communications media will be SATCOM or HF data link. ADS-B requires the aircraft to be fitted with an FMS and CDU, and with some means of displaying message alerts and annunciation

Automatic dependent surveillance – broadcast mode (ADS-B). ADS-B will be used to transmit 4-D position and flight plan intent based upon GPS position using line-of-sight VHF communications. Either Mode S or digital VHF radio will be used to transmit the data. ABS-B requires a cockpit display of traffic information.

The CNS triad offers a suite of methods of addressing the FANS objectives. The ultimate aim of FANS is to provide a 'free flight' capability, as shown in Figure 46.

Under prevailing navigational means, an aircraft flying from a departure airfield to a destination airfield would need to file a flight plan entailing flying down prescribed airways from VOR/DME beacon to DME beacon. Upon arrival in the vicinity of the destination airfield, the aircraft may have to enter a holding pattern while waiting for clearance to land. The approach may necessitate an ILS approach that would involve over-flight of the outer marker some 5 to 7nm on the extended runway centre-line. Many inefficiencies exist in this type of routeing, which is typical of many flights carried out over land today.

Free flight, on the other hand, would enable a more direct routeing from the departure to the destination airfield, clearly yielding savings in time and fuel. Furthermore, upon arrival at the destination airfield the aircraft could execute a more direct and efficient approach and landing; most probably enabled by phasing its approach with that of other aircraft by means of 4-D, required time of arrival navigation techniques. An FMS embodying a full performance aircraft model could make further savings by optimising the use of fuel during flight to enable the aircraft to meet the necessary 4-D way-points at the appropriate time with a minimum fuel burn.

The universal application of the principles of free flight is some way off. Free flight may already be used on specific flight segments where air traffic density is relatively low, but universal application depends upon the proven maturity of the technologies described. Complex political and financial factors compound the technical issues. Free flight in Europe will probably not be fully available until the partitioning of the airspace is simplified and rationalised compared with the complex structure existing today. The cost of new equipment, both air and ground based, is not trivial. Airlines will wish to be assured that the investment in new equipment will provide an adequate return as new FANS capabilities are introduced. Nevertheless, in the medium and long term the pressures of air traffic density and fuel economy will ensure that many FANS features will attain maturity.

Bibliography

Galotti Jr, V P, *The Future Air Navigation System (FANS)* (Ashgate Publishing, 1998). A book addressing FANS from the air traffic controller's viewpoint.

Kayton, M, and Fried, W R, *Avionics Navigation Systems* (2nd edition, John Wiley & Sons, 1997). A comprehensive reference addressing navigation issues.

Moir, I, and Seabridge, A, *Aircraft Systems* (Professional Engineering Publishing (IMechE) in association with the AIAA, May 2001). A unique reference addressing aircraft systems.

Pallett, E H J, *Aircraft Instruments and Integrated Systems* (Longmans, 1992). A basic reference addressing the principles and practice of older avionics systems.

Pratt, R, *Flight Control Systems: Practical Issues in Design and Implementation* (IEE Publishing, 2000). This reference addresses the design, development and certification of aircraft flight control systems

Spitzer, C, *Digital Avionics Systems, Principles and Practice* (2nd edition, McGraw-Hill, USA, 1993). An authoritative reference laying out the guidelines for military and civil avionics systems.

Stimson, G W, *Introduction to Airborne Radar* (2nd edition, SciTech Publishing, USA, 1998).

8
Exploring the Limits
Ian Goold

On the early morning of 18 February 1977, at Dryden Flight Research Center (FRC), Edwards Air Force Base, California, USA, Fitzhugh Fulton completed his pre-take-off checks at the end of Runway 04. Overhead a pair of Northrop T-38s circled, ready to accompany him on another Edwards maiden flight. Previously a senior USAF test pilot, Fulton was now flying for the US National Aeronautics and Space Administration (NASA); he had flown the Boeing B-29 and B-50 from which many NASA research aircraft had been launched, and the mighty North American XB-70A bomber. This day he would fly a Boeing 747-123 on its first flight with the Space Shuttle Orbiter riding piggy-back, the method chosen to return Orbiters to Florida's Cape Canaveral should they ever 'land away' in California when returning from space. But Fulton, his crew, the attendant Dryden personnel, and hundreds of media representatives had to wait. Take-off would be impossible: a skein of geese had got there first, and was flying right down the centre of the runway, blissfully unaware that they were delaying departure of the 747/Shuttle *Enterprise* combination.

Nature, always unforgiving of those who do not understand its rules, took precedence. As US Air Force historian Richard Hallion notes in his account of Dryden research: 'Geese and jet engines do not mix'. Man must avoid the birds while trying to emulate them, and to outfly them in as many ways as possible – faster, further, higher. Each of those superlatives can be applied to pilots and aircraft who have flown the skies over the Mojave Desert to the east and north of Edwards. This chapter cannot recount all the successes (and failures) that have attended experimental flying since the Second World War, but it can provide snapshots of aviation history as the aerospace industry has conducted half a century's worth of research and development (R&D) flight that has led aeronautical progress in speed, range, and altitude.

'Flight testing of new concepts, designs, and systems is fundamental to aeronautics. Laboratory data, and theories based on these data, cannot give all the answers,' NASA administrator James Webb told the US Congress in 1967. 'Flight research plays the essential role in assuring that all the elements of an aircraft can be integrated into a satisfactory system.' Formal testing of experimental aircraft after the Second World War was to analyse the requirements of control and the contribution of alternative configurations to ultimate performance; designers and pilots wanted to understand how representative theoretical and windtunnel-derived data were of actual behaviour in the air. According to a UK Aeronautical Research Council (ARC) paper: 'The behaviour of aircraft may be accurately predicted, by numerical analysis or by computer, provided the aerodynamic derivatives and the moments of inertia are known. In the early stages of design this information is necessarily supplied by windtunnel tests or estimation, but it is very desirable that it should checked by flight measurement, so that confidence can be placed in these procedures.'

Assessment of the Airbus A310's performance under icy conditions, if accretions formed on those parts of the wing and empennage unprotected by anti-ice systems, was carried out by making 'simulated ice shapes' of metal and conducting flight tests with them attached to the appropriate areas. The upper picture shows one of the simulated shapes attached to the port wing root leading edge; the lower view depicts a selection of the shapes tested, compared with the 'clean' leading-edge skin in the foreground.

During the early 1940s fighter pilots had begun to experience the 'compressibility' effects of airflow in terminal-velocity dives. In the UK a Spitfire PR.XI flown at the Royal Aircraft Establishment (RAE) was thought to have approached 90 per cent of the local speed of sound (Mach 0.90), and a new generation of jet and rocket engines was seen as the key to obtaining sufficient sustained power to explore sonic performance. The challenge was to develop engines powerful enough to overcome the increased drag that was generated at near-sonic speeds, as well as to delay that increase for as long as possible. The related reduction in lift and movement of the centre of pressure (as shock waves moved back over the aerofoil) raised the required control forces, while also diluting the effectiveness of hinged control surfaces.

As high-speed lift was reduced, many designs exhibited a tendency to pitch down into an outside loop (or 'bunt'). And the standard solution to delaying the drag onset – swept wings – brought an attendant challenge for the operation of aircraft at slow speed. 'Introduction of large angles of sweepback has led to many problems at the lift co-efficients used during take-off and landing', says another ARC paper. 'Many of the difficulties stem directly from changes in aerodynamic characteristics produced by quite small areas of separated flow even at moderate lift co-efficients. Some of these problems cannot be resolved in the windtunnel, and available tunnel information must be checked by flight measurement.'

In the USA, pioneering sonic work took place over the Mojave Desert, which met government requirements for security and space in which to test experimental aircraft. The Bell XP-59A Airacomet first flew there in October 1942, followed by the Lockheed XP-80 Shooting Star in January 1944 at what was to become (from December 1949) Edwards Air Force Base. In 1946 NASA predecessor the National Advisory Committee for Aeronautics (NACA) set up a formal flight-test unit there to begin a process of research that continues today as NASA's Dryden Flight Research Center.

While the pace of change during the war had produced new ideas about the shape of wings to come, and the Mojave had hosted a wide variety of designs – from the Heinkel He 162 through radio-controlled Boeing B-17 and the mixed-powerplant Convair XP-81 and Ryan XF2R-1 to the Douglas XB-42 and Northrop's XB-35 and YB-49 flying wings – the industry became increasingly preoccupied with the question of supersonic flight. Was it even possible, and, if it could be achieved, what design characteristics were required?

The need for such research had been acknowledged at the 1935 fifth Volta Congress on High Speeds in Aviation, which produced two important principles – the requirement for supersonic windtunnels, and discussion of swept wings to delay the onset of increased drag as airflows became sonic. Four years later a US Army Air Corps board considering future military-aircraft requirements received an aeronautical R&D paper that proposed extensive transonic work to correlate theoretical data with actual performance, using gas-turbine or rocket propulsion. In Germany and the UK such propulsion development was already under way.

From 1940 NACA began to recognise the value of a full-scale transonic-research craft to address compressibility phenomena, and it is worth summarising the events that led to such a programme, since it established a procedure that would be successfully pursued by governments, military services, and the aerospace industry in subsequent generations. Compressibility problems had been experienced in Europe with the Messerschmitt Bf 109 in 1937 and later in the USA with the Lockheed P-38 Lightning (which flew in early 1939). Subsequently, Republic P-47 Thunderbolt, Curtiss SB2C Helldiver and Bell P-39 Airacobra combat aircraft began to suffer airframe failures during high-speed dives.

New NACA high-speed aerofoil sections were thought capable of allowing aircraft to reach sonic speed, but there was no immediate consensus on the best configuration, power unit, wing form, and undercarriage. A major frustration for engineers was the inability of contemporary windtunnels to provide accurate data between Mach 0.7 and 1.3 because of 'choking' as shockwaves compromised transonic airflow. In the early 1940s, availability of turbine engines that would overcome the limitations of propeller performance at high speed was seen by US officials as the key to permitting exploration into the supersonic regime.

Delivering the US Institute of Aeronautical Sciences Wright Brothers' 40th anniversary lecture in December 1943, RAE director William Farren discussed the planning and application of research for aeronautics. The next day NACA's jet-propulsion committee, considering how to apply British turbojet experience, heard Bell Aircraft suggest that military services, NACA, and manufacturers develop a research aircraft. The industry would build it for the Services, and NACA would conduct flight research and share data. (Farren did not disclose that the UK had issued a specification covering development of an experimental machine, which became the Miles M.52 project, to be capable of attaining 1,000mph (1,609km/h, or about Mach 1.5) at 36,000ft (11,000m). And plans were afoot in Germany for the DFS 346 swept-wing supersonic rocket aeroplane.)

Simultaneously, the US Army Air Forces (USAAF) became convinced that a rocket aircraft was needed to study transonic flight and NACA and RAE were sharing thoughts on falling bodies for research, tracked by radar. Early transonic data was becoming available from NACA experiments with aerofoil shapes mounted on a North American P-51 Mustang, which generated supersonic overwing airflow when dived at around Mach 0.75. Since falling-body, wing-flow, and NACA rocket-propelled

FASTER, FURTHER, HIGHER

In the early postwar years, high-speed piston-engine aircraft were used to conduct tests at transonic speeds. An example was North American P-51H Mustang 44-64192, which was assigned to the US Navy in 1948 and went to Langley Field, Virginia, under Project NAV-4. After being specially instrumented it was given 'unique aerofoil sections which could create transonic airspeed' on each wing, as seen here. With aerofoils, assemblies, etc, attached, flights were conducted under the desired test conditions. The aircraft was subsequently used by Grumman until 1952, logging 659 hours on the swing-wing XF10F Jaguar programme.

research provided only limited potential because of the cost and effort required and the absence of suitable wind-tunnels, a dedicated research craft was increasingly seen as the way forward. Military officials could not agree on a single solution, with strong divergence of opinion between the US Navy's choice of a Mach 0.85 turbojet machine and USAAF preferences for a Mach 1.2 rocket-powered aircraft, but both were to be satisfied: rocket work was conducted with the Bell XS-1, and the Douglas D-558-1 provided initial turbojet data.

Together, the projects 'could serve as models for research-aircraft procurement', according to Hallion. He cites designers' recognition of the requirement for research aircraft well before government and industry, and customers' stipulations that manufacturers achieve a single goal as simply as possible so that research could be introduced on subsequent projects. The two aircraft remained essentially unmodified, could be used for other work, and exhibited normal flying characteristics despite being used to explore uncharted territory. And the ways in which the research was managed yielded valuable principles that were applied later to leadership of larger space exploration programmes.

Given the lack of knowledge in the mid-1940s about high-speed flight and the implications of operating close to the speed of sound, it is worth considering some of the early post-Second World War US research that contributed to advanced aircraft design. Sonic work can be regarded as having three phases. Initially, research was to prove experimental configurations or to understand supersonic behaviour; then, explore hypersonic flight; and, third, develop an orbital spacecraft. Subsequent research has studied individual areas such as paraglide vehicles, supercritical wing aerodynamics, tilt-rotor technology, oblique wing configuration, highly manoeuvrable aircraft technology, and the Space Shuttle Orbiter.

The sonic programme began in 1946 with glide and powered trials with the XS-1, and by May 1947 20 powered flights had been made by two aircraft. The first machine, with a thin, 8% thickness:chord-ratio wing, was used to reach Mach 1.1 as quickly as possible, while studies of transonic control and stability were conducted with the second example, which had a 10% ratio aerofoil.

Until the first supersonic flights, available data came from windtunnels running at around Mach 0.85 or from Mustang wing-flow trials, which achieved a speed of Mach 0.93.

The XS-1 provided all-important stability and control information via telemetry as test pilot Charles Yeager flew increasingly fast. When longitudinal control was lost at around Mach 0.94 because shock waves inhibited the effect of elevators, NACA and military engineers decided to use the adjustable tailplane for control. (Subsequently, it was concluded that disturbed airflow had compromised instrument readings, the actual speed having been Mach 0.997.) Once Mach 1 had been achieved – Mach 1.35 was attained within three months – additional X-1A, -1B, and -1D variants were ordered.

The second XS-1 explored the implications (for example, 30%-higher transonic drag) of a thicker wing, and speed data from both aircraft was correlated with windtunnel information. The discovery that the adjustable tailplane improved transonic controllability led directly to 'all-moving' surfaces used on the North American F-100 Super Sabre, the first production aircraft to exceed Mach 1.0 in level flight. Other results included improved airspeed-measurement techniques.

Flying in the wake of the Bell XS-1 had been the Douglas D-558-1 Skystreak, which was not truly supersonic (despite having a similarly thin straight wing and tail) but did not require an air launch. For five years until 1952 it provided sustained service in Mach 0.80+ turbojet research. NACA research-aeroplane projects leader Hartley Soulé reported that the X-1 and D-558-1 provided 'very complete coverage of design information for high-speed straight-wing airplanes [up] to transonic speed'. Together, they demonstrated the legitimacy of theoretical techniques by offering actual in-the-air comparative results.

Initially, the D-558-1 exhibited unsatisfactory handling above Mach 0.82, with longitudinal-control stick forces increasing 500% by Mach 0.87, accompanied by a tendency to wallow. But it made a positive input with the discovery that vortex generators introduced to speed up boundary-layer flow stabilised shock waves: trim changes were reduced and the limiting Mach number increased. Ultimately, the modification enabled the Skystreak to reach a maximum speed of Mach 0.99 and was applied in development of the Boeing B-47 Stratojet.

The losses in May 1948 of Skystreak No 2 following a compressor burst and a NACA P-51 Mustang that broke up while diving underlined the inherent danger in the research work. The swept-wing Douglas D-558-2 Skyrocket investigated the pitch-up from which many early swept-wing aircraft suffered, in this case during a Mach 0.60, 4g turn that confirmed wind tunnel predictions and increased airframe loading by 2g. The event was followed by a further incident near the stall in 'dirty' configuration (with gear and flaps down) that resulted in a 6,900ft (2,100m) spin before recovery.

By mid-1948 the principle of dedicated research aircraft had been established, and NACA expanded research to other areas. The Bell XS-2 would explore swept-wing technology for speeds as high as Mach 3.0; the stub-wing Douglas XS-3 was to be for Mach 2.0+ turbojet work; transonic research would be conducted with the semi-tailless Northrop XS-4; transonic delta-wing studies would occupy the Convair XF-92A; and variable-geometry wings would be a feature of the Bell X-5. (From June 1948 the 'S' was dropped from formal designations.)

In modifying aircraft to offset inherent shortcomings, NACA developed concepts that generally could be applied to subsequent designs to overcome problems. To correct characteristics such as stall or spin tendencies or unstable 'hunting' motions, NACA came up with wing modifications such as saw-tooth leading-edge extensions, slats, chordwise fences, or thicker trailing edges.

One machine with a particularly vicious problem was the variable-geometry X-5, but despite the tendency toward stall-spin instability it provided valuable insights and proved the 'swing-wing' philosophy that combined low and high speed performance. Its great utility (and economy!) value to NACA was that different degrees of sweep could be investigated in a single airframe. Variable geometry had been planned for the Messerschmitt P.1101 research aircraft before the end of the Second World War. Probably unaware of trials performed in the very earliest days of aviation, Grumman claims the 'world's first' epithet for its XF10F-1 Jaguar, which was developed for the US Navy during 1946-52 and made 32 flights with 13.5-42.5-degree wing-sweep. Although that mechanism proved reliable and effective, the aircraft was let down by airframe, engine, and stability and control problems.

Northrop's twinjet X-4 was similar to Britain's experimental de Havilland D.H.108 in having a swept wing, no tailplane, and pronounced instability at near-sonic velocity. NACA was called in to help Northrop to analyse initial flight-test information before the X-4 embarked on dynamic stability trials. Its inherent instability yielded a mountain of data concerning the interaction of motion about all three axes. One line of research involved thickening the trailing edge by the simple expedient of wedging open the split speed brake to improve roll control. So doing, NACA inadvertently provided a means to adjust the lift:drag ratio that would help to predict the behaviour of the much-later North American X-15.

The XF-92A was a leftover from a failed fighter design, but NACA wanted the lightweight, thin-wing, low-aspect ratio delta for supersonic research: its contribution to US designs such as Convair's F-102, F-106 and B-58 was echoed by French manufacturer Dassault, which extrapolated from the small Mirage I delta through the Mirage III fighter (a Fairey Delta 2 lookalike) to the Mirage IV high-speed bomber.

The extremely underpowered Douglas X-3 appeared a research failure, but was redeemed by virtue of its propensity for inertial coupling in flight and 'throwing a tread' on the ground. Wild gyrations generated when the X-3 was subjected to a snap roll at very high speed mimicked theoretical predictions just as early-production North American F-100 Super Sabres were being lost under similar circumstances of roll divergence. Increased wing and tail area was the solution, but the X-3 had provided the hard demonstration. The aircraft's huge appetite for tyres (a result of small-diameter wheels and high-speed airfield performance) led to revised design specifications for aircraft tyres.

An area of experimental NACA research with a decidedly low glamour quotient involved investigation of the 'dragworthiness' of external stores such as drop tanks and bombs which gave rise to distinct problems of airflow interference when attached to otherwise efficient wings.

Reviewing the first generation of experimental aircraft in the light of criticism that such activity was a 'make-work' exercise, NACA dismissed the charge. Critics claimed that research was expensive and diverted resources from 'real' aircraft, did not produce improved engines, and did not offer solutions because 'X' aircraft were unrepresentative. NACA was adamant that its work on aircraft configuration was the only such activity on which 'real' aircraft could be based and contributed directly to the USAF 'century-series' fighters from the F-100 onwards. Since research was essentially about trans- and supersonic aerodynamics, stability, and control, engine development was not the point. And because the configuration of just about all of the initial X aircraft was adopted by industry, the designs anticipated subsequent development and ensured that characteristics such as high tailplanes, small fins, and configurations likely to exhibit roll coupling or pitch-up were avoided.

NACA provided an impressive list of the areas in which high-speed experimental research aircraft had contributed to the data available to industry for future developments (see page 38).

The great contrast between the USA and the former Soviet Union in aerospace development was the latter's circumspection about boasting of its achievements. Much that is known about developments in the old USSR is due to dedicated analysis by a few observers, augmented by personal relationships established over many years. (Western, 'technicians' were only slightly less conspicuous with their tape measures at 1960s/70s Paris air shows than their Soviet counterparts as each caught up on the physical details of competing hardware.) One senior British commentator, invited in the early 1980s to write the definitive history of Russian aerospace, said simply: 'We don't have the information'. Accordingly, very much is not known about historic Soviet experimental research and development, despite an open attitude to contemporary matters. The same observer said in

The Grumman XF10F-1 Jaguar of 1953, the world's first variable-sweep fighter, suffered from numerous stability and control problems and a troublesome Westinghouse J40 engine. Although the project was abruptly halted, the manufacturer retained faith in the concept.

EXPLORING THE LIMITS

Republic's XF-91 Thunderceptor experimental very-high-speed, high-altitude interceptor had a variable-incidence wings of inverse taper and thickness. While the variable incidence permitted a high angle of attack for take-off and landing, the inverse taper and thickness, combined with leading-edge slots, delayed wingtip stalling at slow speeds, as the wing produced most of its lift on its outboard area. In high-speed flight the power of its 15,200lb-thrust, afterburner equipped General Electric J47 turbojet was supplemented by a 6,000lb-thrust Reaction Motors XLR11 auxiliary rocket motor, and in December 1952 this combination enabled the XF-91 to become the 'first American combat-type aircraft' to exceed the speed of sound in level flight.

1995: 'It is a cinch to find out what the Russians are trying to *do*, but still difficult to discover what they *did*.'

But some details have emerged about experimental aircraft, particularly from fighter manufacturer MiG. The MiG I-270 Zh of 1946 was a straight-winged single-seater that gave way to the MiG-9 first-generation turbojet fighter. The suffix defined its ZhRd (*Zhidkostniy Raketniy Dvigatyel* liquid-propellant rocket engine) powerplant, which the aircraft was designed to test. Two aircraft were built, with fuselages based on the Junkers Ju 248, but were written-off in landing accidents. Hallion has said that the design's potential as a rival to the Bell XS-1 'cannot be totally discounted', despite the I-270 Zh probably being limited to high subsonic speed.

The late-1940s MiG-9L/FK was an airborne laboratory to assist development of air-to-surface missiles, particularly the KS-1 Komet to be launched from Tupolev Tu-16s used for anti-shipping patrol. An unpressurised rear cockpit was introduced for a missile-guidance system operator, roughly above the wing trailing edge of the stretched MiG-9 airframe. Two radar antennae were fitted; a target illuminator above the nose intake with leading-edge receivers to detect reflected signals, and a fin-mounted transmitter/receiver unit that was used to develop both launch and missile guidance systems.

Two MiG-21I 'Analog-144' aircraft were built, based on the MiG-21S but fitted with delta wings based on scaled developments of that for the Tupolev Tu-144 super-

FASTER, FURTHER, HIGHER

sonic airliner. The aircraft were intended to prove airflow characteristics and flight controls, and to train Tu-144 test pilots. The MiG-21I/1 featured inboard leading edges with a 78-degree sweepback, reverting to 55 degrees for the outboard two-thirds, supplemented by inboard flaps and outboard elevons. The MiG-21I/2 sported a larger wing with a double-ogee planform, with four elevons on each side. Neither machine had a tailplane.

The aircraft were flown at speeds of up to Mach 2.06 and down to 212km/h (about 133mph) and up to altitudes of 19,000m (62,300ft). Unfortunately, delays in defining wing and elevon specifics held up development and the type did not fly until April 1968, less than eight months before the Tu-144 itself. The MiG-21I/1 crashed in 1969 as basic tests were completed, but the second machine flew for several years.

The MiG-23PD (*Podyomnye Dviatyeli*, or lift jet), also known as the MiG-23-01, was intended to meet a specification for a Mach 2.0+ short-take-off-and-landing aircraft in competition against the MiG-23-11/1. The PD adopted the MiG-21's tailed-delta layout and incorporated two 5,700lb-thrust Kolyesov Rd-36-35 lift engines. These were mounted in the mid-fuselage with a slight forward-thrust component, and used only for take-off and landing, when a louvred door was raised to supply air. Both lift exhausts could be directed through a rotating grid that acted as a thrust vector, which allowed forward propulsion to be augmented at take-off or opposed on landing in a reverse-thrust mode. Take-off roll reportedly was about 200m (655ft), with landings requiring about 250m (820ft). The aircraft appeared at the 1967 Domodyedovo display, but tests ended soon after.

Three MiG-23-11/1s were built. The design adopted a shoulder-mounted variable-geometry wing providing sweep between 16 and 72 degrees about swivel points 1.5m (59in) either side of the centreline. The wings had leading and trailing edge flaps, plus spoilers that provided roll control in conjunction with tailerons. Because the wing pivots provided no redundancy, load factor was limited to 3.1g. Take-off distance was 320m (1,050ft) and landing with tail 'chute 440m (1,445ft). Following 97 test flights, the factory report included judgements that the MiG-23-11/1 offered significant reduction of field-length requirements; 'great ease' of handling throughout the flight envelope; and low g forces in rough air at maximum wing sweep. The MiG-23-11 provided the basis for the production MiG-23S and much-stronger subsequent MiG-23M, MF and MS variants.

Two experimental MiG Ye-8s were built for a requirement for an all-weather day/night interceptor, based on the MiG-23PF, but with nose intake replaced by a ventral intake below cockpit to feed an uprated R-21f engine. A radome housed the Sapfir 21 radar in the nose, which carried 2.6m (102in)-span destabilising foreplanes just below the centreline. The foreplanes moved freely in the airstream at subsonic speeds, but above Mach 1.0 were locked. Five months after the April 1962 maiden flight, the project was halted after engine disintegration to Ye-8/1.

Another canard design was the MiG Ye-152P (similar to the non-canard Ye-152M planned interceptor), but the project was dropped in favour of the Ye-155, which was intended to intercept fast, high-flying aircraft such as the Lockheed A-12 and North American XB-70. The Ye-155 gave designers a demanding requirement: to provide a

In an attempt to combine Mach 2 capability with STOL performance, MiG produced the 23-01, which had an R-27-300 turbojet as its main powerplant and two Kolyesov RD-36-35 lift jets in the mid-fuselage, inclined slightly forward and used only during take-off and landing. A hinged louvred door opened above them to admit a supply of air to the lift engines, and they had rotating grids on their nozzle throats to allow the pilot to change the direction of the thrust vector. The 23-01 first flew on 3 April 1967, but flight tests were abandoned that same year when MiG turned its attention to variable-geometry wings.

vehicle from which to launch missiles against winged targets at any speed and height up to 25,000m (almost 82,000ft) when continuous high speeds will have produced a 300°C heat soak for the airframe and systems. Mikoyan selected welded steel for primary structure, with heat-resistant 11% light alloy elsewhere and 9% titanium alloy in hot areas. An original configuration was chosen for the twin-R-14D-powered aircraft, with a thin shoulder-mounted wing and pairs of dorsal fins and outward-canted ventral fins.

Work with ten development Ye-155P and R prototypes led to the MiG-25, but modifications generated during flight test demonstrate the value of experimental aircraft. Foreplanes were discarded; fin area was increased 'significantly'; ventral fin chord was reduced (to avoid ground strikes on landing); winglets were removed (but replaced by anti-flutter fairings); five-degree wing anhedral was introduced (in place of triangular wing endplates); and differential control for the all-flying tailplane introduced a taileron facility. The sheer performance of the aircraft provided the basis for three airframes designated Ye-266 for absolute speed and climb-to-height record purposes.

Experimental Sukhoi aircraft have included the T-37 that was designed to provide data and proved to be the manufacturer's fastest test interceptor. Based on the PT-8 produced for radar exploration, this aircraft used Sukhoi's standard tailed-delta configuration with a more-powerful integral engine with intake-mounted radar. Air-data heads were fitted on a radar-centrebody-mounted pitot tube. Airframe novelties included a removable rear fuselage for engine access and a stringerless monocoque forward fuselage that included integral fuel tanks and air ducts. Performance was enhanced by a low-profile cockpit canopy, but the powerplant was its Achilles' heel and unreliability led to scrapping of the programme.

The first Soviet vertical/short-take-off and landing jet aircraft was the Yakovlev Yak-36, which used vectored thrust instead of a combination of lift and cruise engines. With outrigger legs to balance the tandem centreline landing gear, the Yak-36 was powered by two R-11v engines fed by a broad nose intake. Jet efflux was by a single rotary nozzle on each side below the c.g., augmented by reaction control outlets on wingtips, nose probe, and tailcone. Yakovlev spent two years ground-testing the systems and engines while completing some ten examples, of which at least two flew.

Problems encountered during flight test included severe pitch instability while transitioning from vertical to forward flight. As a result, Yakovlev reverted to separate power sources on the Yak-36M, which was more reminiscent of the Hawker P.1127, with large intakes below and behind the cockpit to feed a single lift/cruise engine. Air for the tandem lift jets was provided through inlets on top of the fuselage, exhausting through 30-degree vectoring nozzles. A novelty was the safety system by which the pilot

Soviet Russia's first jet-lift V/STOL aircraft was the Yakovlev Yak-36, which had a pair of R-11v turbojets with single rotary nozzles beneath the wings on the aircraft's c.g. Reaction control outlets at the extremities of the wings and fuselage gave stability and control in hovering flight. Three aircraft were built; the third is depicted, fitted with UB-16-57 rocket launchers on underwing pylons. Testing began in 1962, and the first free hover was made on 23 June 1963.

was automatically ejected from the aircraft if combined height and descent rate was outside limits. Subsequent modifications to develop the operational Yak-38 included revised cheek inlets, underwing pylons, and strakes alongside the upper air inlets. Operational evaluation of the production aircraft proved critical to performance in high ambient temperatures as a result of insufficient installed thrust, which was increased in later models.

While NACA and the US military was pushing on with experimental high-speed research in southern California, designers and engineers in Europe were planning their own assault on the 'sound barrier'. Indeed, one British test pilot, Roland Beamont, had become the first supersonic European when he flew the North American XP-86A Sabre in early 1948, several months before the first supersonic flight in Old World skies, accomplished by John Derry in the third D.H.108 in September 1948 – almost a year after the Bell XS-1.

The world's first venture into supersonic aircraft performance might have been the British Miles M.52, a research design responding to a 1943 requirement for a 36,000ft (11,000m) speed of 1,000mph (1,600km/h – when a level 500mph/800km/h had not yet been attained). The project was cancelled less than six months before first flight amid other economic cuts in early 1946, ostensibly because it did not feature a swept wing, by then regarded as *de rigeur* – despite the M.52's having been intended for supersonic performance. (Future UK research was planned to be conducted with unpiloted, rocket-powered craft, and trials with a 30% scale model M.52 proved the basic design with a Mach 1.4 flight in October 1948, a few weeks after Derry's flight.)

Dedicated high-speed research aircraft – De Havilland D.H.108

The wooden de Havilland D.H.108 was intended to investigate swept-wing behaviour and generate half-scale data for application to tailless configurations being considered for the D.H.106 jet transport. Standard Vampire fuselages were used for three prototypes, the first being fitted with a 15% larger wing sporting 43 degrees of sweep. A conventional rudder was complemented by elevons outboard of the flaps. Fixed Handley Page leading-edge slots were incorporated following warnings from the RAE Aero Department that at low speed the configuration could lead to 'Dutch rolling' (oscillations around the longitudinal axis) or wing-drop with attendant loss of control. After more than two years of data-gathering following its first flight in May 1946, the initial subsonic D.H.108 went to Farnborough for RAE control, stability, and landing trials and was destroyed after spinning during stall tests.

High-speed trials of the layout were conducted on a second aircraft fitted with redesigned canopy, adjustable slots, powered controls and a larger 45-degree sweepback. Flying soon after the first machine, it immediately demonstrated faster speeds than the current 616mph (991km/h) world record, but during a rehearsal for a record attempts the aircraft broke up in a dive at Mach 0.875.

The third D.H.108, with a stronger fuselage, longer nose and automatic slots, flew in July 1947 to gather additional research data, including pressure plotting and investigation of pitch oscillations above Mach 0.88. Never-exceed (Vne) speeds were adopted and tests showed that pitch control was insufficient at Mach 0.97 and roll control was lost at Mach 0.99. During a dive on September 9, 1948, during which the D.H.108 generated -3g and pitched over beyond the vertical after loss of longitudinal control, the aircraft became the first in Europe (and the world's first jet) to exceed the speed of sound. It later joined RAE for longitudinal stability and control trials and was lost in a high-speed dive in February 1950.

English Electric P.1

Although this was a very different aircraft once it reached its final operational form as the Lightning F.6, the English-Electric P.1 design arose from post-Second World War recommendations that the UK develop a Mach 1.4 supersonic fighter. The P.1 had a 60-degree swept wing (virtually a notched delta) with ailerons on the wingtips; preferred tailplane position was low on the rear fuselage, although this was not confirmed until after the alternative T-tail configuration had been tried on the Short SB.5, built to test the overall configuration, including other degrees of wing sweep.

The P.1A first flew in 1954, and almost immediately exceeded the speed of sound in level flight. Early changes included provision of increased nose-down trim, locking of leading-edge flaps for landing, and non-use (except in emergency) of forward-opening airbrakes (which caused severe pitch-up and buffet). Poorly designed locks led to the loss of three cockpit canopies during high-speed flight, including one at Mach 1.0+. A notch cut in the outboard leading edge overcame uneven lateral control forces at low speed. The design target of Mach 1.2 was achieved, but ultimate speed was never proved because of reservations about directional stability. Such considerations were emphasised with the loss of the Lightning T.4 and T.5 prototypes after directional stability led to structural failure at high Mach numbers. The fin was made larger and stronger on Lightning F.2A and subsequent models.

A second P.1A, which flew in 1955, introduced twin Aden 30mm cannon armament, a modified equipment hatch, a much-needed 250gal (1,136 litre) belly fuel tank (standard on all Lightnings), and toe brakes (deemed by military pilots to be too sensitive).

The addition of fixed-nozzle afterburners designed for maximum performance severely reduced available dry thrust to less than that required for level flight should one engine be lost in landing configuration. Other military objections involved the uncomfortable ejection seat, and low canopy/helmet clearance. To reduce subsonic drag (by 20%), the first P.1A was modified with extended tip chord, resulting in a kinked leading edge and slightly higher supersonic drag (overcome by increased thrust). The two aircraft, which accumulated over 500 flights, were powered by Armstrong Siddeley Sapphire engines that were supplanted by Rolls-Royce Avons in the P.1B and Lightning. Further development was performed on 20 pre-production Lightnings, including the conical centre body in a now-circular nose intake first seen on the Mach 1.7 P.1B (which became Britain's first, and the world's third, Mach 2.0 aircraft).

Fairey F.D.2

The experimental F.D.2 represented Fairey's last throw of the dice in fixed-wing aircraft when the research machine flew in late 1954, but its genesis can be traced from 1947, when the UK government asked if Fairey's projected tail-standing vertical take-off F.D.1 could be developed for supersonic operation. After a false start with a twin-jet featuring very-swept leading and trailing edges, conventional tailplane, and all-moving tip ailerons, Fairey had by the end of 1949 settled on a single-engined delta configuration that comprised the minimum weight and smallest possible frontal and surface areas. The wing had a very low thickness/chord ratio, which required a slim undercarriage. To ensure sufficient pilot vision on the approach and ground, the nose including cockpit was articulated downward 10 degrees.

Limited windtunnel work preceded the first flight, with some supersonic test results not being analysed until after the aircraft had achieved Mach 1.0 in October

1955. Thoughts soon turned to records, and the aircraft set an absolute speed record of 1,132mph (Mach 1.73) in March 1956.

The aircraft, whose design was almost exactly duplicated in the Dassault Mirage III, gained a reputation among pilots for being easier to fly at supersonic speed than in slower regimes, where constant attention was required. The two F.D.2s were used for RAE control and stability tests (including Dutch-roll assessment) and low-speed drag measurement, and the design proved sufficiently adaptable to be converted for use as the BAC 221 in high-speed delta research. Plans for a developed F.D.2 design including rocket power and capable of reaching Mach 2.5 and 90,000ft (27,400m) did not proceed.

Bristol Type 188

Two examples of the Bristol Type 188 were built for sustained Mach 2.0 flight to study the effects of kinetic airframe heating, but the programme was primarily a failure since the aircraft singularly failed even to achieve that speed. The extremely futuristic configuration suggested in a 1952 RAE report – minimum-diameter circular fuselage, with engines almost as fat located halfway along each wing – was fabricated from stainless steel, which posed significant problems since materials compatibility was required between the airframe and different attachment fasteners (such as bolts, rivets, and screws). Another problem for the 188, Bristol's first turbojet (but the last design to carry the manufacturer's name), was manufacturing technique: Bristol developed new argon arc-welding procedures known as 'puddle-welding'.

The original specification called for an airframe to withstand Mach 2.0 at sea level, with a drag limit to accommodate shallow dives through the maximum Mach number achievable in level flight, but required speed was later raised to Mach 2.5. After windtunnel and flying-model tests a semi-delta shaped wing was used outboard of the engines, which were attached to the fuselage via a straight wing section. Longitudinal control was provided by an all-flying tailplane atop the fin, which had an extended chord to accommodate loss of power of one of the de Havilland Gyron Junior engines.

The oval fuselage section was just enough to house the pilot in his ejection seat, and much of the main body was used for fuel tankage. Behind the cockpit an equipment bay housed electronic paper and magnetic recording facilities and telemetry instruments for acceleration, strain, temperature, and vibration, and cockpit-refrigeration system for the cockpit. Bristol developed a ground station to monitor aircraft performance and behaviour in 'real' time and communicate information to the pilot. Orders for three of the original six 188s were cancelled, while one other became a static-test sample at the RAE.

First flight in April 1962 came just a year after initial roll-out, following resolution of air-intake problems that became apparent in ground tests, and a year ahead of the second 188. But flight-testing was to prove an anti-climax, since (ironically) Bristol was never able to wring more than Mach 1.88 from the aircraft. Heat-soak time was limited to two minutes (compared with the ten minutes specified) because of high fuel consumption from the underpowered engines. These were also prone to supersonic surge, leading to yaw and pitch problems. After rejection of the cost of redesigned air inlets and selection of an alternative powerplant, the project was abandoned in 1964. The two Bristol 188s had accumulated some 40 hr of flying in fewer than 80 flights. A positive product of the programme was the first British use of flight-data telemetry.

Supersonic research – Handley Page H.P.115

Once again underlining the value of flown research aircraft to confirm windtunnel derived data, as well as to perform original experimental work, the Anglo-French Concorde supersonic airliner project required two such vehicles: the BAC 221 (a modified Fairey Delta 2) to explore wing performance and the Handley Page H.P.115 to provide information about slow-speed handling and stability of slender delta wings.

Initially, it had been thought that two quarter-scale low-speed aircraft (one unpowered, for which Slingsby Aviation outlined a possible design) were needed, but the work was consolidated without reversion to a glider contribution. Nevertheless, the likely flight characteristics arising from the H.P.115's basic design (particularly control about the longitudinal axis in crosswinds or turbulence) suggested the machine would prove to be a handful for test pilots and there would be no reserve vehicle if the aircraft was lost. The aircraft first flew in August 1961, and proved not to have any tricks up its sleeve, being enjoyed by all pilots. They reported good longitudinal handling with precise response even in turbulence.

The Handley Page H.P.115 was built to test the low-speed characteristics of slender-delta wings, its plywood wing leading edges being easily removable to enable alternative units with conical camber to be substituted. Smoke generators beneath the inner wing leading edges fed coloured smoke over the upper surfaces, enabling the wing vortex to be seen and photographed to ascertain the point of vortex burst.

FASTER, FURTHER, HIGHER

Slow-speed handling showed that at a 30-degree angle of attack and 60kt indicated the H.P.115 was well within control. In other areas, rudder-induced roll was large and the aircraft was said to be directionally 'stiff'. Lateral control fell with speed and dutch-roll damping was neutralised, eventually becoming divergent as speed fell further. Oscillations as high as +/-30 degrees could be corrected at will through large helpings of left or right stick, or by lowering the nose. A maximum ground angle of 14 degrees endowed the H.P.115 with a minimum touchdown speed of 90kt. After five years' testing the aircraft was modified with microphones (to detect vortex noise), strain gauges, pressure-plotting transducers, and a wingtip parachute before providing five further years of research, including vortex generation and analysis.

BAC 221

Following agreement for development of the Anglo-French Concorde supersonic airliner in the late 1950s, it was recognised that data was needed on the behaviour of such a slender wing and recourse was made to the Fairey Delta 2. This aircraft already had been considered for possible further research into alternative delta planforms tied to Hawker Siddeley supersonic project work. Rationalisation of the UK aerospace industry saw Fairey absorbed into Westland, which passed the proposal to Hunting, which in turn was merging with Bristol, English-Electric and Vickers. When the music stopped the parcel was with the Bristol division of the new British Aircraft Corporation (BAC).

Modification of the F.D.2 as the BAC 221 for high-speed tests of the ogee wing for Concorde was carried out from 1960 and also included new systems and installation of telemetry equipment. BAC proposed a 6ft (1.8m) fuselage stretch and taller undercarriage. The middle section of the new wing's leading edge was swept by 65 degrees, being blended fore and aft to the fuselage and wingtips. The lower wing skin was bulged to accommodate the retracted undercarriage, which was adopted from the Lightning. A long chine extended the leading edge almost to the point of the nosecone, with appropriate mechanical flexibility to accommodate lowering of the (retained) drooping nose, whose position was now variable.

Other changes included incorporation of a camera atop the fin to record airflow on the tufted starboard wing (the port wing was drilled for pressure, boundary layer, and temperature plotting). A rocket 'lateral-thrust' unit was fitted to the fin and above and below each wingtip to provide impulses for flutter investigation. To save costs, the underwing air intakes were fixed rather than adjustable, being optimised for a speed just below the Mach 1.6 design speed. Because there was to be no system ground-test rig, this work was completed on the aircraft, delaying first flight until May 1964.

Trials led to an 8in (20.3cm) fin extension (to improve directional stability) and adjustable autostabilisation, before the aircraft was delivered to the RAE Aero Flight at Bedford in May 1966. This timing was now so far into the Concorde programme that the project's ability to influence the SST was already compromised. Nor was the 221's wing an exact scale replica, since it lacked the requisite camber, droop, thickness:chord ratio, and twist, and the flying-control layout was different. Accordingly, its work alongside the H.P.115 low-speed development aircraft was largely taken up with investigating approach and landing phases of flight and research connected with Concorde's automatic throttle system. The 221 retired in 1973; four years after Concorde's first flight, but still almost three years short of the airliner's entry into service.

Proof-of-concept vehicles – Avro 707

To meet a 1947 UK requirement for a high-altitude bomber to carry a nuclear load at high subsonic speed, Avro chose a delta configuration, which permitted bombs, engines and undercarriage to be contained in the wing. Information on the bomber's likely performance and flight characteristics was based on data collected

The high-speed behaviour of sharp-edged, slender ogival-planform wings was assessed on the BAC 221, the highly modified first Fairey F.D.2, WG774. First flown 1 May 1964, it joined the slender delta research programme at RAE Bedford in 1966 – too late to have any significance on Concorde, but it yielded much useful data.

EXPLORING THE LIMITS

The first of two Avro 707A high-subsonic test aircraft, WD280, displays its Vulcan-style wing-root intakes.

Avro 707B case study

According to UK Aeronautical Research Council papers based on RAE technical reports, the Avro 707B suffered reduced static longitudinal stability above a lift co-efficient of 0.5, although this was recovered in cruise flight above a Cl of 0.7. There was also a loss of elevator power 'from cruising to landing configuration'. Lateral behaviour had become more important as the availability of instrument approach aids required pilots to be much more precise in maintaining the optimum flight path, while simultaneously new aerodynamic configurations commensurate with high-speed flight were degrading aircraft lateral characteristics in this regime. For example, tests of the 707B showed that, as approach speeds fell below 115kt, drag increasingly compromised both speed and flight-path control, although this had been preceded by aileron 'snatching' as flow began to separate at the wingtips from about 118kt. Pilot reports that low-speed aileron response was good were supported by flight measurements, which showed that, even when flow was separated over most of the wing ahead of the ailerons, roll-control had fallen by only 25%. Low-speed rudder capacity was limited; at speeds as low as 90kt directional control was 'very poor'. The sea-level rate of roll with full aileron deflection during approach was almost 15 per cent faster than that of the contemporary Gloster Meteor NF.11 nightfighter, but neither compared well with the North American F-86E Sabre or, more particularly, the Hawker Hunter (which, at 65 degree/sec, rolled twice as fast as the 707B). To understand 707B wing airflow characteristics, standard flight-test procedures such as tufting, smoke generation and pressure plotting were used. Analysis showed 'quite sudden' flow separation to a 'considerable' depth (indicated by tufts), spreading inboard from the wingtips as speed reduced and angle of attack increased. Smoke generation illustrated the separated flow. Pressure plotting was useful in checking effectiveness of various modifications such as wing fences (to resist spanwise movement of airflow) or a notch cut in the leading edge (to speed up the local boundary layer), although both changes led to impaired handling qualities attributed to aileron buffet and overbalance. The modifications did little to improve performance by reducing airflow separation. Accordingly, low-speed control stability (also a consideration with the Boulton Paul P.111) came to be seen as a greater limiting factor to touchdown speed than the simple maintenance of an adequate margin above the conventional stall speed.

FASTER, FURTHER, HIGHER

from research conducted with three Model 707 one-third-scale experimental aircraft.

The first 707 crashed less than a month after its first flight in September 1949 (possibly due to jammed airbrakes producing a high rate of descent, aggravated by a rearward c.g.). The second aircraft, the 12ft (3.6m)-longer 707B, flew a year later to gather 80-350kt low-speed data relating to stability with a 51-degree swept leading edge (44.9% at quarter chord), compared with the 60 degrees leading edge of a 'pure' delta that was used on the Convair XF-92A. It logged 100hr before going to the Aircraft & Armament Experimental Establishment at Boscombe Down in September 1951, finished its flying days at Farnborough after a crash during crew training at the Empire Test Pilots' School, and was later scrapped at RAE Bedford.

Dubbed 707A, a third machine was used to research behaviour at high subsonic speed from June 1951. Following problems with turbulence around the 707B's dorsal air intake, the 707A was fed via wing-root intakes as would be used for the full-size bomber (produced as the Avro 698 Vulcan). A second 707A accelerated testing from February 1953, to be joined five months later by the 707C, which sported side-by-side seats for pilot training. The 707C carried out supersonic tests at Bedford in 1958, but four further 707Cs were cancelled. A half-scale 60,000ft, Mach 0.95 twin-engine Avro 710 was cancelled, along with a full-size aerodynamic prototype.

Lack of powered controls on the 707As initially produced stability and control problems. After being used to prove a compound leading-edge sweepback modification, the original 707A was used for delta-wing low-speed airflow investigation at the Aeronautical Research Laboratories in Australia. The second 707A took part in auto-throttle trials to offset drag changes at higher angles of attack when landing.

A product of 707B testing was a longer nosewheel strut to improve runway incidence, which was introduced to the emerging Vulcan design. The 707 trials also endowed the Vulcan with a smaller fin, and tilted and angled jet pipes to enhance longitudinal stability and overcome thrust-related trim changes.

Handley Page H.P.88

To assist its understanding of the crescent planform chosen for the wing of the Victor strategic jet bomber, Handley Page tested a 36% scale model wing on a Supermarine 310-type fuselage. The resulting H.P.88 flew in 1951 with a wing shape heavily influenced by Arado 234 bomber studies in Germany. (An earlier one-third scale H.P.87 glider was abandoned.) The wing incorporated a two-stage sweep to leading and trailing edges, although changes to the H.P.80 design rendered the planform unrepresentative even before the aircraft flew. Also, the low wing mounting to match the airframe did not reflect that chosen for the Victor. Prototype manufacture was

The crescent wing for the Handley Page H.P.80 Victor bomber was tested on the Blackburn-built Handley Page H.P.88, a Supermarine Type 510 fuselage with a new wing. Completed in 1951, it exhibited pitching problems that developed into porpoising and, after only 14 hours' flying, broke up in flight, killing its pilot.

EXPLORING THE LIMITS

Possible sweep angles and relative wing/tailplane positions for the English Electric P.1/Lightning supersonic fighter were evaluated using the Short SB.5 low-speed testbed, WG768. Different wing-root attachments enabled three different sweep angles to be tested, and the air-to-air photograph on the left, taken in 1952, shows the aircraft with its wings set at 50 degrees sweep, their lowest angle, and with a high-set variable-incidence tailplane. The view on the right depicts the SB.5 in 1960, with the maximum wing sweep of 69 degrees and the tailplane and elevator repositioned at the bottom of the rear fuselage.

contracted to Blackburn, but initial flight trials exhibited pitch sensitivity from about 230kt and extreme control difficulties above 255kt. A Blackburn-modified tailplane trailing edge permitted speeds of up to Mach 0.82, but subsequent Handley Page trials ended when the aircraft broke up during its fastest-ever low (300ft/90m) run just 36 days after its maiden flight. Investigation showed that the airframe had suffered +12/-5g loading at 525kt (i.e. the accelerometer needle reached both 'stops'!) and structural failure occurred behind the wing.

Short SB.5

The Short SB.5 was built to investigate low-speed behaviour of the highly-swept layout selected for the English Electric P.1, itself an experimental design expected to pave the way for the UK's first truly supersonic fighter. Despite the S.B.5's relatively light weight, it embodied a stiff wing (with permanent 20-degree leading-edge droop) capable of being mounted at 50-, 60-, or 69-degree sweep, and carrying the fixed main undercarriage (with adjustable rake to accommodate the varying c.g.). Other flexibility inherent in the design (which was essentially a seven-eighths-scale P.1) included a movable tailplane (that initially was placed atop the fin but later moved to the lower rear-fuselage position adopted for the P.1) and an inboard leading edge drooped to modify the camber of the inner wing.

After a first flight in late 1952 and seven months of tests, the wing was moved to the intermediate position. During 1958-60 the aircraft returned to the manufacturer for replacement of the Derwent engine by a more powerful Orpheus to permit acceleration from near-stall speeds, remounting of the wing in its extreme location, installation of a zero-altitude ejection seat and revised instruments, and repainting. Altogether eight combinations of wing sweep and leading-edge droop were tested. The SB.5 was then used for basic RAE handling trials until the mid-1960s, after which time it joined the Empire Test Pilots' School.

Vertical take-off research – Fairey F.D.1

In 1946 Fairey began to study possible tail-standing vertical-take-off (VTO) but conventional landing fighters, and proposed formal research at a time when pilotless models were seen as the best method for a vehicle involving complex controls. Accordingly, the company started work on rocket-powered, delta-winged models launched from short near-vertical ramps, alongside studies of a full-scale piloted aircraft. A delta wing section which offered no tip-stalling at slow speed and good lateral behaviour was chosen for tests of slow/medium control and stability with a rocket/jet-engined prototype that would be tested first as a glider. Initial model trials indicated control and stability problems, accompanied by dangerous engine incidents, and although the conventional Fairey Delta 1 was flown in 1951 (then the smallest delta design to fly), official interest was withdrawn. The aircraft was retained for investigation of delta-wing lateral and longitudinal stability as well as braking-parachute research, and provided a stepping-stone to the later Fairey Delta 2.

Short SC.1
In the early 1950s jet engines were being developed offering power sufficiently greater than installed weight to stimulate thoughts of direct lift. The Rolls-Royce thrust measuring rig (dubbed 'flying bedstead') made the first successful demonstration in 1953. Two Rolls-Royce Nene engines were installed back-to-back with exhausts diverted downwards from points close to the c.g.; balance was achieved through the use of 'puffers' directing air tapped from the compressors. The rig had a thrust:weight margin of 16%, somewhat short of the specified 25% target. Almost 250 flights were completed by Rolls-Royce at up to 50ft (15m) altitude and 10kt before the rig was passed to the RAE, in whose hands it crashed in September 1957 after failure of autostabilisation. A second machine crashed fatally two months later.

Nevertheless, the UK government sponsored a small Rolls-Royce lift engine (the 2,000lb-thrust RB.108) and issued a specification for a VTOL research aircraft that would transition to wing-borne flight with lift engines inoperative. Since Short Brothers believed (in 1953) that future supersonic airliners would require vertical-lift engines, the company successfully proposed a small tailless delta-winged aircraft (later dubbed SC.1) using four centrally located RB.108s for lift and a further tail-mounted unit for forward thrust. After two machines were ordered Short Brothers began to develop requisite automatic stability-control equipment, power controls and undercarriage: the new precision-engineering division used its analogue computer to design hydraulic and electronic mechanisms, including magnetic amplifiers for maximum reliability.

To meet RAE requirements for a quick transition between manual and auto control, Short introduced triple-channel servo controls. In most respects the airframe was conventional, while levers for up to 35 degrees thrust 'tilt' were added to the pilot's normal power controls. The SC.1 used puffers, with compressor air blown through half-open valves slaved to aileron and elevator circuits with nose and tail valve adjustable to also provide lateral thrust to change heading.

First flights with the two aircraft took place conventionally (with no lift engines installed) in April 1957 and in tethered hover in May 1958, but not before Bell had hovered the X-14 vectored-thrust research vehicle in February 1957. An early modification was the introduction of a dorsal extension from the fin to improve directional stability. Initial public demonstrations at the 1959 Farnborough Air Show were compromised when grass cuttings clogged the engine-intake debris guard, leading to an expeditious descent. The first transition from level flight to a vertical landing occurred in April 1960 after three years of test flying and little more than six months before Hawker Aircraft began to hover its single-engined P.1127 experimental vectored-thrust aircraft. Soon rolling take-off techniques were being developed to reduce ground erosion and increase take-off weight for steep, non-vertical departure. A proposed two-seat SC.8 trainer variant using a fifth lift engine did not proceed.

Extended trials through 1960-63 saw the introduction of improved autostabilisation, inlet louvres (which closed during forward flight), and 'puffer' valves. Faulty autostab gyros led to a fatal accident in October 1963, which was followed by the introduction, three years later, of heading and sideslip controls, horizon instrument, and navigation display, later supplemented by a head-up display and ground/air data link ahead of all-weather and night-flying trials at Bedford. Short's plans for further development included military and supersonic commercial transports, and ground-attack and naval-strike fighters. The company also worked hard but in vain to promote VTOL launching platforms from which fighters and bombers might be deployed.

Hawker P.1127
Hawker Aircraft's private-venture P.1127 experimental aircraft combined vertical (or very short) take-off and landing capability with the performance of contemporary ground-attack fighters. A key to this was its use of a single engine to provide all required power in whatever mode of flight. Accordingly, the design did not suffer the penalty of other VTOL projects which had dedicated lift engines that provided only a deadweight once the aircraft was wing-borne. Developing a principle originated by French engineer Michel Wibault, Bristol Siddeley's Pegasus engine discharged thrust through four adjustable nozzles; two 'cold', using air from the fan, and two 'hot' from the exhaust, that permitted the P.1127 (it was never named) to take-off and land vertically or conventionally, and 'fly' sideways (via the appropriate use of balancing 'puffer' jets) and even backwards (since the nozzles rotated past the vertical to provide a slight rearward thrust

EXPLORING THE LIMITS

With test pilot Bill Bedford at its controls, the first flight prototype of Hawker's revolutionary P.1127 vertical take-off-and-landing fighter, XP831, makes its first free hovering flight, at Dunsfold, on 19 November 1960. The cable trailing from the starboard wingtip is a communications link.

A major drawback of many of the V/STOL systems tried was that the lift engines might be nothing more than an additional burden when the aircraft was in level flight. Germany's first such aircraft, the EWR/SUD VJ 101C of the 1960s, had four reheated 3,650lb-thrust Rolls-Royce RB.145 engines mounted in pair in swivelling wingtip pods, which resolved that problem, but the two 2,750lb-thrust RB.145s in the fuselage were dead weight when the machine was hovering.

component). Since the aircraft could lift itself off the ground the implicit thrust:weight ratio was greater than unity, endowing the P.1127 with spare power – though it suffered the usual fight against weight gain, which was more critical than in conventionally configured aircraft.

The first vertical take-off at Dunsfold in October 1960 was followed five months later by the commencement of conventional operations at RAE Bedford. Thereafter, as four development P.1127s joined the two prototypes, Hawker worked to prove the techniques needed to obtain transition between wing-borne and reactive-lift modes. The development machines were used for engine, performance and handling trials, pilot conversion, and aerodynamic modifications.

An indication of the design's sensitivity to the engine configuration came in 1961, when the second prototype crashed after a thrust nozzle broke off in flight (rendering the machine uncontrollable), and the first prototype crashed at the 1963 Paris Air Show after a system failure sent all nozzles into the aft-thrust position while it was manoeuvring at slow speed.

NATO interest was such that a 'tripartite' evaluation squadron representing UK, US and German interest was formed to operate nine development machines, dubbed Kestrel FGA.1s. German interest did not proceed, but six Kestrels (AV-6As) went to the USA. The aircraft was subsequently acquired as the Harrier by the RAF and the US Marine Corps (AV-8B). The concept was further developed as the basis for the Harrier II in co-operation with McDonnell Douglas. A projected supersonic P.1154 to suit RAF/RN needs did not proceed despite years of discussion, because it became too expensive and too late. But in the 1970s naval interest was rekindled with the advent of through-deck cruisers and the Sea Harrier was announced in 1975.

German interest in the concept was confirmed later with the VFW-Fokker VAK-191B, which incorporated two Rolls-Royce lift engines and a third vectored-thrust unit as a proposed Fiat G.91 replacement. Germany had earlier conducted trials in the early 1960s with the Entwicklungsring Sud (Bolkow/Heinkel/Messerschmitt) VJ.101C, a VTOL research aircraft powered by two integral lifting and four wingtip tilting engines. Making its first transition from vertical to horizontal flight in September 1963 (but abandoned a year later), this was intended as a precursor to a Mach 2.0 V/STOL project. The Dornier Do 31E VTOL transport project flew in 1967 with wingtip liftjets and a pair of podded Pegasus vectored-thrust units.

Other experimental V/Stol craft have included the Dassault Balzac and Mirage III-V, each with eight liftjets; the SNCA du Nord tilt-duct project; Lockheed's XV-4A Hummingbird using diverted thrust from fuselage-mounted P&W JT12As; Ryan's XV-5A with 'fans in the wing' for vertical lift; the Ling-Temco-Vought XC-142A and Boeing-Vertol/Canadair CL-84 tilt-wing projects; and the Curtiss-Wright X-19 tilt-propeller research aircraft.

Pure research – Boulton Paul P.111

With near-sonic experimental flying allotted to the D.H.108, and high-speed swept-wing behaviour being investigated with the Hawker P.1052 and Supermarine Type 510, early delta tests were the function of Boulton Paul's P.111 and Fairey's F.D.1. These were designed for academic work, rather than tied to particular projects. Deltas were tested for their potential to reduce weight and

FASTER, FURTHER, HIGHER

Avro Lancaster PA474 was used as an aerodynamics testbed for some years. Here it is seen with a suction wing protruding from its fuselage for laminar-flow tests in connection with Handley Page jet airliner projects exploiting boundary layer control. A pair of 60hp Budworth marine gas turbines connected to the wing by six flexible suction pipes were fed with air drawn through manifolds, needle valves and venturi-meters from numerous slits in the wing's surfaces, and a similar but separate system collected air from the wing's glassfibre leading edge. Tests began at the College of Aeronautics in October 1962 and were continued until late 1965, by which time the installation, now improved, had been transferred to a Lincoln.

To ensure that a swept wing retained its angle of attack and remained free from buffet regardless of its deflection under load, Geoffrey Hill and Short Brothers' project designer David Keith-Lucas devised the 'aeroisoclinic' wing, in which the torsional and flexural axes were coincident and the lower skin of the wing's leading edge was separated from the bottom edge of the box spar to avoid buckling. Originally fitted to a glider, the wing was then tested on the Short S.B.4 Sherpa, seen here, which was powered by two 350lb-thrust Turbomeca Palas turbojets and first flew in 1953.

Built for a British-government-funded research programme into delta-wing aircraft, Boulton Paul P.111 VT935 underwent extensive tests from 1950 until 1958. Three alternative wingtips could be fitted for comparative testing. The bulge on the rear fuselage is a housing for an anti-spin parachute.

drag (with no tailplane), as well as to provide better air penetration at higher speeds. The P.111 was built to a 1946 specification covering pure research into stability and control of delta wings, but did not fly until 1950, when longitudinal control was found to be very sensitive; one factor was a nose-up trim while the undercarriage was retracting. Control about all axes was compromised by the behaviour of powered systems and related artificial 'feel'.

Changes made during repairs following a forced landing included a nose-mounted pitot tube, new undercarriage doors, and mid-fuselage petal airbrakes (mainly to increase drag on approach so that engine speed was retained for electric power generation), the machine being redesignated P.111A. Variable gearing was introduced to reduce control sensitivity (with automatic trimming available above 400kt). There was no conventional stall, so lateral control became the minimum-speed criterion, and the P.111 achieved a maximum speed of Mach 0.93.

Short Sherpa

The Short S.B.4 Sherpa ('Short & Harland Experimental and Research Prototype Aircraft') was based on the earlier SB.1, the first aircraft designed and built by Short Brothers in Belfast after its move from Rochester and designed to maintain the manufacturer's acquaintance with swept-wing research. The S.B.1 had been a one-third-scale glider designed to test a so-called aero-isoclinic wing originally intended for the Short PD.1 that competed for the UK V-bomber requirement. The wing, inspired by Geoffrey Hill's ideas for perfect inherent stability and control, had a 42.25-degree leading-edge sweep and all-flying tips that were operated together as elevators or differentially as ailerons; there was no tailplane.

EXPLORING THE LIMITS

The principle involved 'torsional and flexural axes [that] were coincident'; the leading edge was separated from the lower wing skin to avoid local buckling. Short wanted to use the wingtips to apply loads that would cause the main wing to flex in such a way that incidence remained constant (and hence buffet-free) as speed increased. In the event, the glider crashed and work continued as the Turbomeca Palas-powered S.B.4 Sherpa version, which first flew in October 1953. Extensive use of flight recorders provided information about the expected behaviour of such a full-size variable-camber wing when operated at high speed and high altitude. Short tried unsuccessfully to use the wing to meet the requirement that led to the Blackburn Buccaneer.

Hunting H.126

Although the UK National Gas Turbine Establishment had been since the early 1950s investigating the potential for a 'jet flap', using diverted efflux to augment take-off and landing characteristics, it was not until 1963 that full-scale flight tests began. The chosen vehicle was the purpose-built Hunting H.126, which sported a high-aspect-ratio shoulder wing and was powered by a Bristol Siddeley Orpheus engine. The plan had been for all the exhaust air to be expelled from the wing trailing edge, providing thrust and contributing to aerodynamic efficiency over the lifting surface of the wing. In practice it was found that the use of relatively cold air, blown round a rounded trailing edge and ultimately exhausted forward, provided the best gains in lift coefficient. Despite the inherent problems of transferring the exhaust gas through the airframe structure and wings, construction of the H.126 used standard alloys. Flap construction allowed for a greater expansion of the (heated) upper surface than of the lower surface, which was in ambient temperatures.

The aircraft carried extensive measuring equipment, particularly to monitor structural temperatures. Conventional controls were augmented by 'puffer' valves below each wingtip, below the rear tailcone, and on each side of the rear fuselage. Pitch and yaw inputs were commanded by the pilot, but lateral augmentation was governed by independent autostabilisation. Only a small amount of the exhaust was discharged through conventional outlets (which were mounted on the fuselage underside, to offset the nose-down trim change from the 'jet' flap when power was increased). After completion of UK trials, the H.126 was taken to the USA for full-scale windtunnel tests at Ames FRC in California in 1969.

Built to investigate the principle of augmenting lift by use of a jet flap, the Hunting H.126 low-speed research aircraft had a Bristol Siddeley Orpheus engine, the efflux of which was ducted through the wing and discharged as a thin sheet of gas from the trailing edge, producing both lift and thrust over the full-span flaps. The outer ends of these flaps also worked as ailerons. Two small direct-thrust nozzles provided additional thrust, and further nozzles in the rear fuselage provided pitch-trim and yaw control. First flown on 26 March 1963, the H.126 was able to take-off and land at much lower speeds than conventional jets.

FASTER, FURTHER, HIGHER

McDonnell Phantom II 62-12200 was going to be a standard F-4B for the US Navy, but became a testbed for future models. It was used to test manoeuvring leading-edge slats and a beryllium rudder. After being tested with a fly-by-wire control system in 1972, it was given canard winglets and revised intake bodies, as shown.

Twelve years later, when Space Shuttle Orbiter *Columbia* alighted at nearby Dryden FRC in April 1981, it was watched by thousands of spectators there and millions more on live television around the world. This first re-entry of a winged, piloted spacecraft from Earth orbit was accompanied by an appropriate signature tune: 'the rumbling of its own sonic booms', according to Hallion, a reminder of the late 1940s' exploratory flights that 'pushed back the frontiers of supersonic and hypersonic flight' in those same skies.

Bibliography

Barnes, C H, *Bristol Aircraft since 1910* (Putnam, London, 1964). One of the publisher's classic company histories, several more of which appear below.

——, *Shorts Aircraft since 1900* (Putnam, London, 1967).

Brown, D L, *Miles Aircraft since 1925* (Putnam, London, 1970).

Buttler, A L, *British Secret Projects: jet fighters since 1950* (Midland Publishing, Hinckley, 2000).

Gunston, W T, *Back to the Drawing Board* (Airlife, Shrewsbury, 1996).

——, *World Encyclopaedia of Aircraft Manufacturers* (Patrick Stephens, Sparkford, 1993).

——, *The Osprey Encyclopedia of Russian Aircraft, 1875-1995* (Osprey Aerospace/Motorbooks International, Osceola, Wisconsin, USA, 1995). Extensive coverage of civil and military aircraft built in the Russia from the earliest times; an invaluable English-language reference.

——, (Ed), *Chronicle of Aviation* (Chronicle, London, 1992).

——, *Tupolev Aircraft since 1922* (Putnam, London, 1995).

——, & Gordon, Y, *Yakovlev Aircraft since 1924* (Putnam, London, 1997).

——, *MiG Aircraft since 1937* (Putnam, London, 1998).

Francillon, R J, *Grumman Aircraft since 1929* (Putnam, London, 1989).

——, *McDonnell Douglas Aircraft since 1920* (Putnam, London, first edition 1979).

Hallion, R P, *On the Frontier: Flight research at Dryden 1946-81* (NASA, Washington DC, 1984).

——, *Supersonic Flight: Breaking the Sound Barrier and Beyond* (Brassey's, London, revised English edition, 1997).

——, (& Time-Life editors) *Designers and Test Pilots* (Time-Life, Alexandria, Virginia, USA, 1983). A volume in the publisher's series entitled 'The Epic of Flight'.

Hygate, B, *British Experimental Jet Aircraft* (Argus, Hemel Hempstead, 1990).

Jackson, A J, *Avro Aircraft since 1908* (Putnam, London, first edition 1965).

——, *De Havilland Aircraft since 1915* (Putnam, London, first edition 1962).

Mason, F K, *Hawker Aircraft since 1920* (Putnam, London, first edition 1961).

Mondey, D, (Ed) *The International Encyclopaedia of Aviation* (Octopus, London, 1977).

Rawlings, J D R, and Sedgwock, H, *Learn to Test, Test to Learn: the History of the Empire Test Pilots' School* (Airlife, Shrewsbury, 1991).

Sturtivant, R, *British Research and Development Aircraft: Seventy Years at the Leading Edge* (Haynes, Sparkford, 1990).

Taylor, H A, *Fairey Aircraft since 1915* (Putnam, London, first edition 1974).

9
Advanced Aircraft in Production
Mike Hirst

There was a time when aircraft were fashioned from metal by craftsmen, whose skills were a blend of traditional and modern engineering etiquette. They were traditional because they used tools that their forefathers would have recognised. They were modern too, because they accepted that their workmanship often had to be so accurate that a ruler would not suffice to measure any dimension on their work. They used micrometers as easily as they handled a rasping file, or set a countersunk rivet without so much as a tell-tale scratch on the aluminium skins that it fastened. Production-line mechanics who rose to the top of the field had to be exceptionally skilled.

Good mechanics have to be sensitive to the material they are using, and the requirements it fulfils. That much was proven when many so-called unskilled workers were used to build, service and repair aeroplanes in the Second World War. But with the war over, and jobs in aviation businesses on the decline, a reduced band of individuals took over almost all the production and servicing jobs again. Almost all major postwar aircraft production work was in metal, and largely alloys of aluminium. After that, the business changed little for many years, although the attention to detail became ever more important. Those who had always aimed for the best were the best to do the job. Safety was enhanced by having natural-born perfectionists. Production engineers' skills were paramount, prided as well as practised. But today's skills, while still founded on an evolutionary trail from rudimentary material-forming skills, are often based on layers of computer and intellectual knowledge that would have been regarded as misplaced in a latter-day production engineer.

Aircraft are nearly always symmetrically shaped, which seems to appeal to the human senses that associate symmetry with beauty. There is an odd kind of psychology at play when people murmur about a 'beautiful' aeroplane, but streamlining that mimics the grace of a well-toned body seems sure to win human approval. Shown a curvaceous silhouette, even aero-philistines will concede at least a degree of approval.

These attributes were ingrained, too, in designers. While graceful lines were hard to conceive when most aircraft were biplanes, and stiffness was achieved with cross-bracing and struts, the most revered designers were those who strived to combine beauty and utility. This required an understanding of materials on two counts. First, in terms of the relationships between material and aerodynamics in the airframe; and second, in terms of the advances in materials that affected the efficiency of aero engines.

Achieving that natural beauty was a knack that came slowly, but natural designing talents emerged in some extraordinarily clean-lined designs in the 1920s and 1930s. Almost all were expensive in terms of time and effort – and therefore cost – to manufacture. Given that a consequence of a graceful design was an escalation of cost, most designs manufactured were utilitarian. This quest has become almost never-ending. Keep it affordable, but make it perform as if it is the best. Striving to get faster and higher has simply compounded the problem.

An early example in which the balance was magnificent was the de Havilland Mosquito. Especially relevant, from the perspectives being considered in this chapter, is how the Mosquito performed so splendidly and yet was so easy to build that it appeared readily, in quantity. It had a splendid configuration. The shape was not just aerodynamically good, but lent itself to ease of manufacture – in wood. There were many metal components (engine bays, undercarriage, flying controls and so on), but the aircraft's success was attributable largely to the fact that it could be manufactured by production staff who had virtually no innate knowledge of aeroplanes. The best skilled woodworkers in Britain were cabinet makers, and the furniture manufacturing businesses of Britain supplied most of the input to the Mosquito assembly lines, where the aeronautical production engineers slotted, bolted, screwed and glued the remarkable 'wooden wonder' together.

Much of what they did was almost 'lost', in that their techniques were eclipsed by the art of machining great slabs of metal, but they have returned, in a refined sense, today. What is important to record about the Mosquito is that the design was assembled from sets of subassemblies that had well-defined physical interfaces. The subassemblies were usually major airframe components – a fuselage or a mainplane, for example – where the wood was arranged so that its grain was aligned to the best direction to bestow stiffness as well as strength. There were places where several criss-crossed layers were necessary, and these were glued together, using what were remarkable adhesives at the time. Joints were often screwed and bolted too, but distributed loads were almost all absorbed in large sheets of high-quality plywood.

There was plenty of room for inaccuracies, and they resulted in imperfect aircraft. One of the last checks on a brand new Mosquito would have been to attach a cord to the nose cone and use it to measure the distance to a wingtip. The tolerance, over some 35-40ft (11-13m) was an inch or two (3-5cm). But that was the easy bit of the test. When the cord was taken to the other wingtip, the difference in the measurement between tips had to be within about half an inch (1cm). Given that the aircraft was a conglomeration of thousands of parts, once this test confirmed an acceptable assembly, the purist instincts of the traditional mechanic were delighted.

The world moved on, and jet aircraft were almost invariably made of aluminium alloy. This material can be forged, milled, etched, hacked, pressed, bent, even cast into the desired shape. The more robust the structure, the more massive the work involved, and in some cases a stiffer, and stronger, material was specified. The popular, but not oft used, substitutes were steel or titanium. Steel is heavy and corrodes, unless one uses stainless steel, which is difficult to machine and to attach to other metals. Ordinary steel has its uses, but it has to be given exemplary protection, and frequent checks are essential.

Titanium is even heavier, but it is even stronger. It is used especially where space is at a premium, as in undercarriage reinforcement and the attachment regions of wings. Titanium is mined, and the ore refined in an electrolytic process, so it is expensive, and production per year, world wide, is meagre compared with the massive tonnage of steel or aluminium.

Aluminium alloys were the first choice because pure aluminium is very soft. Certainly it is lightweight, but it has relatively little strength. Alloyed (mixed) with a few per cent of additives like manganese, copper or zinc and, more recently, lithium, aluminium is suddenly much stronger, although it is still light. The strength-to-mass ratio is exemplary, and the material is naturally occurring and available in abundance. There are many aluminium alloy specifications, and they are so specialised that a designer will choose different specifications for the top and bottom skin of an aircraft wing. The deciding crite-

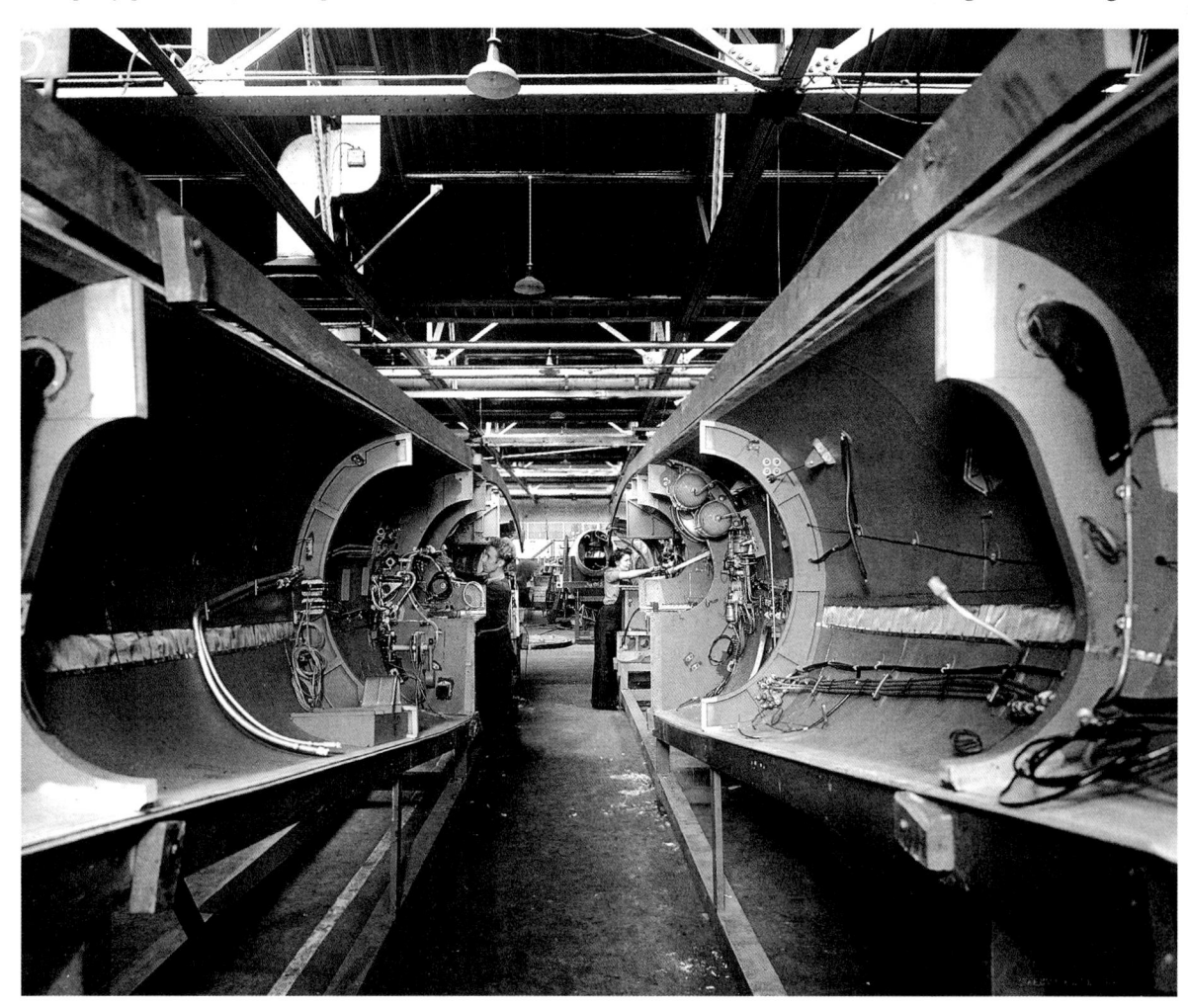

De Havilland's versatile Mosquito incorporated many structural features that simplified assembly. Here, the two halves of the moulded plywood fuselage are being fitted out with services and equipment before being joined together.

ADVANCED AIRCRAFT IN PRODUCTION

Although it was built in very limited numbers, Concorde did go into production. In this photograph, taken in Aerospatiale's Toulouse St Martin factory in 1973, aircraft No 209, the ninth production aircraft, is being assembled in the foreground, while No 201 is in its finishing stages in the distance. The latter aircraft first flew on 6 December that year. Limiting Concorde's cruising speed to Mach 2.0 meant that its airframe could be built of aluminium alloys, thereby avoiding the manufacturing problems encountered with metals such as steel or titanium.

ria in such a case is to ensure that they are suited to being in tension (on the bottom) or in compression (on the top), when the aircraft is flying. Different specifications have evolved around needs for alloys that will flow easily when cast, or to be crack resistant, or able to withstand long periods at high temperatures. One reason that Concorde is a Mach 2.0 aircraft (it was designed to cruise at Mach 2.2 in fact), is that at the cruising speed the leading edges of the wing experience 'heat soak' at a temperature that the special alloy used can just withstand without losing its strength and stiffness. Even so, the airframe life is finite, and like most aircraft Concorde is checked frequently, and when signs of ageing are found, repairs are made. The airframe of an aluminium alloy aircraft eventually has to be withdrawn from service. Many aircraft are scrapped because, although they still look fit to fly, their most highly-stressed structural members have reached their 'fatigue life'. Fatigue – the propensity to crack when repeatedly stressed, relieved and stressed again – is a weakness of all metals, and aluminium alloys are particularly susceptible.

When carbon fibre appeared in the late 1960s, it was viewed very suspiciously by production engineers. The idea harboured by the scientists who created carbon fibre was simple. As carbon molecules will form long strings that are extremely tough, if minute strands of carbon (a very common and therefore inexpensive element) are created, the hair-like material is, per unit weight, many times stronger than steel. But while fibres are strong in tension, they are hopeless in bending, and will flop about. The solution is to create directional layers of carbon fibres in a matrix (the name for any substance that will stick them together). The most common modern matrix compounds are epoxy plastics. If layers of carbon fibre

FASTER, FURTHER, HIGHER

are laid together and their strand directions adjusted to run predominantly in different directions, a thick layer can be built up that has directional strength properties; it is high-tech plywood. The designer can tailor the strength of the component, and instead of taking a great slab of material and milling it to size, the component is now built up until it has the desired strength, and stiffness. Traditional engineering acumen suddenly seemed less important in aircraft production. The layers were laid in a mould of the desired size and shape, and when the mould was full its contents were 'cured' in an oven. This usually required the mould to be tightly packed, and care had to be taken to exclude all air. The matrix became viscous, flowed together between layers, and then hardened, leaving a solid article.

This production process has proliferated as people have wanted aircraft that will fly higher and faster, because the aircraft perform better if they are lighter, and carbon fibre is the lightest airframe material of all.

Creating a wing skin is no longer a job for the milling and etching department. The large and expensive machines used for such work are replaced by large tables, on which the moulds are placed, and ovens, where the moulds are stacked before air is evacuated and the articles heated. What emerges can be relatively precisely shaped, just needing the edges to be trimmed. This can be done by high-power lasers or concentrated water-jets. It is often necessary to do more, almost traditional, work to create holes for fasteners and so on. But overall, what took many people many hours is now a job that can be completed by a few people in many fewer hours. Furthermore, the airframe can be assembled from fewer parts, which means less assembly time. The fact that the raw material is expensive is now traded off against reduced production effort, giving an immediate benefit. Further benefits accrue from the increased aircraft performance attributed to the carbon-fibre construction.

This material had become a dominant contributor to

When this photograph of the Bristol Type 188 high-speed research aircraft was taken at Filton in 1960, it was optimistically expected that it would attain Mach 3. To cope with kinetic heating at this speed, stainless steel was used as the primary structural material, causing difficult manufacturing problems. After all the effort, the aircraft proved unable to attain the anticipated speeds.

ADVANCED AIRCRAFT IN PRODUCTION

modern designs, but the way that designers got there led production engineers through hell and back, because the route was not obvious, so there were some peculiar decisions in the period 1960-1980, roughly. Given that production staff dominate the payroll at an aircraft factory, it is vital to know the techniques and skills that future generations will need, and to invest often vast sums of money to get the production effort right. This investment has to be researched and committed many years before the aircraft that will exhibit the benefits become public items. The quest for aircraft that will meet superlative claims has been a driving factor in the many turns on the tortuous road that has been trodden.

Around 1960 a team at the Bristol Aircraft Company was tasked to design and build the Bristol 188 high-speed research aircraft. This was a single-seat aircraft, loosely based on the configuration favoured for a Mach 3.0 (2,000mph; 3,218km/h) cruising speed V-bomber replacement for the RAF. This speed was regarded as suited to an aircraft cruising above 70,000ft (21,300m), and thus being able to outrun air or ground launched missiles. Such an aircraft would be a very difficult target indeed. Aluminium alloy would melt at the skin temperatures that the passing air would generate from friction upon the airframe, and there were not many alternatives. Bristol chose to make the aircraft from stainless steel. This was diabolical to work, as it was harder than many production tools. They did it, using stainless steel rivets and learning much, too, about welding in an inert atmosphere (argon arc welding). The riveting was hell, as the tough material could not be formed if it was not treated to tremendous hammer blows, and yet the structure that was being manufactured was delicate to the touch. Argon-arc welding excluded oxygen and prevented bubbles forming, and could preserve the integrity of joints, and choosing how to close a joint (violent riveting, or tempestuous welding) was not an easy decision.

At last the Bristol 188 flew. It looked futuristic, but the makers then discovered that the design had a chronic deficiency of thrust; or was it an excess of drag? It flew but little, and never reached a speed that the front-line English Electric Lightning fighters of the day could already exceed in RAF service, and had been designed to flight conditions that the emerging generation of surface-to-air missiles could combat. The gleaming 188 was an expensive one-off, but in the minds of Britain's aeronau-

Although it began life painted black, the North American X-15 turned white in 1967 when, in preparation for an attempt on the absolute world speed record, NASA coated it with an ablative material that dissipated heat and protected the aircraft and its pilot from the high temperatures generated at hypersonic speeds. On 3 October 1967, piloted by Major Pete Knight, the aircraft set a record of Mach 6.72 (4,534mph; 7,297km/h).

FASTER, FURTHER, HIGHER

tical production engineers it was a milestone, because they had learned that building in stainless steel was going to be expensive. It is not impossible, just improbable, and in over 40 years not a single stainless steel design has been considered in Britain.

In the USA they were exploring the limits with similar optimism. For a start, they had accumulated a few hours of experience with hypersonic research aircraft such as the North American X-15, which was built of titanium and nickel-chromium based steel components. The Inconel X material used was not dissimilar to stainless steel, but owed much to the knowledge that had been accumulated in the specification of material properties for aero-engine turbine blades. On the X-15 (of which only three were flown) the structure was fabricated with ingenuity and a degree of pragmatism. Not least was the specification of ablative coatings, which were in their infancy but were needed for recoverable space vehicles.

These are applied rather like a viscous paint. They harden on the external surface and seem hard at normal temperatures, but when the temperature rises above a few hundred degrees Celsius they soften, and 'run', driven aft by the air stream and streaming away as an incandescent mess within the aircraft's wake. This conveys away a lot of the excessive heat, which relieves the designer and builder of worrying about the effects of internal structure becoming overheated. While this was fine for a research aircraft, which after each high-speed test flight had to be 'painted' all over again, its unsuitably for a production aircraft was pretty obvious.

For example, the X-15's windscreen could be completely misted by ablative material streaming from the nosecone, so at launch the V-shaped windscreen had the port window covered by 'eyelids'. The aircraft would scoot to Mach 7 or so at sub-orbital altitudes, then, as it returned to the thicker air of the ionosphere, the clear

The Americans encountered the problems of working with stainless steel with the North American XB-70 Valkyrie, a prototype Mach 3 bomber, which first flew on 21 September 1964. Sustained flights in excess of Mach 3 were made by the second of the two Valkyries built, which continued to perform valuable research work after the first aircraft was lost in a mid-air collision on 8 June 1966.

194

window would mist. Near subsonic speed the eyelids would be opened, and although crews reported detectable pitch and roll effects as they opened horizontally, they restored a view ahead. Ablative materials are still used on very-high-speed air vehicles today, notably the US Space Shuttle Orbiter. The vehicle is a relatively conventional substructure coated with 'tiles' that lose their outer surface, and after each flight each tile is inspected and badly ablated tiles are removed and replaced. It is a relatively crude solution, but a typical production engineer's way of turning good science into a manageable operational solution.

Aircraft that were going to fly faster, higher and further in the atmosphere were not given to such a pragmatic solution. But that did not stop the Americans launching into a project, at the same time as the Bristol 188, that would also be a Mach 3 cruising speed aircraft. But the Americans went the whole hog and built a prototype Mach 3.0 supersonic bomber, the North American B-70 Valkyrie. It was an awesome project. Two aircraft were flown. Conscious of having to sustain volume production within a finite (but still lavish) budget, they also chose to use stainless steel, and had to endure hours of frustration in developing new metal cutting and joining techniques. They did this well enough to fly their aircraft up to the top end of the flight envelope they were designed to achieve, carpeting parts of the US desert with the most powerful sonic boom ever created by an aircraft. But it was almost a pipe dream. Integral fuel tanks revealed the porosity of welds, and while solutions were found (welding techniques similar to those used in the UK were developed again), one tank proved just too troublesome to tackle. Again, because missile technology was advancing fast enough to degrade the perceived invulnerability of this class of aircraft, it became another shelved project. Any designers considering the merits of a Mach 3-cruise supersonic airliner should remember these two projects and the heartaches they caused, before putting any conceptual description to a production engineer.

The Bristol 188 first flew on 14 April 1961, and the B-70 on 21 September 1964. This was the heyday of heady prognoses that forecast higher, faster and further as a matter of course, and these two projects were at the head of so much hype and publicity that their demise was a sad blow to pride, both in Britain and the USA. France had attempted similarly ambitious but less spectacular projects, such as the Nord Griffon turbo-ramjet single-engine fighter and the Leduc ramjet-powered research aircraft. After these excursions into the unknown, and in the early 1960s, the French embarked on a technical liaison with the UK and, with the team at Bristol especially, began to design and build Concorde.

Concorde is a rare aircraft, with 20 examples (prototype, pre-production and production) built and flown. Nevertheless, it was in production, and that meant that all the issues that had to be addressed to build thousands (which wild optimism had expected at one stage) were taken into account. First, the production engineer's most persistent nightmare was avoided; the aircraft had an aluminium alloy airframe. Existing tooling technology, while still representing a substantial investment, was applicable to the Concorde programme. The Anglo-French constructors invested in milling, chemical etching, stretch forming, and so on, that was traditional world-wide. The materials were compatible with the experience of existing airframe mechanics, so inspection provision in service was conventional too. The only problem that designers had was coping with the great extent to which the airframe, and the wing in particular, would change shape as components expanded in the heat soak at Mach 2 cruising speed. They solved it through an ingenious combination of material specifications and pin-jointed ribs. The latter allowed the wing to expand, for fuel tanks to remain secure, and for the aerodynamic profile to be within very strictly defined dimensional tolerances. The engineers who built Concorde airframes needed to use all their accumulated knowledge to construct them, which supports the adage that 'what is right, looks right'. In terms of its structural design, and the way it was designed to be producable and maintainable, Concorde did more than is ever credited to it.

A decade or so before Concorde, while the Bristol 188 and North American B-70 were mesmerising the public's eyes, the Americans had already flown the Lockheed A-11 Mach 3 aircraft at Indian Springs Air Force Base, Nevada, on 26 April 1962. The A-11, and its better-known derivative, the SR-71, punched holes in the record books, yet they were unknown to the public at large until American President Lyndon B. Johnson revealed not just its existence, but also the extent of its performance, in 1964. The 'Blackbird' set many challenges in terms of its production.

Before taking up that thread, imagine the reaction in the Soviet Union in the late 1950s as these astonishing projects were being created in the West. A sober, and really very impressive outcome was the Mikoyan MiG-25 fighter. This flew in 1964, and was a Mach 3 aircraft that did live up to its promise. The design had been spawned by the Soviet Union's desire to have a counter to the B-70, and while that was cancelled in the mid-1960s, by 1976 the Soviet Air Force was believed to have over 400 MiG-25s in service. No doubt the Soviets were aware of the presence of the Mach 3 SR-71 as it boomed its way across their territory and filmed all that lay beneath it. That there was never a clash of these East and West titans was probably down to the poor capability of the MiG-25's interception radar, rather than to lack of performance alone. Certainly the USA took to using spy satellites as soon as it could, minimising the need to fly hazardous atmospheric missions.

FASTER, FURTHER, HIGHER

The MiG-25 design was almost pure simplicity. To withstand short periods at high speed, a substantial proportion of its airframe, including the wing main spars and fuselage frames, was manufactured in steel. It was an enigmatic aircraft to the West, until one was flown to Japan by a defecting pilot in 1976. Western intelligence descended on it, and succeeded in convincing gullible press correspondents that it was clearly inferior to anything in the West because of the steel (implied 'rusty') components. In fact, the aircraft confirmed many fears. An incredibly rugged and reliable airframe, it was aerodynamically adept, structurally sound, and easy to build and maintain. The propulsion system was no more complex than it needed to be. In fact, all that let it down was the inferior quality, compared with what was available in the West, of the aircraft's electronic systems.

The MiG-25 was a product of the race for speed and altitude waged between the USA and the Soviet Union during the Cold War. The pacemakers in this race emerged from the Lockheed Skunk Works in California. Within the factory Lockheed let innovative designers try the unthinkable, and the roll-call of types has been brief but legendary. The first acknowledged product, the U-2, was almost a subsonic Lockheed F-104 Starfighter. It had an extra-large engine, with no afterburner, and a wing very unlike that of the F-104; it was more like that of a glider. The horizontal and vertical tail surfaces had to be extended, and the cockpit was a less streamlined affair than that on the F-104. With so much lifting surface, and a powerful turbojet engine, the aircraft, even with a full fuel load, would climb like a homesick angel. Coaxing it back down was another matter. It was designed and built in 1954-55, and did not really challenge production engineers because the airframe was largely conventional, in aluminium alloy. That should be looked at in perspective, however, as the long periods at high altitudes required special lubricants to be developed and introduced into service, even grease for flying control surface hinges. The U-2s were built in batches, and were maintained and operated by specialist crews, who over 40 years later say little about their work. On 1 May 1960 USAF Major Gary Powers was shot down over the Soviet Union and

A change to titanium was made when Lockheed built the astounding A-11 (later SR-71). Its milled and chemically etched skins expanded faster than the aircraft's internal structure at high speeds. Consequently, they had to be able to slide over one another.

ADVANCED AIRCRAFT IN PRODUCTION

used as a bartering tool by the Soviets. This effectively ended the U-2's initial deployment life, although it was reborn in the late 1980s as the TR-1, fitted with huge underwing pods and operated as a tactical reconnaissance aircraft.

The A-11 (later, with a role change, the SR-71), the second major design from the Skunk Works, was much more challenging to produce. As stainless steel had been such a nuisance, Lockheed chose to use titanium. This is an expensive material, but it is a natural element, and given the metallurgical knowledge of the day it was much more predictable than stainless steel. The aircraft had many complex panels, and the titanium skins were milled and chemically etched to make the most of the high specific strength (ratio of ultimate tensile strength-to-weight) of the material, and yet to leave a pattern which would provide stiffness too. The processes that were used to coax the titanium into shape were almost conventional metal-forming processes, implemented using specially strengthened machine beds and special cutting tools. Like Concorde, the SR-71 is plagued by thermal expansion. If the whole structure could have 'soaked' to the same temperature they could have neglected the problem, but it was clear that the skins would expand faster than the internal structure, which would be 'cooled' by fuel. They had to attach the skins to the wing ribs and fuselage frames at a limited number of points, and let the skin panel edges ride over one another. As the aircraft accelerated and the skin friction effect heated the airframe, the skins would expand, slipping over one another. Eventually the internal structure would warm up, and the skins would return approximately to their original relative positions. It was simple and ingenious, albeit far from perfect. The overlapping joints were a source of extra drag, but the edge dimension – subject to not making the skins impossibly thin, or prone to vibrating – was minimised to reduce this. Then there was the problem that the integral wing contained fuel, and fuel leaks out of overlapping joints. The solution was pragmatic. One fuelled the aircraft up, expecting fuel to seep out at about a steady rate. It could have been as high as 10 per cent per hour. It was traditional to keep SR-71s in special shelters, where fuel could drip, nay, trickle, on to the hangar floor and be collected in integral trays that kept tyres protected. The places stank of kerosene. It was traditional to take off with only as much fuel as was necessary to rendezvous with a tanker close to the point where the aircraft would be ready to climb rapidly and go supersonic. At this point it would drink to its limit (even to a gross weight that exceeded maximum take-off weight) before departing on its covert mission. It was still seeping fuel, as well as guzzling it, but, by all accounts, as the aircraft got hotter the seepage rate reduced. On exceptional flights another rendezvous would be necessary to ensure sufficient fuel to reach a suitable base.

Super-plastic forming and diffusion bonding is used to make small but complex structural components lightweight. This is a section through a manhole cover made by British Aerospace for the Airbus A310-300, which had no fewer than 26 of them. The weight-saving benefits totalled 33kg (73lb) on these components alone.

Even the SR-71s were not really built in quantity. They were batch built and almost hand-crafted in the Skunk Works, workers being sworn to secrecy so successfully that little has been revealed about how the aircraft were really manufactured. We can be sure that many of the techniques we have inherited today were dreamed of, and research into them started, in those days. An example would be superplastic forming and diffusion bonding (SPF/DB), which is used almost routinely, but only on small articles, to build complex forms in tough and difficult-to-machine materials, on modern combat aircraft. The Skunk Works also fostered interest in laser (and water-jet) cutting technology.

In today's technical teams the production engineers needed to work up design solutions relying on these methods are as important as designers. They have to be adept at the traditional engineering skills, and they need to be well qualified in the sense of understanding the issues involved in managing a vastly expensive piece of plant. In the case of SPF/DB technology, the plant has the potential to go wrong, even catastrophically, but it has equally a propensity to yield defective products to an engineer who does not understand how to use it well. Inserting slabs of expensive material into a process that takes time and exhibits a 0 per cent or 100 per cent success rate depending on the qualities of the engineer in charge, is one of today's production challenges in titanium.

One of the best examples of an SPF/DB-formed component today is the heat exchanger exhaust ducting in the Panavia Tornado's dorsal fin. When fabricated from sheet it was a difficult and time-consuming article to build, not least because rivets had to be set in small spaces. In service these rivets, immersed in a high-speed and hot airflow just above the engine bays, would shear because of heat and vibration, and fall out. The SPF/DB equivalent component is created by loading pre-cut and pre-bent sheets of material into the SPF/DB unit, where they are held on specially shaped dies. The temperature is raised, air is excluded, and the forging process is accelerated with the hammer blow that drives the sheets against the dies being created by a small and controlled explosion. In a split second the sheets are slammed together so completely that there is no metallurgical evidence of a joint. The grains of metal actually fuse together. Once material has been formed like this, there is no turning back. The Tornado ducting unit is created to dimensional tolerances on the air passages and overall size that a hand-fabricated unit cannot equal, and, perhaps best of all, it is just one part, not a dozen sheets with several dozen rivets within it. It is much more reliable and far less prone to fail in service. Although it costs more to build, it costs less to maintain. Because it fails less frequently it is safer, and the overall conclusion is that it is a cheaper solution. The knowledge that a modern production engineer needs to have at hand to contribute to the design, and eventually to manufacture such items, is formidable.

In fact, recent generations of aircraft have revealed that dimensional tolerances are the biggest challenge of all, and the production engineer has had to learn to handle them adeptly. The driving force has not been to make an aircraft that is faster or higher in terms of performance, but one that is less easy to detect, or 'stealthy'.

One unexpected benefit of the extremely futuristically-shaped SR-71, whose configuration had been predicated solely by aerodynamic considerations initially, was that it proved (in certain directions) to be almost undetectable. Some simple theory would give one an inkling of why, and perhaps this was a deliberate yet unacknowledged policy during design. But Lockheed spokesmen stoutly deflect too many probing questions. They tend to remind all who hypothesise on the topic that the aircraft was designed in the era of slide rules, and thus claim that its radar cross-section (effective target size to a radar) was difficult to predict.

In the mid-1970s the Soviets published some academic papers that predicted how radar energy was reflected, absorbed or diffracted by objects. The investigations were so penetrating in terms of mathematical rigour that they astonished a group of young engineers at Lockheed. Ben Rich, the new design chief in the Skunk

While much of the Lockheed Martin F-117's airframe is of conventional aluminium alloy, carbon fibre is used extensively in its external structure. The aircraft is built to dimensional tolerances several orders of magnitude finer than any previous one, with basic overall dimensions critical to a millimetre. The dimensions, calculated using computer programs, were applied to computer generated drawings.

ADVANCED AIRCRAFT IN PRODUCTION

Works, listened to a couple of young engineers who had been reading the Soviet research. They concluded that Soviet mathematical capability was in advance of their own, but that Americans had superior computers, and that if they wrote radar cross-section evaluation programs based on the Soviet analytical techniques they would be able to predict the radar cross-section of an aircraft in three dimensions. Rich gave them a number of months to create the data for the SR-71. He had the actual aircraft radar cross-section data, and was sceptical of their ability to succeed, but within two months he had convincing proof that the theory was producing a superb fit with reality. A few more aircraft had to be modelled to gain full confidence in the techniques used, but apart from that there was no denying the fact that radar cross-section could be predicted.

As the team doing these calculations grew, and as their confidence in the methods they were using grew, they moved away from predicting the radar cross-section of existing aircraft to predicting the radar cross-section of unbuilt projects. Eventually they were able to write overall aircraft configuration rules, and this is where they horrified production engineers (and aerodynamicists). They realised that the perfect aircraft would have all its radar-reflecting surfaces perfectly aligned to be parallel to a minimum number of directions. There were other rules, to do with corners and radar frequency, and the tendency to diffract, rather than reflect or refract, electromagnetic energy, so like so many mathematically derived theories, the concept is easier to describe than it is to implement. But the gauntlet was down for production engineers. The aircraft had to have perfectly aligned surfaces.

An associated evolutionary track was that, although these aircraft could be drawn by hand, the vertices at important component junctions were so critical that dimensions tended to be computed. Thus the move towards computer-aided design (CAD – charted in more detail in chapter 10) was given extra impetus. Overnight, production-jig tolerances were reduced from millimetres to microns, and at first there was concern that the tight tolerances would mean effort spent building them with such precision was likely to lead to nought if the imperfect aircraft were as detectable as conventional designs. In the late 1970s some rough and ready test aircraft were flown; the Have Blue projects. As flying machines they were frankly lethal, but as they roared across the Nevada desert, virtually undetected by the radar units that were trying to track them, they proved the point, so the Skunk Works created the world's first stealth aircraft to go into production. The F-117 Stealth fighter (in essence a light-attack aircraft), was designed in the late 1970s and first flew in the early 1980s.

The point proven in the secret trials with the Have Blue designs was that, given radar detection ability of an inverse-fourth power, if one could reduce the radar cross-section by an order of 10^4, the detection range in any circumstances would be reduced by 10. If a ground-based radar was able to detect a conventional aircraft at 50 miles (80km) range, the 'stealthy' aircraft would only be detectable at 5 miles (8km) range. If it flew at over 5 miles altitude it was virtually undetectable. This was a whole new era, in that the superlatives that had been attached to aircraft – higher, faster and further – were now joined by another – stealthier. The F-117 was immediately compromised by long-standing concerns about production. Could it be produced at an affordable cost, and still be an effective air vehicle?

Although it was a Skunk Works aircraft, this model was built in substantial quantities, in the order of 100 aircraft, within a decade. Acknowledging that it was subsonic and it had to be affordable, much of the airframe was designed in conventional aluminium alloy, and major internal structures were manufactured in Lockheed factories, or by subcontractors, using conventional tools and practices. Even so, the design tolerances imposed were very high, and the number of suitable sources was low. It was difficult in that respect. As the major subassemblies were combined, Lockheed's engineers took responsibility for maintaining dimensional tolerances over the complete airframe. Because it was a 'facetted' aircraft there was plenty of opportunity for defining reference points, and it was the first time that aeronautical engineers used a technique that had developed in building surveying. They measured between various points on the aircraft structure using a laser. It was also the first aircraft on which the panel edges were all aligned to be parallel in a limited number of directions. Where a production joint or a door (bomb bay or undercarriage door, for example) deviated from these lines the panel edge was zigzagged, aligning the edges with other edges. Having parallel edges meant that any reflected radar energy was aligned to a common direction, and just rarely would it reflect back towards the illuminating source. This minimised detection through reflection. The dimension of the zigzag pattern, and indeed the overall dimensions of the aircraft, were determined by the need to ensure that other electromagentic propagation effects were minimised; notably diffraction. Again, the production engineers were held to dimensional tolerances that were several orders of magnitude different to previous aircraft. Exact figures are not bandied about, but whereas the nose-to-wingtip dimension on a similar-size conventional aircraft, such as the F-18, would possibly be 0.25in (6mm), on the F-117 the length was critical to a millimetre or so. The designers calculated these dimensions using computer programs, and they converted the product of applied physics to computer-generated drawings. Thus the design began to evolve, within the minds of a design team, but with reality growing out of CAD processes.

Carbon-fibre production, as expected, had established

The advanced wing being mounted on a Harrier fuselage during the building of the prototype of the Advanced Harrier AV-8B at Mc Donnell Douglas in St Louis, USA, in September 1978. At that time the supercritical wing structure was the largest aircraft part ever made from composite material. Including the control surfaces it weighed 1,374lb (623.6kg), and by making all the spars and the upper and lower surfaces from graphite epoxy a saving of 330lb (150kg) over conventional materials was achieved.

itself in the 1970s. Initially the adherents of this new material were too ambitious. The vision was that aircraft could be moulded, almost like a toy plastic kit aircraft, but this required curing processes and dimensional tolerances that were difficult to achieve, and, above all, structural integrity was not always assured. The idea that one laid layers of fibre 'cloth' into a matrix was borrowed from the glass fibre industry, where these techniques were well established. Aviation applications grew in the 1970s in the high-performance sailplane market too, but these were relatively expensive, limited-production aircraft, and the hand-crafted wings and fuselages were of variable quality, subsequently tested individually to ensure that integrity was acceptable. The proportion of units manufactured and failed was often high, again contributing to cost.

Established manufacturers wanted the value of carbon-fibre reinforced plastic (CFRP), and made their moves less boldly. One leading user, in mass production, was McDonnell Douglas. In the late 1970s the company researched and implemented CFRP construction of wings, basically mixing conventional metal-aircraft manufacturing technology and CFRP techniques. It did this on the F-18 and the AV-8B by having wings which had metal spars and ribs and CFRP skins. The manufacture of the skins was done in moulds, where layers of carbon-fibre 'rovings' (simply fibres lightly impregnated with uncured matrix, so that it held together like a fragile mat) were built up by hand. The mould was bagged, the air evacuated, and the whole mould placed in a modest-size autoclave, where several hours of heating cured the matrix and created a very dense carbon-fibre skin panel. McDonnell Douglas was not averse to placing metal fixtures in the skins, and when the panel was cured it therefore had inset metal flanges onto which spars could attach. It also drilled the skins, riveting them on to the metal substructure. Technologist engineers had a field day as they devised methods of running drills at variable speeds by using resistance as a feedback. As the skin was drilled in the wing jigs, the drill speed would compensate automatically for the material's toughness as it penetrated layers of CFRP and aluminium alloy. The result was

ADVANCED AIRCRAFT IN PRODUCTION

A weight saving of 22% was achieved by making the fin of the Airbus A310-300 from carbon fibre-reinforced plastic instead of aluminium alloy, and, excluding nuts and bolts, the component had one-twentieth the number of parts. Here, the fin for the first A310-300, built at MBB's Stade factory, is prepared at Toulouse in February 1985 for attachment to the fuselage. Comprising a fin-box, leading edge, rudder and access panels, the component was claimed to be the 'first major piece of CFRP committed to airliner series production'.

smooth, burr-free holes, and a reduced chance of fibre edges being rough and causing fatigue cracks when the structure was under load.

With time, confidence with CFRP was growing. One clear indicator was the increasing size of autoclaves that aircraft manufacturers were buying. In essence, these seemed likely to replace milling machines and chemical-etching baths. A rule of thumb was that 50ft (16m) was about as big a structural unit as could be cured. But consistency over the whole piece was assured, and the one-piece structure often replaced many hundreds of smaller items in a pre-CFRP design.

As confidence grew, production specialists were driven harder by the advantage of lower component counts, and they began to encourage the design and production of whole 'box' structures. Sometimes these were 'closed' with rivets – this implying that the whole unit was CFRP; skins, ribs and spars – with the spars and ribs co-bonded to one skin, and the last skin lowered on, in a jig, and riveted in place. This technique is now extensively used in the commercial airliner businesses to build aircraft fins, and sections of flaps and control surfaces. Some fully co-bonded CFRP is in production too, such as the Eurofighter Typhoon wing, where the whole box is cured in one operation, requiring no jig placement for final closure of the whole wing box. In general, the military have used CFRP more extensively than the commercial sector, as they have clawed for best performance, but the manufacturing techniques are steadily gravitating into civil production.

The leading edge of carbon-fibre applications has been on stealth aircraft. Carbon will conduct electricity, but when it is bonded into its plastic matrix and suitably treated it can be a radar-impervious 'shield' too. A large proportion of the F-117 external structure is carbon fibre, and it has to be manufactured to dimensionally tight tolerances. Most of it is flat, or single curvature, which helped a great deal on this initial stealth design.

Gradually the US scientists were gathering enough information to feel confident about building an even bigger aircraft, and in the late 1980s, after a lot of speculation, the Northrop B-2 stealth bomber was rolled out. Exactly what CFRP production processes are used, and how the aircraft is built up, has to remain a tale that is partly speculative. This is because the US military acknowledges that the defining attributes of this aircraft are determined in these processes, and they are coy to be open about a technological advance that has cost them a great deal to establish. It is well known that the aircraft is dimensionally symmetric, from nose to wingtips, to an accuracy that needs laser to measure it. The aircraft has many metal innards, but CFRP is used as extensively as they dared, to minimise radar reflecting properties. The latter attribute is enhanced by having a coating, suggested to include a large proportion of depleted uranium, tailored to absorb electromagnetic energy in certain wavebands. How this shape is defined and then manufactured depends greatly on the fusion of CAD and modern manufacturing technologist knowledge. The traditional engineer has little contact with the conception, definition and production of the large-scale items on the B-2, which costs approximately a billion dollars per copy. But we can be sure that their deftness of touch, traditional skills and micrometer-sharp 'feel' will still have a part to play in assembling these CFRP leviathans. The search for

The Eurofighter Typhoon's wing torsion box is made from carbon fibre composites, using co-bonding techniques to eliminate the use of mechanical fasteners on the lower skin to substructure. After the lower wing skin has been laid up on the co-bonding tool, as seen here, to accept the pre-formed uncured spars, the upper skin and tool are fitted and the whole assembly is placed in an autoclave, where the spars are simultaneously cured and bonded to the lower skin.

FASTER, FURTHER, HIGHER

Built to laser-determined accuracy, the Northrop B-2 stealth bomber makes extensive use of CFRP to minimise radar reflections from its airframe. The incorporation of parallel edges in the design ensured that reflected radar energy was aligned to a common direction and only rarely reflected back to its illuminating source, and a coating also helped to absorb electromagnetic energy.

superlative aircraft attributes has changed the production business, and the trend to use composite materials, such as glass fibre and carbon-fibre, now it is established, will accelerate to embrace all practical possibilities.

Interestingly, the heyday of innovative production development, the 1960s, has proved to be a watershed, and the skills of the best production engineers today are somewhat different from those anticipated at that time. It is still precision engineering on a big scale; genuinely more precision than ever, and with electronic connivance that was unheralded barely 30 years ago. However, it would be unwise to believe that production techniques have again reached a plateau. The age of 'intelligent' materials is here, and this means that a designer can almost dial up a specification, and a concoction of elements will be used to produce the perfect match. There is talk (indeed, there has been for many years) of ceramics that are chemically related to the ceramics most people would know of in their household, but their potential has yet to be proven. Overall, aviation factories throughout the world still depend on men and women with that blend of traditional and modern – nowadays combining the 'eye' with the 'keyboard' more and more – and the aircraft that meet the most superlative claims are still a challenge to build in quantities.

Bibliography

Alexander, W, and Street, A, *Metals in the Service of Man* (Pelican, 1976). A readable introduction to the properties of metals, which also studies the most widely used alloys of metal.

Barnes, C H, *Bristol Aircraft since 1910* (Putnam, London, 1988). A comprehensive company history with individual type histories of its aircraft.

Francillon, R, *Lockheed Aircraft since 1913* (Putnam, London, 1982). A complete history of this innovative US aerospace company and its products.

Gordon, J E, *The New Science of Strong Materials* (Pelican, 1976). The first authoritative book on carbon-fibre technology, by the principal researcher at RAE Farnborough.

Gunston, W T, *Bombers of the West* (Ian Allan, Shepperton, 1973). A survey of the principal bombers produced by the leading powers of the western nations since the Second World War, with development and operational histories.

——, with Gilchrist, P, *Jet Bombers from the Messerschmitt Me 262 to the Stealth B-2* (Osprey, London, 1993). A type-by-type study of the world's principal jet bombers, concentrating on their technical development.

——, *The Osprey Encyclopedia of Russian Aircraft, 1875 to 1995* (Osprey, London, 1995). Extensive coverage of civil and military aircraft built in the Russia from the earliest times; an invaluable English-language reference.

Miller, J, *The X-Planes* (Midland Publishing, Hinckley, 2001). The ultimate book on the extraordinary range of advanced experimental aircraft tested in the USA since the Bell X-1 of 1945.

10
Demise of the Drawing Board
Mike Hirst

Seeing an aeroplane, especially as it soars up gracefully, is a gratifying sight to any aerophile. But for the person (or team) that designed it the visceral joy is almost intoxicating. As aircraft grow in complexity, it is easy to forget that each is the physical manifestation of what would have started as a set of ideas. The design process is driven by needs – the customer's needs. The customer needs an aircraft that can do this, or do that, and, on the basis of personal knowledge, and the legacy within their environs, designers can work up a proposal to meet the aspirations of all those who wait to see their product.

At first the design is a concept. It does not matter whether it is a high-performance fighter or a simple light aircraft. To appreciate what goes into the design of a modern production aeroplane it is as well to start with a simple picture, and to let it grow. If one considers the conceptual design processes of home-builders who have the courage to design an air vehicle, they will start with sketches, perhaps of monoplanes, or biplanes, of high-wing or low-wing, of aircraft with nose gear or tail gear, etc. They may have chosen an engine, and be designing around it. They may have decided on how many occupants it will carry, and even if it is only two they have to considered how they would be accommodated; side-by-side or in tandem. Initially, the wing might be a 'gut-feel' design, of approximately the area expected, but the span might not be certain, the planform a 'favourite' in terms of taper and aspect ratio, and the flaps (if any) may not yet be well defined. There are many decisions to make. Most designers are happy to have a number of configurations to which they can refer, and of which they will know their individual strengths and weaknesses. This is true of all aircraft designers, whether they are considering something for weekend recreation or to punch holes in the stratosphere.

Eventually, after a few days or a few weeks for a light

The detail on low-speed windtunnel models is quite remarkable. On this one-eleventh-scale model of the Airbus A321 in a tunnel near Toulouse, France, the patches show where instrumentation is embedded, to collect pressure tapping information for computer-based analysis.

aircraft, but maybe after many months or years of deliberation for a complex high-investment project, the day comes when the actual configuration has to be chosen, and refined. It is a day when everyone involved knows that they are committed now to a design that is deficient, compared with other possible configurations, in some ways. But they will have chosen it for its strengths, and will be setting out their stall to preserve those vital assets that they have decided will make it a better choice than any of its competitors. If you want to know what is good and bad about a design, talk to a designer. Talk to a marketing person and the chances are that you will hear especially of the weaknesses of competitors. They live off the tales that come out of design offices when disagreements have been aired, and this vital staging point in the product's history has been passed.

By this time the aircraft is a shape, and it might have been drawn by draughtsmen. The shape will be subject to refinement, as aerodynamic tests are performed in laboratories and in windtunnels. Most importantly, it will be a sheaf, or many volumes, of data. These will define the aircraft's performance, and critical especially among that data will be the aircraft's mass estimate. It will have related the mass of every item to either the catalogue from which it is derived (if it is a bought-in item) or to its shape and loading, if it has been designed in situ. A small company designer might still work out these data manually, but increasingly the data is derived by formulae buried in the hearts of computers. The design process, even at the conceptual stage, can be computerised.

After the conceptual phase has resulted in a justifiable configuration, the preliminary design stage begins. The approximate calculations of critical loads on structure are replaced with case studies that will define exceptional conditions. A light aircraft designer will consider different payloads, and different fuel loads. A military aircraft designer might do the same, but the combinations of loads – fuel tanks, bombs, external sensor pods, etc – can be huge. In every case the cautious designer has a wind-tunnel model constructed and measures pressure distributions so that actual structural loads and the distribution of loads about the centre of gravity can be determined, although for a light aircraft designer this might be impossibly expensive. Those involved in simple projects are well advised to stick to well-documented, proven, solutions to critical issues.

At the other end of the scale, it is not uncommon for a modern military aircraft to enter service with a very limited approval, compared with the many weapons configurations in the sales brochure. Just as loads are being determined, so models may be used – again, immersed in windtunnel airflow – to see how stores will fall away from an aircraft. If they become caught in swirling flows, bombs can rise and strike an aircraft's tailplane, disabling the aeroplane instead of the intended target. Such tests reveal enough for a design to be refined, perhaps growing vital strakes, or retractable air 'dams', that will lessen such risks, or the designer might have to resort to using tiny but powerful explosive charges to push the ordnance out of the airflow disturbed by the aircraft. These are critical design decisions, and they are taken often long before the aircraft will ever take to the skies. In recent times the need for such tests has diminished, however, as designers have refined the calculating methods they use to predict the micro-scale behaviour of air. They can now 'model' the airflow in regions around an aircraft with enough confidence to make reliable predictions of how objects will behave in such circumstances. But the need to ensure that a high-value project is not jeopardised by a poor decision means that some cases will be tested using models, and the experimental data compared with theoretical results to verify that the modelling techniques are correct.

As well as the airframe there are the engines and the aircraft's systems. The engine is a vital item, and the fuel system that meters energy from the often-cavernous tanks within the vehicle to the powerplant have to be regarded as part and parcel of the propulsion system. The light-aircraft designer will almost certainly choose an existing engine and design around it; but a high-technology project can be based on a new powerplant, or use novel installations, and need a lot of attention devoted to defining key issues in the preliminary design stage.

When it comes to the on-board systems, a light-aircraft builder will have decided what kind of flights are likely to be undertaken, and will have chosen navigation and radio communication sets, and associated instruments and control units, accordingly. It would be usual to consult sup-

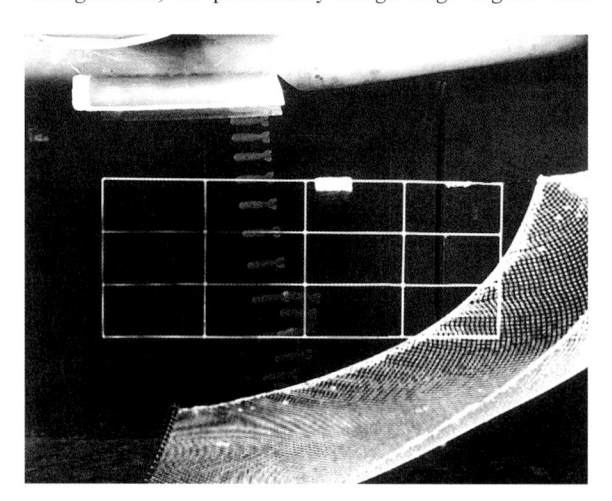

Some aspects of a design are very difficult to predict. This time-lapse film of a model bomb being released from a Boeing B-47 bomb bay provides information on trajectory and turbulence. On the basis of test at different attitudes and speeds the designer will determine if operational limits need to be imposed in service.

DEMISE OF THE DRAWING BOARD

pliers' catalogues and to use the components that they guarantee will integrate. An airliner designer does not do things much differently, because airlines like their equipment to be similar to that used by everyone else. This ensures that replacements for failed items can be found readily, and the integration task is made easier by 'standards' defined in great numbers of thick manuals. In a military aircraft, up to the present time, at least, things are done rather differently. The conceptual design stage will have involved operational analysis, where the expected performance of the aircraft, and its systems will have been expressed in computer programs, and mock combats fought, or 'virtual' targets (complete with 'defences') attacked. The evaluation stages may have even used flight simulators to let aircrew get a 'feel' for the concepts being considered, so crews might be 'flying' a new aircraft over a decade before it actually takes to the air.

In this case, the systems design teams have to get suppliers to agree to equipment specifications that operational analysis work has shown to be critical to combat success. An example might be an airborne interception radar, which might be requested, for example, to have a declared minimum detection range against a specified size of target. It will have been agreed what power it would use, and have its reliability postulated, and much more. Specifications can be enormously long documents, boring to read, but vital in terms of their detail to ensure that the right kind of product is determined; and they will contain clauses declaring that it has to be delivered on such a date at such a cost. In the systems area many hundreds, if not thousands, of electrical, electronic, hydraulic or pneumatic components can be involved in this tedious process. In the airframe area the same is true, but the components are more often mechanical.

Once an aircraft is in service, crews soon tell the designer if they cannot gain access to equipment, or if they find internal configurations intolerable. The ease of access to equipment on the F/A-18 Hornet will contribute greatly to its turnround and mission availability.

When design progresses from preliminary to detail design stages it has been traditional to develop an engineering mock-up. This is a full-scale replica of the Airbus A330/340 structure, used to check assembly details and wire and pipe runs in the wing leading and trailing edges. In the future this might be done entirely using computer-generated images.

While systems and airframe designers might seem a million miles removed from each other, they actually need to be very closely associated. If a radar, for example, is heavier than expected, it will affect the aircraft's mass and balance, perhaps significantly enough to have a measurable effect on performance. If it is deficient in performance, the onus might fall upon the airframe performance engineer to find some extra sparkle, such as climb rate or turn rate, that will restore parity in the battle success equations. So the systems designers often dig into quite detailed design, ahead of their airframe colleagues, so that they can have confidence in the mass, volume and installation assumptions that affect them. An example of the physical design conflicts that can occur comes from looking at how equipment is reached by technician. There will be target times set for the rectification, or the removal, and the replacement of equipment. Poor access to equipment that needs frequent attention could be the Achilles' heel of an aircraft, causing flight delays and reducing the annual aircraft utilisation of an airliner, or reducing the combat readiness of a military type.

So, to the third phase of design. Once the preliminary design is frozen, detailed design can begin. It is rare to find a date when this happens, whether the design is large or small. Designers try to identify components that will be relatively unaffected by continuing refinement of the preliminary design, and press legions of draughtsmen into action, turning out drawings that will lead to items being machined or fabricated, and prepared for assembly.

For example, most airliner designers will 'freeze' the fuselage cross-section at the conceptual stage. Soon after entering the preliminary design stage, and conducting critical flight phase loading evaluations, they will have a good idea of the most critical stresses that the fuselage structure will have to withstand. They will have knowledge, too, of the configuration details that will allow them to define components that make up the whole. This is not always straightforward, as they strive to keep the parts count low (essential for cost control), and yet sufficiently

FASTER, FURTHER, HIGHER

numerous to allow stresses to dissipate safely if one item should be damaged, or fail, between maintenance checks (called 'structural redundancy'). Compromise is the art of design, and simplicity is the key to good design.

Lofting might have been used, until recent times, to determine many component dimensions. It is a drawing process that aviation designers adopted from maritime practice, where it was first recorded around 1795. Lofting consists of drawing the full-size shape of major aircraft components on a large flat surface. The section is drawn at defined stations along the component, such as the wing or fuselage, and geometric interpolation or extrapolation is used to determine sections at other critical stations, such as production breaks, rib or frame locations, and so on. These sections can just as easily be determined by mathematical derivations, but this takes time, and humans are prone to getting the answers wrong because we are adept at displacing numbers in columns and rows, and so on. Nowadays computer-aided design (CAD) programs will take the datum station ordinates and deduce other station ordinates by calculation. The loftsman's skills have been transformed into lines of digital code that await activation in a computer; testimony indeed to the march of technology into what were the domains of skilled artisans.

While weight is crucial, a designer will have contingencies too, knowing, for example, that skin panels can be thickened if stresses rise above expectations. Confidence comes from familiarity with the job, and the experienced designer/ draughtsman is a much-sought-after, but increasingly rare, component of the modern aircraft factory. Over a period of time the fuselage, wings, tail surfaces, etc, will be committed to detailed design. By now irreversible design commitments have been made, and the sins of the initial concept, while they might have been easily ignored earlier, will begin to show how much of a risk the designer has taken.

The number of drawings will mount up, and once each is approved they are passed to the build team, which creates each item and supplies it to the assembly team. Many other specialists are exercising their skills by this stage, too. The airworthiness engineer is double-checking assessments of failure effects, and seeking confirmation that consequences of failures considered by the designer fall into accepted failure mode categories. Weights engineers are monitoring overall mass, calculating where the aircraft centre of gravity is located, the moments of inertia about each aircraft axis, and so on. Electrical power engineers are considering the power requirements of the equipment. They 'audit' the way that electrical load will vary throughout the flight, and develop a power generation and distribution philosophy that will meet constraints on space, weight, and so on. Hydraulic power engineers do a similar job, at about the same time.

The actual detail of aircraft structures is remarkable. This F/A-18 Hornet rear fuselage shows how the frames are closely spaced, and each is ribbed with vertical and transverse webs to endow stiffness and achieve exceptional low mass targets.

Systems, meanwhile, are being designed down to a detail level elsewhere. The accepted design process, in most cases, is that one works 'top-down'. Once a top-level (sometimes called 'functional') specification is received, the system is divided into subsystems. A radar system might be subdivided into antenna, transmitter, receiver and signal processor subsystems. Each subsystem team has to agree interfaces with its partner teams. These are not always easy to define. The physical connections, through sockets and structural items, will usually be easier to identify and define than the information-based interfaces, where the onus is on teams to supply data under defined conditions. The latter are often expressed against templates, which will have grown from experience. For example, every digital electronic data exchange will have had its range, resolution and accuracy characteristics defined, in the knowledge that, in doing this for information that passes across subsystems, the design detail is setting store on the capabilities that are essential for the components within them.

Components are the items one step below subsystems. In a radar they might be travelling-wave tubes (used to generate the transmitted energy pulses) or standard electronic items such as microprocessors or random-access memory (RAM) chips. If the project is to be relatively risk-free, the subsystem designer will choose commercial off-the-shelf components. But if the performance needs of the system are exceptional, the technology in the components might also have to be exceptional. Somewhere a component designer, who might be thousands of miles away from the original aircraft design team, and indeed might be designing a new electronic 'chip' without even knowing that it will be used in an aircraft, will have the ultimate design task to perform.

In the perfect world, each of the levels in the design process – system, subsystem and component – will have a detailed specification, that has flowed down from the level above, and thus they will have some guidelines on what performance capability is required of their contribution. The components are tested, and when they work as desired they are combined to form subsystems. These are tested against their specifications, and when these work as desired they are combined to form the system, which is then tested to see if it works as desired. This is called a V-diagram design process, after the shape of the diagram that is used to describe it.

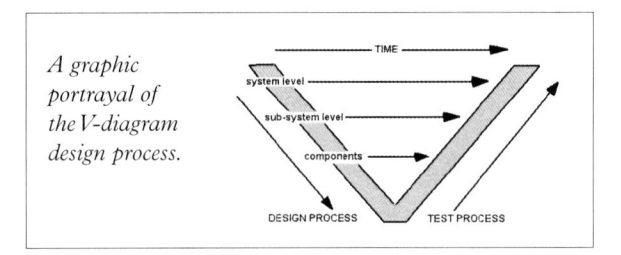

A graphic portrayal of the V-diagram design process.

Given that all this is happening manifold – in radar, electrical, display, control, hydraulics, fuel, even aero-engine, supply chains – it is sometimes an event little short of miraculous in nature that everything is brought together on time, on budget and meeting needs. The aircraft design authority has to hope, if communication lines are limited, that the strengths they perceived in conceptual design, and that they have set out to preserve, will still be realisable when a device comes in too late, too heavy, to expensive, unreliable, virtually unmaintainable, or whatever.

Indeed, if one plots the people (or the hours) involved in design, one finds that the quantities leap significantly as one proceeds from conceptual, to preliminary, to detailed design stages. One finds, too, that the degree to which a design is well managed, or not, is driven by the confidence that the design team has at each of the points where they move up a stage on the scale. Cutting down on the conceptual-stage effort can seem to save time and money at first, but if poor assumptions get carried through the design process, the chances of a error somewhere down the supply chain increases, and the rectification cost, in terns of labour, money and time, rises inexorably. A rule of thumb is that an error that goes uncorrected costs ten times as much to rectify at each stage, and that applies right through conceptual, preliminary and detailed design stages, manufacture, assembly, test and operations. An extra pound spent in conceptual design can lead to many thousands of pounds, even millions of pounds, being saved later in the project.

Aeroplane designers are not alone in facing these issues. They are the components of the greatest challenges facing designers in many high-technology-based industries. But no manager can afford to keep prolonging a development stage, even knowing it is going to save money, if the consequence is that the design process becomes so prolonged that the product is facing more competition in the market when it is ready for delivery. That can mean less market share, less revenue, lower production rates, higher unit costs. The economic case spirals into the failure zone, often never to be recovered. When projects cost billions, this is not an acceptable situation for any business to face.

So technology has been recruited to help the designer be a better manager. While 'back to the drawing board' was the oft-quoted retort when a failure was discovered (and until recent times, and in many cases, it has been the appropriate response), drawing boards are hard to find in aircraft factories nowadays.

The changes began several decades ago. It all started as electronic systems began to play an increasingly significant role in the design of aircraft, around the 1970 period. Up to that time, aircraft and their systems were designed, their intricacies were studied in reams of calculations, and when there was sufficient evidence to pin-

FASTER, FURTHER, HIGHER

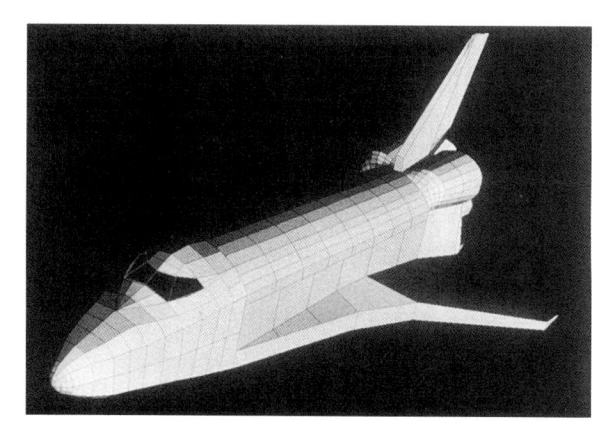

This deceptively simple-looking 'panel' model of a Space Shuttle is an example of the way that a modern product is described in computing terms. The three-dimensional shape is immersed in computation fluid evaluations that reveal the pressure distribution over the shape, and in turn the information is then used in finite-element modes, so that the stiffness, deflection and mass of different structural solutions can be evaluated before the structural configuration is committed.

point specific values of specified items, an overall design evolved stage-by-stage, on paper.

This osmosis of electronic computation meant that arithmetic data that was considered on a slide rule in the 1960s gravitated on to the desk calculator around 1970, eventually on to an electronic calculator and, finally, in the early 1980s, on to the desktop computer. Throughout this period it was still commonplace, at every stage, that where risk was identified as a major issue, the critical components within the design were built and tested, and if they failed to perform as planned they were modified. The details were recorded, and the next test done. This required time, so risk was accepted only if there was time to do the development necessary. Keeping track of decisions was difficult in big design offices, and when the design process involved a vast number of teams spread globally, the idea that all decisions would be known in a timely manner, at any location in the organisation, was academic. Sometimes batches of changes were made before anyone got round to modifying the original drawings, which showed their maturity as they spawned growing tables of drawing modification (mod) numbers, all

As the aircraft is flight-tested the stress office staff have to endure the ignominy of seeing one example of their latest creation tested to destruction on the ground. This is the Airbus A320 structural test specimen, with its wings subjected to their ultimate load.

DEMISE OF THE DRAWING BOARD

Sometimes the designer might not care to trust calculations and intuition alone. If an unconventional layout like this Burt Rutan designed Scaled Composites 151 is envisaged, it is as well to build a flying scale model, to check out where doubts might have remained. This Pratt & Whitney JT15D-1 powered proof-of-concept aircraft ws built in 1991.

duly dated and signed. Legions of people strived to ensure that the recording process had managed to catch every change that had been made, and that as one compared drawings they were still all current, not chalk and cheese.

Around 1980 electronic system designers realised they could 'automate' some or all of the stages in the design process. The way this all developed was rather unheralded, but magnificent. On choosing to design an electronic component to do a particular job, designers could express their needs – a circuit that would oscillate at a give frequency, perhaps – and the computer would invoke a legacy stored within programs held inside it. These would recognise the kind of circuit that was best suited to the purpose, and would evaluate the values of unknown parameters, until it was certain that the component would function as desired. On some systems the 'design' could be 'tested' while it was still a picture on a computer screen. It would respond just as if it was the test article on a laboratory bench, and using a 'virtual' oscilloscope or probe the engineers could look into the properties of the system. They got time-histories of volts or amps drawn more precisely than a hundred test points could have been plotted, and achieved in a fraction of the time.

As systems grew bigger, and they were broken down with logical design rules applied through the V-diagram process, it was reasoned that any system could be designed in a staged manner. Each component, and then each subsystem, as it was developed in a computer-based simulation, could be tested in consort with other parts emerging from parallel design processes. In the end, all the necessary data being known, the system could not just be drawn by the computer, unaided by human hands, but its many components could be checked for cost, for delivery schedule, and so on. Computer-aided design had come of age.

This was hardly evident to the public, whose recognition of what was afoot only came about when mechanical design engineers acquired CAD systems of equivalent fidelity. Elementary mechanical CAD systems appeared in the 1970s. The idea was simple. Instead of doing calculations on paper and recording the results so that they could be transcribed to a set of engineering drawings, the computer could do the calculations, then draw the resultant article. This was fine for designing a component whose loading was easy to define, but it was a chore to enter complex loading data.

A good example is a wing. A wing usually has several spars, these being the crucial limbs that extend from root to tip, and on to which are connected ribs that extend from leading edge to trailing edge. The spars and ribs are then enclosed in the skins. The wing has mass (an airliner wing can weigh several tonnes), but it is designed to generate 'lift', the aerodynamic force that will hold the aircraft aloft. Hence, a wing will droop from root to tip because of its own weight when on the ground, and will bend upwards when in flight, acting as the aeroplane's support. The amount it will bend has to be known precisely for the overall performance of the aircraft to be understood, but getting to this data is a long process.

Before CAD came along, the aerodynamicist would analyse windtunnel data and determine how the lift force was distributed along the span and the chord (from leading edge to trailing edge). Now that we have computers that will do repetitive and complex sums without pausing, or making blunders, because they do not tire throughout the day, these evaluations are done in what is called computational fluid dynamic (CFD) evaluations. This yields loading data which can be loaded straight into the stressman's wing design program. Told what configurations are acceptable, the CAD system will calculate the vertical and horizontal loads across a grid-like mesh, and transfer this data into shear, bending and torsion loads on spars, ribs or skins at various positions.

Sometimes the only way to get such calculations started is to make a guess! In his historic autobiography Slide Rule (Heinemann, London, 1954) the novelist Nevil Shute (N S Norway) described the process to design hoop frames for the R101 airship in the late 1920s. He related how he would make an educated guess, calculate data at all the 'nodes' around the multi-pointed frame, then arriving back at the place he started from he wanted to find a value that approximated well to the initial guess. This took many iterations, and occupied many days for each frame, but he said that getting a suitable solution was 'akin to a religious experience'.

The modern CAD system does this across the complete mesh description of all wing components, and it is testimony to the complexity of modern design that an airliner wing can require the most capable modern 'supercomputer' hardware to work away for twenty-four hours or so to yield satisfactorily iterated results. This amount of calculating, in Nevil Shute's day, was unimaginable – possibly equivalent to the world's population working away for a decade or more, without a stop to eat, drink or sleep.

While one can contend, quite genuinely, that these computer-based processes take the designer so far away from the design process that they lose the 'feel' for the data, there are joys, too. Like the electronic engineer's analytical capabilities described above, the mechanical engineer's CAD system will track many properties too. Given material specifications, each wing skin thickness will have been deduced, and its mass taken into account in determining the stresses in the wing. The local loading points where deformation might be difficult to assess, or from where fatigue cracks might be expected to propagate, will have been identified. Each component's mass, the location of its centre of gravity, and assessments of its moments of inertia about various aircraft-related axes will also have been evaluated. In some systems the production processes needed to create each component will have been determined, and the person-hours and machine-hours needed to achieve the desired result calculated. Consequently, as the aircraft design proceeds, the actual factory capacity, and the skills and techniques

The latest airliners exist in lifelike form long before metal is cut and an aircraft emerges from the development process. Even before adopting the A380 designation, Airbus Industrie had realistic computer-generated images of the A3XX. This semi-complete image reveals the computer-generated 'framework' on which the aircraft's exterior is applied.

needed, will be known. If the design requires a massive production investment that the manufacturer cannot sanction, the investment constraint can be expressed as a limitation that the design process must observe, and the design will be revised. On the Northrop B-2 stealth bomber the acceptable dimensional tolerances, even for large components, are minute compared with those attained on any earlier-generation aircraft. The shape is drafted to great accuracy, and assembled largely robotically, illustrating how sometimes the technology that has only just emerged has been crucial, too, in allowing designers to make significant progress in some areas.

Among the more esoteric properties of the wing that are considered are the susceptibility to vibration, which can change dramatically as a function of aircraft height and speed; the configuration of flaps and slats; and the distribution of fuel among the wing fuel tanks. These were evaluations that were often made at absolute circumstances (which were believed to be critical), and then the aircraft was flown at close to the theoretical conditions to see if the observed vibration levels agreed with predictions. A flight-test team would approach the critical condition with great caution, and often the speed and height were changed in increments, and data retrieved and analysed between flights. It therefore took many days, even weeks or months of flight-testing, to get these critical data. The modern CAD system can provide data to a computer in real time, and compare data that is being collected from transducers on the wing, and sent back to the flight-test centre on a 'tele-metery' link. Thus sets of conditions can be 'cleared' in a matter of seconds, reducing the flight-test burden allocated to this task to a few hours of flight time. When it brings such benefits, the CAD system is no longer a luxury but a vital part of the aircraft creation system.

So, from wiremen who used to estimate bend allowances on electrical wire looms, to mathematicians who used predict performance data, the CAD system has usurped them, or mechanised their daily chores. The CAD concept is being embraced especially in those project domains where esoteric and advanced aircraft are designed. These are high-risk programmes, so when the go-ahead is given to build a new type it has to be sanctioned with more confidence than a computer-less design process can provide. Today's test pilots can be flying higher and faster and further than any of their predecessors, and yet might never leave the ground. They do it in simulators that meter every drop of fuel, account for every eddy of the air in the aircraft's wake, and provide impeccable renditions of the handling qualities that they can expect when, or if, the vehicle is ever constructed.

Even so, the first flight, when it comes, is still the landmark event it always was. No computer can replace the pilot's senses and knowledge when it comes to rolling down the asphalt on the first takeoff. The human ability

The designer's delight is to see the final product being assembled in quantity. This was the UK assembly plant at Warton for the Panavia Tornado.

to let feelings meet emotions, as they never can in calculations, is still a vital flight-test attribute. A test pilot's wariness for the unexpected might represent the latest breakthrough in our meagre knowledge of life at aviation's leading edge. Computers may have replaced drawing boards, virtually with no exclusions, but they will not replace the pilot so easily.

Bibliography

Blanchard and Fabrycky, *Systems Engineering and Analysis* (Prentice-Hall, many editions, to 2000). The best-known reference for modern design engineers seeking a fundamental grounding in all the disciplines they have to balance.

Eatas and Jones, *The Engineering Design Process* (Wiley, 1996). A readable textbook on the latest design belief systems. Provides an overview of essential disciplines, skills and analytical techniques in several well-integrated chapters.

Foster, R N, *Innovation: The Attacker's Advantage* (Summit Books, 1986). An ageing but still compelling account of the fine line that designers tread between success and failure, and a reminder to those who slave over computers that the mind is still where the designer's seeds are sown and germinated.

Womack, James P., Jones, Daniel T, and Roos, Daniel, *The Machine that Changed the World* (Rawson Associates, 1990). No, not the aeroplane. A fascinating account of a study conducted by MIT scientists that compares and contrasts the design methods in different automobile manufacturing organisations. It showed US firms where they had to go to catch up with Asian builders.

11
Flight Simulation
L F E Coombs

The aircraft in its three-dimensional environment is a far more difficult machine to control than those that manoeuvre in fewer dimensions, such as land vehicles and ships. From the earliest days of powered, heavier-than-air flight the ability to change or correct aircraft attitude was only mastered quickly by a few. The majority needed about ten hours in a dual-control trainer aircraft before they could be let loose on their own. Advertisements of the early 1900s sometimes referred to the liability of pupil pilots for 'breakages'. Therefore thought was given to the use of 'ground' training devices, which might enable pupils to acquire the skills of control co-ordination without leaving the ground and without incurring 'breakages'.

In the *ab initio* stage of a pupil pilot's progress a simulator provides freedom from the limitations of flight in poor visibility. Even in real flight there are practical limitations on the extent to which pupil pilots can experience the control effects and aircraft responses at the boundaries of the flight envelope. Apart from cost there is the overall need for safety that makes the simulator a useful alternative to actual flight. At the other end of the pilot scale, a simulator enables the experienced pilot to complete his Instrument Rating check or gain a new aircraft type in his licence without the need to burn thousands of litres of fuel flying a real aircraft at a time when it should be earning money for the airline. The modern simulator enables the military pilot to acquire the skills needed to make the most effective use of the aircraft as a weapon system, particularly when new weapons and systems are introduced into service.

Early days
Early examples of flight simulators included the simple arrangement of a seat on an unstable mounting, on which a pupil sat and endeavoured to keep the seat upright by co-ordinating the aileron and elevator controls. The task was made more difficult by the instructor, who could make changes to the system. Because one of these early flight simulators was specific to learning to fly the French Antoinette monoplane of 1909, it anticipated the time when modern simulators represent the flight characteristics of a particular type of aircraft. The Sanders Teacher of 1910 had wings and was slightly more realistic in that it could be turned into the wind to generate lift. There were also the 'Penguins'; aircraft with reduced wing area that could be taxied at speed but were unable to take off. Other early 'simulator names' are Anderson, Billing, Gabardini, Ruggles and Walters.

Despite the tremendous increase in the number of pilots to be trained during the First World War, there does not seem to have been a corresponding increase in simulator numbers and type. The majority of training aids were aptitude testers such as the Ruggles Orientor. This may have been the first to make use of electric motors to effect attitude changes in the replica aircraft. Another early simulator of the effects of aileron, rudder and elevator movements consisted of a model aircraft suspended in a glass windtunnel. This provided much entertainment at exhibitions.

The civil and military simulators of today are a long way, in terms of technology, from the devices used in the early years of powered flight to assess a pupil pilot's ability to co-ordinate the control of roll, pitch and yaw.

The essential Link
Link is the name most readily associated with simulators. Although it was not the first, it was undoubtedly the first mass-produced, all-purpose, affordable, flight training aid. It was also a moving-base device that provided some of the sensations experienced when applying bank, pitch and yaw. But the early Link Trainers could not be used to simulate extreme flight conditions; even though they could be stalled and spun.

The movement of the small fuselage, with its single-seat cockpit, was effected by air motors whose design owed much to the church and theatre organs made by the

Eardley Billing's 'Oscillator' was available 'for trial by flying aspirants' at Brooklands in 1910. Designed to teach pupils 'the essential points in regard to the manipulation of an aeroplane', it could be made to climb, dive and bank in light winds.

A pilot makes a 'cross-country' flight in British Airways' Link Trainer at Gatwick Airport, Sussex, in November 1937. The school was mainly being used to train pilots for the South Atlantic service.

Link family business. For example, the training motor for turning the cockpit about its vertical axis in response to aileron and rudder control demands consisted of a number of bellows connected to a multi-throw crankshaft. Similar technology was used for aileron and elevator. It was all very simple and could be mended using simple tools and by unskilled personnel (such as the writer). In some ways Link's design was anticipated by the 1918 patents of Lender and Heidelberg.

The potential value as a training aid of Edwin Link's 'Blue Box' of 1929 was not immediately recognised by the aviation world. It became an amusement arcade attraction, and Link even added a coin-in-the-slot feature. In some way this turn of events anticipated the present day 'desk-top' PC flight simulators, which are neither toys nor completely realistic training devices.

Link's 'Blue Box' Trainer arrived on the aviation scene just at the right time. Aviation in general was expanding and figuratively pushing against the 'visibility barrier'. If pilots inadvertently or foolishly attempted to fly in cloud for more than a few minutes they usually ended up striving to recover the aircraft from a dangerous situation. The Link Trainer with the hood down taught the basic rule: 'Believe and follow what the instruments tell you. Ignore the sensations of apparent movement and do not try to fly by the seat of your pants.'

A significant milestone in simulator history was the decision made in 1930 by the Pioneer Instrument Co to use a Link Trainer to demonstrate the importance of its instruments in helping a pilot to avoid loss of control in cloud. It is important to note that Link did not advertise his invention as a 'simulator'. It was an Aviation Trainer. Not until after the Second World War did 'simulator' become common usage. Although not of great importance, it might be of interest to note that some argue that you should not call a training device a 'flight' simulator unless it at least moves around the three axis of roll, pitch and yaw. As Humpty Dumpty says to Alice in *Through the Looking-Glass*: 'When I use a word, it means just what I choose it to mean …'.

In 1934 the US Army was persuaded to buy six of Link's Pilot Makers. This was the year in which the US Army Air Corps had lost eight aircraft with five pilots killed and six injured in crashes while attempting to carry the US Mail. The pilots' lack of training in flying on instruments when confronted by poor visibility was the major cause of the disasters. At about this time Link tried to interest the US Navy in his 'Aqua Trainer'. This sat in the water and was powered by a small petrol engine driving a propeller. The Aqua Trainer could rise off the surface and respond to the pilot's control inputs. It provided a pupil pilot with the essential 'feel' of a seaplane during the critical flight phases of take-off and touchdown on water. It was an innovative idea, but failed to interest enough potential customers.

By 1936 Link had added an instructor's table equipped with duplicate instruments and an electro-mechanical device that moved a 'crab' carrying a marker pen across the chart on the instructor's table.

Once under the hood in the Link, the pilot was isolated from the world. It was used by the airlines and training schools for navigation to radio waypoints, let-down procedures and approaches to an airfield. It became recognised as a way of teaching pilots in 'flying the range'. In other words, navigating across the USA among the directional and non-directional radio beacons and following the aural and visual indications of the radio ranges. It is important to note that there has always been a strong aversion to using a flight simulator as the only means of teaching someone how to control an aircraft. Except in low-visibility conditions, there is nothing like the real thing. It was argued that the basic skills of controlling an aircraft should first and foremost be taught in a real aircraft, but on those days when even the birds were grounded the 'Link' stepped in.

Over the years Link trainers became increasingly more sophisticated, and later models were no longer 'general purpose' because their flight characteristics were programmed to replicate those of a specific aircraft type. A notable example was the AT-6 Harvard Link trainer. The Second World War increased significantly the operation of long-range over-water flights, such as the transatlantic ferry service. In the absence of long-range radio navigational aids, crews had to rely on navigating by the stars. Link, in association with Captain P V H Weems, USN, the authority on celestial navigation, developed the massive Celestial Navigation Trainer (CNT). The 'fuselage' of the CNT was suspended within a dome on to which were projected lights representing the stars. A bomber

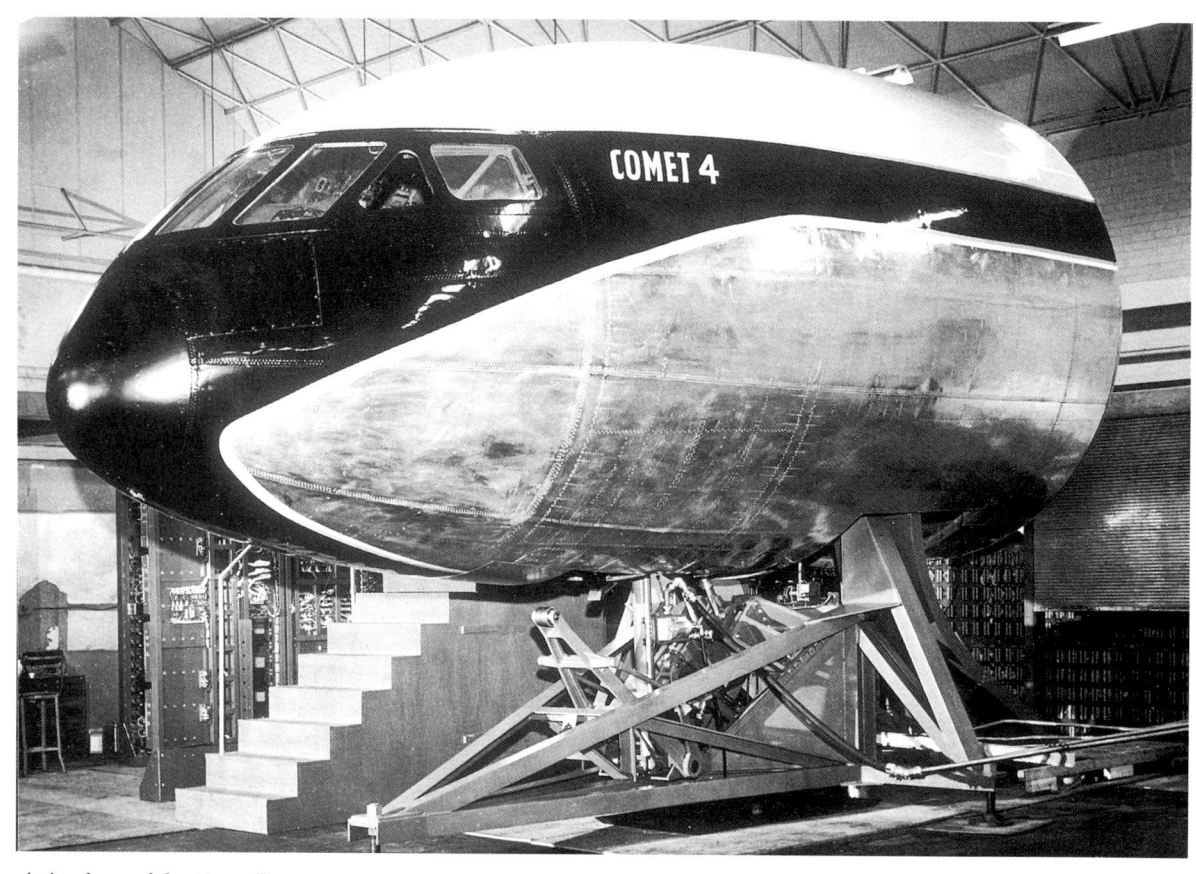

A simulator of the 1950s. This de Havilland Comet 4 one-axis simulator for British Overseas Airways Corporation had movement only in pitch, and no simulation of the world outside the cockpit.

crew could complete a long-range flight, navigating by sextant shots of the stars. A reproduction of the terrain was projected below so that the bomb-aimer could use the bombsight. The RAF, the original 'customer', calculated that by the end of the war it had saved 50 per cent of its training time and costs by using the CNT. The importance of the CNT to the history of this subject is that it was the first production-standard, moving-base simulator with positions for more than just a 'solo' pilot.

Developed in parallel with Link's Pilot Maker were examples of the more basic types of simulators used to test the visual, tactile and control co-ordination and responses. In the UK the RAF started to use the Reid Reaction testing device in 1932. (Its primary purpose was to eliminate, at an early stage, pupils who were likely to prove bad pilots. It also preselected pupils who were more fitted by temperament to make good fighter pilots. The Reid apparatus tested the speed of reaction in making restoring movements of the controls.) Link trainers were also used by the RAF in the Second World War for the aptitude assessment of potential pupil pilots. During the immediate prewar years and the war years themselves there was a massive increase in the production of Link Trainers. At the end of the war units were coming off the assembly line every 45 minutes.

Other than the Link Trainer and the CNT, the US air services and the RAF had few, if any, moving-base simulators during the Second World War. Among about 30 types of synthetic training devices used by the RAF were: a fixed-base bomb-aiming, training device that used a moving-map display to simulate the movement of an aircraft relative to target; the complete fuselage of a bomber in which pilots, navigators and flight engineers, but not air gunners, could make a simulated sortie, during which their skills were monitored by an instructor who could inject faults and errors into the different systems (the Silloth Trainers); and a fixed-base Spitfire cockpit used for teaching the correct techniques for identifying and intercepting target aircraft in different visibility conditions. In the last device the 'targets' were model aircraft suspended from an overhead moving chain that simulated different ranging situations. Air gunners were trained to aim at a simulated target that was projected by an optical system to the inside surface of a dome which had a gun turret at its centre. A dome was also used by the Royal Aircraft Establishment (RAE) in the late 1940s for its air combat simulator. This used a Spitfire cockpit and the optical projection of an enemy aircraft on to the surface of the dome. In 1973 the RAF made a 'homemade' fixed-base simulator for one of its Avro Shackleton squadrons. This particular example met the peacetime need for realistic training in searching and attacking when there were insufficient real targets to represent wartime conditions.

Just as Link based his technology on that of the pneumatic organ, so did the UK simulator builders. As an example, the Automatic Player Piano company, which made training devices for the RAF, was already familiar with the levers, linkages and bellows that made up so much of the simulator technology of the time. Although the 'organ and piano builder' simulators contributed significantly to training in the 1940s, their mechanisms were adversely affected by climatic conditions and mechanical problems. This encouraged the development of simulators based on an electronic computer. Bell, in the USA, delivered the first production operational electronic simulator in 1944.

The importance of training

The upsurge of civil aviation after the Second World War, and the significant increases in aircraft performance, introduced many new problems. The basic training method of acquiring experience by sitting in the right-hand seat of an airliner was no longer sufficient to ensure a safe standard of ability. It was realised that the new generation of civil aircraft needed comprehensive training equipment for the flight crews. In 1948 Curtiss-Wright developed a flight simulator for Pan American's Boeing Stratocruisers, the first full aircraft simulator to be owned by an airline.

Jet Training

The jet airliners that entered service in the late 1950s, such as the Boeing 707, required an accurate landing approach path and needed to be established on the extended runway centreline and the descent path at least five miles out from the touchdown point. If they were well off the required approach path and within the last mile or two of the runway threshold they could not be

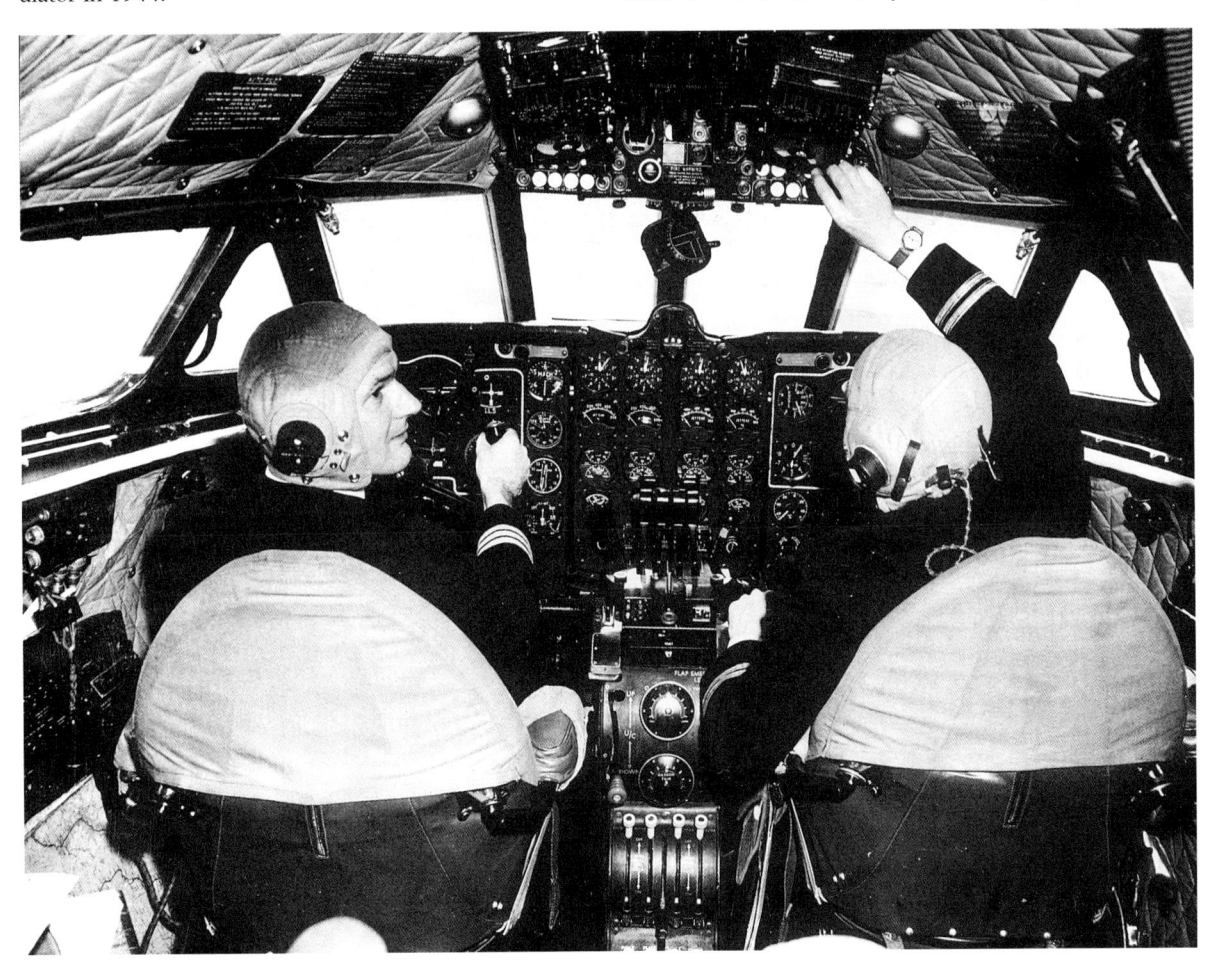

The interior of the Comet simulator of 1959.

brought back easily and safely on to the 'glide' slope and localiser. The high approach and landing speeds, compared with earlier years, of the postwar generation of civil airliners encouraged the development of new approach light patterns. The Calvert systems, developed by the RAE at Farnborough, were evaluated on a simulator consisting of a wide belt on which were small lights to represent the approach pattern. These were viewed from a fixed eye point. Compared with modern simulators these fixed-base and fixed eye-point training and research devices were extremely crude.

Simulators had to be developed that could replicate all phases of flight; particularly the approach and the touchdown phases. However, in the early years of the 707 only fixed-base simulators were available, and even as late as 1958 in jet transport development terms the de Havilland Comet 4 simulator only had motion in pitch. In the 1960s some simulators relied on a small closed-circuit television (CCTV) camera that was 'flown' over a detailed model of terrain. The resulting image was then presented on a large TV monitor mounted in front of the cockpit windows. Compared with modern visual displays, the field of view was limited to about 50 degrees in azimuth and 30 degrees in elevation. At the time this technology was viewed with wonder.

In 1970 a CAE four-axis Boeing 747 simulator was developed for British Airways. A second in 1975 had the Singer Link-Miles Night Vision System. These were part of the airline's £1.5million order for five multi-axis platforms. The night vision system was generated by a computer, and therefore was an important technological advance away from the CCTV systems.

The modern simulator is a combination of a big-memory digital computer, an electro-mechanical or electro-hydraulic motion system and a 'real world' image generator. This provides displays of data and visual representations of the real world along with movement that responds smoothly to control inputs. If cost is not the limiting factor, you can have virtually any function or situation and any effect simulated in full colour and with

An early two-axis simulator platform. Movement in pitch was effected by the crossed support arms and hydraulic mechanism. Roll was effected by the hydraulic ram at the side of the platform.

realistic motion and movement of both the user's cockpit and the 'real world'. The question that sometimes has to be answered in the world of simulators relates to the required degree of realism. Even though a simulator can provide a less-costly alternative to training and research in an actual aircraft and, of course, provide no-risk flight, it is still an expensive item of capital equipment that has to earn is way. The first cost could be about 45% of the simulated aircraft. A typical comparison of operating costs between those of a modern flight simulator and using a real aircraft for training is in the order of 1 to 8 for a Boeing 747, 1 to 3.5 for a 757 and 1 to 2.5 for a 737. The elimination of 'breakages' is an important advantage. A comparison of operating costs can also indicate the point in a training programme at which a trade-off has to be made between the capital and operating costs of a simulator and the costs of 'real' flight training. However, the capital and operating costs of a simulator for business and feeder aircraft may show little advantage to the simulator. One aspect of flight simulator economics is the ability to repeat and analyse critical and important phases of flight. With civil and military transport aircraft take-off techniques, coping with engine and system failures, departure and approach patterns and landing phases take precedence over en-route simulated flight. For long-range navigation and fuel management exercises a fixed-base, less-costly simulator may be acceptable.

The degree of realism required will depend on a number of factors; such as cost and the training and familiarisation roles required by the user. A simple fixed-base procedures trainer may meet a particular training programme in which none of the instruments and controls is more than an illustration mounted at the correct location on a representation of the main instrument and control panels of a cockpit.

The ubiquitous personal computer can provide a 'fixed-base' simulator of the instruments and sounds of an aircraft along with an albeit limited representation of the 'real world' in which the aircraft is assumed to be flying. However there is no feedback to the 'pilot's' senses, particularly of motion and acceleration effects. The computer and software cost around a £2,000. Despite these limitations, PC-based simulators have a number of advantages. They can be easily moved, and in their laptop form provide a training and familiarisation aid that can be accessed at any time, and their simulated instruments are not limited to one type of aircraft.

These PC-based simulators can be used for introducing, explaining and demonstrating the operations of an aircraft and its associated equipment. Animations of instrument responses, the dynamics of mechanical systems and the flow paths of electrical and fluid systems provide familiarisation and training for both air and ground crews. This type of training aid teaches equipment operating practices and diagnostic procedures such as engine run-up and fault detection sequences, inputting navigation key data, such as waypoints, and selecting flight management computer operations.

Between the PC simulator and the 'all-moving, all-systems' flight simulator are numerous degrees of realism and function. The next step up in complexity is the provision of real instruments and controls as well as sound effects integrated with a computer so that a crew can practise the correct operating procedures for all flight modes. The addition of an image generator provides a view ahead in different visibility conditions co-ordinated with the simulated attitude of the 'aircraft'.

To the foregoing can be added simulated aircraft movement and acceleration by inflation and deflation of seat cushions and vibration. With combat simulators the pilot's g-suit inflation can be varied. Reducing the light level can produce the effect of blacking out under high g forces.

The step upwards to the moving-base simulator so as to provide inputs of acceleration, motion and attitude is a big one, and involves anything up to £1 million or more. A 'six-axis' simulator for a Boeing 747, for example, will set you back many millions of pounds. If you cannot afford your own, then you have to buy time at £1,000-plus per hour on someone else's.

An airline training flight in a simulator involves far more than the crew turning up as if it were a classroom. They have to attend a normal briefing session, flight planning routines must be gone through in detail and, at the end of the flight, there is debriefing. An air force crew goes through a similar and rigorous programme that replicates all aspects of a sortie, from briefing to debriefing.

The unexpected

Fifty years ago or more, aircraft were more prone to failures of engines and systems. Aircrew were always on the alert for sudden failures, and in consequence 'on-the-job' training was often very realistic and demanding. Today there is far more redundancy designed into systems, so failures are less frequent and, because of extensive monitoring systems, they can sometimes be anticipated. Therefore one of the functions of the modern flight simulators is to provide realistic problem situations. Coping with some problems cannot be rehearsed safely in a real aircraft. For example, completely and suddenly shutting down an engine during take-off is never attempted for real. But it can be demonstrated in a simulator.

Research

As early as the First World War, devices were being used to assess a pupil pilot's ability to respond to changes in the attitude of an aircraft. In the 1920s Reid, as mentioned, Rougerie and others developed more elaborate devices for assessing a pupil's latent control ability. However, these had none or limited movement and were primarily used to check response times to changes in

FASTER, FURTHER, HIGHER

This Boeing 747-400 flight simulator at the British Airways training centre at London Heathrow is just one of many used by the airline for crew training and familiarisation.

control position and instrument indications.

Cockpit simulators started to be used for research in the 1940s. A notable example was the Cambridge Spitfire Cockpit, used to assess pilot speed and accuracy of response to control demands in a stressful situation and over a long period. From the 1950s onwards, increasing use was made of training simulators as research tools.

Simulation today
The majority of the world's principal air forces acquire sophisticated simulators as part of a defence package. They depend on simulators for type training, system upgrade familiarisation and for simulating combat without risk. Air forces with deep pockets can acquire fighter/attack simulators that enable two or more 'aircraft' to be flown simultaneously, so that each pilot can formate with other 'aircraft' or engage in air-to-air combat and ground attack.

The constant search for ways to reduce costs in a world of ever-restrictive civil and defence budgets is often met by upgrades to existing aircraft types. Certification for airworthiness requirements and the training of aircrew for an upgrade can be helped by the use of simulators. Changes to the aerodynamic characteristics of an aircraft consequent on reconfiguration can be simulated without the costs and safety implications of flying an actual aircraft. The simulator's role can be extended from aircraft handling, performance and navigation to take in the extended environment in which the aircraft is assumed to be flying. Tactical situations can be tried out in a military simulator cockpit, and in transport-type cockpits air traffic control situations can be simulated, including mid-air collision threats.

Turbulence and noise
An example of the way in which a sophisticated flight simulator can both advance safety and reduce training costs is by simulating turbulence, downbursts and windshear effects. These conditions can be elusive, and therefore searching for them is an expensive and unproductive way of training pilots. Simulating them on the weather radar and through the motion system enables pilots to be trained in the correct operating procedures.

Flight simulators can improve an air force's community relations. In general the public, except in wartime, not only resent paying for an air force but object strongly to being kept awake at night or having their peace and quiet shattered by low-flying aircraft. Simulating night and low-flying training benefits both parties.

Simulation demand
At the beginning of 2001 over thirty civil full-flight simulators were on order from the principal manufacturers. These covered a wide range of aircraft types, including corporate and business jets, regional jets and turboprops, as well as helicopters. The trend is now toward centres established by the simulator manufacturers and available

A Boeing 737-800 simulator used by Lufthansa.

FLIGHT SIMULATION

British Midland pilots undergo training in an Airbus A320 full-flight simulator.

In this study of an Airbus A320 simulator for United Airlines, the instructor's monitoring position and displays are visible on the left.

FASTER, FURTHER, HIGHER

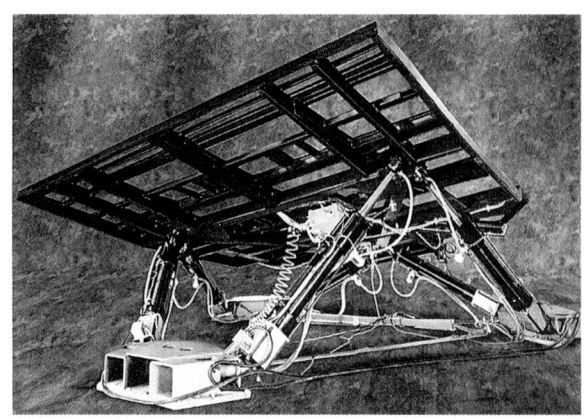

A Moog electric motion platform mounted on six electro-mechanical actuators.

A McFadden six-axis synergistic motion platform (top) and a three-channel system for projecting detailed scenes for take-off, landing, navigation and ground-attack.

to all airlines, whose accountants welcome the opportunity to avoid large capital expenditure.

A typical airline simulator is the CAE full-motion flight type for the Airbus A330-200 acquired by Emirates for its Dubai training centre. This is fitted with the Evans & Sutherland ESIG 3350GT image generating system for reproducing images of all destination airports on its routes, as well as all expected visibility conditions, including fog and sandstorms. The airline also has CAE full-motion simulators for its A310/A300s and Boeing 777s, and a fixed-base system for its A310 training unit.

Although procedures trainers are usually to be found among the fixed-base types, the Emirates airline operates a six-axis full-motion cabin emergency evacuation simulator. For economy, the simulator can be used to represent an emergency in either an A310/300 or in a 777. One half of the 'fuselage' represents the flight deck, cabin and doors of an Airbus, and the other half those of a 777. This very specialised trainer can reproduce the sensations and effects of fire, explosions and decompression. It even has a pool along one side so that a cabin crew can practise all aspects of evacuating passengers into life rafts.

The 'works'
The descriptions of the technology of civil and military flight simulators may be dealt with together. Essentially, the moving-platform mechanisms and the image generating technologies are similar.

A flight simulator is designed to replicate as closely as possible the performance characteristics of a particular aircraft type. Analysis of in-flight performance data by a powerful digital computer provides the basis for a software program. The number of parameters that have to be analysed and correlated runs into thousands. It is obviously not as simple as just arranging the software so that if the pilot advances the throttles the airspeed will increase and the airspeed indicator will show the increase. For example, the fuel flow rate will also increase, the fuel contents will decrease, and many other instruments will also alter their indications and each system will adopt a new datum. It is difficult to think of an element in an aircraft that ignores changes in others. A modern aircraft is made up of many data systems and operating systems, each of which 'talks' in the electronic sense to each of the others. The software program can also integrate a simulator into a simulated air traffic control or air/ground war situation.

The software provides the commands that make the simulator move in response to changes made either by

FLIGHT SIMULATION

the pilot or the flight management system to the control surfaces. The digital signals are converted to analog electrical signals, which in turn are translated into movements of the hydraulic rams or electric-motor-driven legs which support the cockpit platform. The usual arrangement of 'legs' consists of six hydraulic rams arranged as three pairs, each forming a triangle whose apex carries the simulator platform. All six rams work together to make the platform roll, pitch, yaw, sway, surge or heave, or any combination of these.

Sounds, vibrations, and a real-world projection system are added to a moving-base structure containing a replica of the aircraft cockpit. The cockpit and its controls are often the 'real' thing and are readily available, whereas the signals that drive the instruments and the platform motion, and also provide feedback and 'feel' through the controls, have to be generated by software for the flight simulator's computer. Writing the programs that translate real flight data into simulated data is a major task in the development of a simulator.

One way of comparing the quality of different computers is to check the time taken to respond to a demand for a change in 'aircraft' attitude. The modern design goal is 100 milliseconds or less. No system is perfect, and no human pilot is perfect. The flight simulator designer has to reconcile the two. Experienced pilots will comment that a particular simulator 'does not feel like the real thing', or, 'the real aircraft does not fly way the simulator indicated it would'.

Human senses

A particularly important relationship that must be correct to achieve maximum realism is that between the movement of the platform and the human senses. The non-visual human senses that detect movement respond to accelerations. The sensations of movement, or rather accelerations, induced by a simulator must be matched by the simulated display of the outside world. It so happens that the body usually detects movement before the eyes perceive changes to the world. The simulator must

This exterior view of a six-axis system emphasises the size of a complete Airbus A300-600 simulator in relation to the cockpit, in the square house labelled 'Thai', and the substantial mechanism needed to provide roll, pitch, yaw, heave, surge and sway.

FASTER, FURTHER, HIGHER

Realistic night-time effects are provided by this Thales Boeing 777 simulator used by United Airlines.

be designed so that the relationship between the pilot's physical and visual senses are accurately co-ordinated, otherwise the illusion is shattered. In some of the earlier flight simulators the ideal co-ordination was not achieved and they could induce nausea.

The visual projection of the outside world is co-ordinated with the movements of the simulator platform. Under certain conditions of flight the visual scene does not follow the platform. For example, if the physical effects of sideslip are to be induced, the platform is tilted further to one side. However, the visual scene does not show the increased bank angle. In other words, the system deliberately gives false information in order to simulate correctly. If the platform is tilted upward without the visual display or the attitude director moving the horizon downward, as might be expected, then the human body experiences acceleration. Another example is a banked turn. If the simulator platform was left in the banked position the pilot would feel uncomfortable, as no centrifugal force would be keeping the body upright in the seat. Therefore the platform is levelled off, but the attitude displays and the image of the outside scene continue to show that the aircraft is still in a banked turn.

In addition to inducing sensations of movement by altering the angles and attitudes of the platform, motion sensation inducing seats can be used. These have inflatable pads for applying pressure to the body. For example, inflating the pad against the pilot's back gives the sensation of acceleration. The harness can be tightened, thereby forcing the pilot downward as if he were being subjected to positive g. Many and 'devious' are the ways in which the simulator people can deceive.

A realistic image

The requirements for simulators intended for training pilots and other aircrew who operate fixed or rotary wing attack aircraft specify maximum arcs of view. A coverage presenting 200 degrees of simulated 'world' from left round to right simulates the 'over-the-shoulder' scanning pattern needed when searching for and attacking ground targets. Compared with early visual systems, the current visual displays of the external scene are virtually seamless and allow the user freedom of eye point and wide arcs of view. Another essential requirement is to make the external scene replicate as closely as possible the shapes, colours and textures of the ground and any buildings and military equipment. The simulated external scene must be more authentic, for example, than the scenery usually provided with some computer games. The latter is obviously artificial and therefore destroys the viewer's illusion of flying in a real world, particularly as, unlike the professional flight simulator, the view of the outside world is not collimated so as to appear at infinity. The optical system and the methods used to generate the external scene have been the subject of considerable research ever since the first true moving flight simulators were developed.

FLIGHT SIMULATION

The goal has been to provide the cockpit crew with a seamless depiction of the view ahead and to the sides.

The view ahead and to the sides has to be of a quality that matches the realism of the cockpit environment and the movements of the simulator platform. If it fails in this important respect, the illusion is broken. The image generator element of a modern simulator projects a collimated view of the outside world through the visual display system. The design target for image generation and projection is a realistic three-dimensional, variable, all-types-of-weather scene outside the cockpit; along with dynamic representations of other aircraft and ground vehicles. The simulated world in which the aircraft is supposed to be flying has to change its aspect in real time. The generated image is refreshed at a rate of 30Hz for a civil-aircraft simulator. The rate is increased to 50 Hz or 60 Hz (the refresh rate respectively of a British or USA domestic light bulb) for a combat simulator used for low-level flight training, because the pilot needs enhanced detail from which to acquire accurate visual information when attacking a target.

Combat simulation: feeling the g

Fighter/attack aircraft crews are subjected to psychological and physical stresses as they experience the extreme effects of flight: high-g manoeuvring, including 360-degree rolling, pitching up to 90 degrees and looping. These need to be reproduced in a simulator. However, by the very nature of accelerative forces the effects of g cannot be reproduced completely, even in a six-axis simulator platform. Furthermore, the moving-platform mechanism has mechanical limitations that limit the extent to which a movement can be sustained.

The cost and complexity of providing a simulated

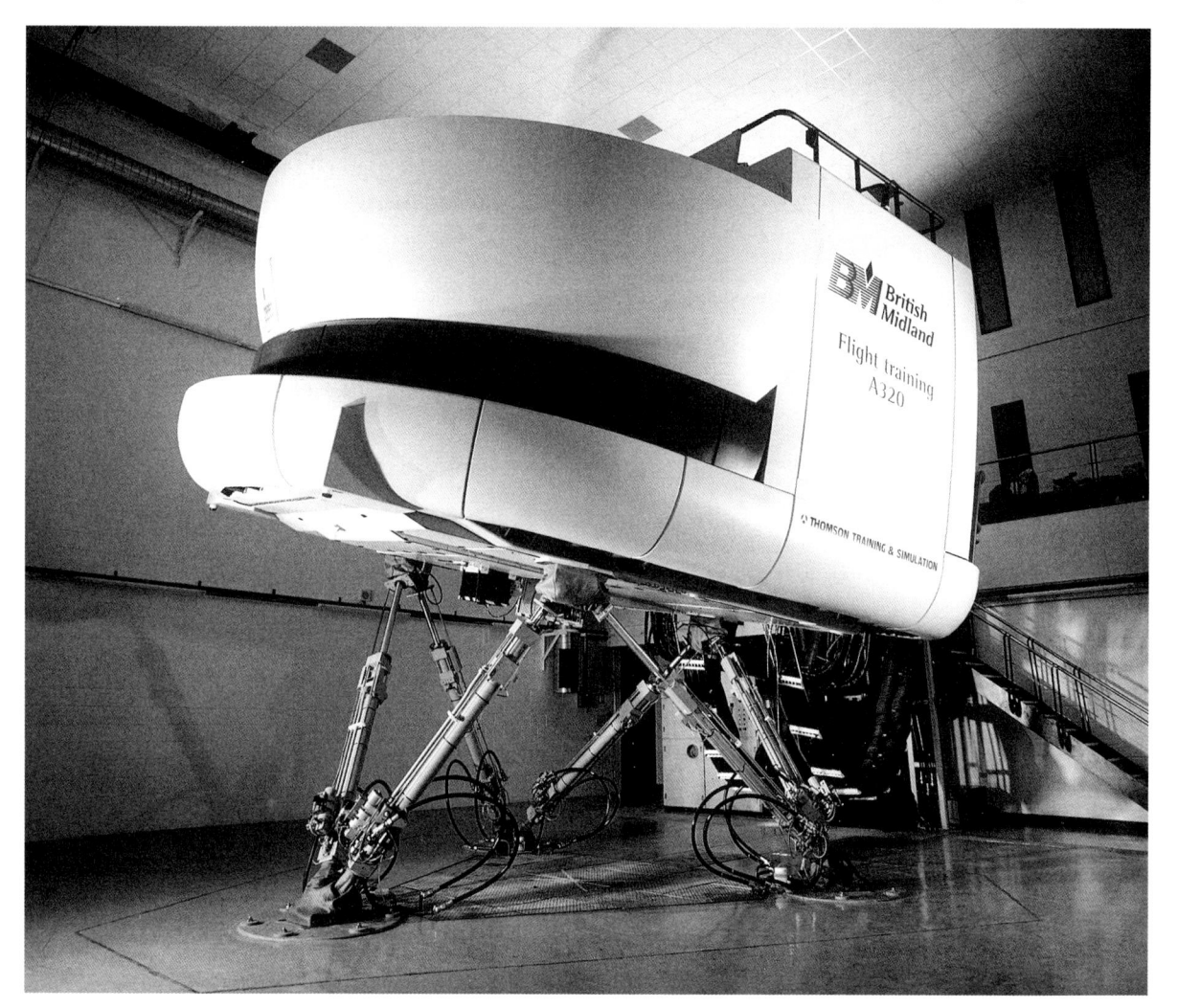

The exterior of British Midland's A320 six-axis platform. This has a 'wrap-around' extension in front of the flight deck windows, on to the interior of which is projected the external visual scene.

FASTER, FURTHER, HIGHER

A BAE Systems wide-screen simulator for carrier landing training.

cockpit environment that can replicate all the manoeuvres and attitudes achievable with a real aircraft have so far set limits to realism. However, a fixed-base flight simulator can be acceptable for combat training because the visual information is more important than the full range of acceleration and attitude sensations. Therefore fixed-base fighter cockpits at the centre of a dome onto which are projected one or more target aircraft may be acceptable for this type of training.

In the 1970s air forces started to make increasing use of combat 'dome' simulators in which two pilots could intercept and attack each other's aircraft. Air-to-air fighting using guns or missiles can be practised at a fraction of the cost of the US Navy's Top Gun 'academy', where real aircraft engage in combat. A typical example used for combat simulation is the twin-dome system introduced by BAe (Now BAE) Warton. Initially this was developed in the 1970s for evaluating the characteristic of different types of fighter/attack aircraft. The twin-dome simulator enabled the aircraft design teams to make comparisons between different types of aircraft and their weapon systems. The flight characteristics and the weapons eventually selected for Eurofighter were the results of 'flying' the aircraft in one dome against known and potential opponents in the other dome. In this type of simulator each dome houses a cockpit and image-projection system. Seated in the cockpits, the pilots see the terrain and sky in different visibility conditions. The spatial reference provided by the scene moves in response to the manoeuvres demanded by each pilot. A typical 'dome' simulator provides large angles of movement about all three axes. The realism is enhanced by noise, vibration and buffeting, as well as the effects of g.

The projection system provides one or more 'enemy' aircraft. The computer works out in microseconds the very complicated mathematics needed to position and move a target on the inside of the dome relative to the control actions of the two pilots. Each dome can also be used individually and its aircraft flown against a target aircraft, or more than one aircraft, flown by the computer. The comprehensive record of every flight provides a means of assessing pilot ability as well as highlighting mistakes that in real combat might prove fatal.

The BAE Wide Screen Visual Simulator is a good example of the sophisticated, multifunction, fixed-base type. An important feature is the group of three projected colour image windows. These are blended to provide a 180-degree horizontal and 48-degree vertical 'window' of the view outside the cockpit. The simulator can be programmed to represent the performance and handling characteristics of different types of fighter/attack aircraft. It was an important element in the development of Eurofighter. In more recent times it is part of the development programme for the next generation of Eurofighter/Typhoon. The image projection system can generate advanced head-up display (HUD) graphics and alphanumerics. These can be evaluated by pilots in different simulated flight conditions in advance of the development of a new type of display hardware.

The twin-dome combat simulator developed by BAE might seem at first sight to be just a very clever way for

Simulated close-in combat in a fixed-base simulator with wide-screen projection of terrain, targets and head-up-display symbolism.

FLIGHT SIMULATION

pilots to engage in aircraft-to-aircraft combat without leaving the ground. It can be used for that, but it has an even more important role to play. This is as a research and development system. The cockpits can be equipped to represent a number of different aircraft types so that, for example, a MiG-29 can fight it out with Eurofighter or with the JSF. Although the cockpits are of the fixed-base type, the realism is enhanced by a wide range of noise effects, the application of vibration and buffeting and dimming and blacking-out of the visual scene when pulling high g.

For air-to-air combat, targets are projected on to the outside scene along with the simulated trajectories of missiles. The complete twin-dome system provides realistic combinations of different types of air vehicles, both 'red' and 'blue'. Inside the cockpits the complete and realistic instruments and controls (because most are the real thing) are complemented by all the necessary communication, weapon system and electronic warfare avionics. Analogous to a chess master playing a computer is the software-generated opponent that can be programmed to be more agile and more aggressive than any known aircraft type.

Alongside BAE's twin-dome system is a single dome housing a Eurofighter-type cockpit. The fully equipped cockpit is mounted on a six-axis synergistic motion platform. The view outside the cockpit is generated by a three-channel projection system which can project highly realistic scenes of the Earth's surface for take-off, landing and for attacking ground and sea targets. Air-to-air combat situations can also be simulated. As with the twin-dome simulator, the system can be used for research into all aspect of military aircraft operation; including deter-

A view showing the relationship between the fixed-base cockpit and the wide-screen visual system.

mining the optimum design of head-up, head-down and helmet-mounted displays. Direct voice input research and evaluation can be undertaken in simulated extreme environments of noise, vibration, g effects and all the other distractions of combat. Even the comparatively mundane research task of the optimum positioning of controls in a cockpit benefits by testing under extreme environmental conditions such as high g. Many simulators have their image-generating systems integral with the motion platform. However, the BAE mission simulator has the image projection system separate from the platform, thereby reducing the dynamic loads on the legs.

These combat simulators hone pilots' skills before they engage in the real thing. For example, RAF pilots used twin-dome simulators for practising combat with Argentinian aircraft before flying out to the Falklands conflict in 1983.

Simulating the virtual

During the design and development of a new aircraft, a virtual version can be simulated by computer, enabling variations in detail to be assessed in relation to expected performance. Simulation includes CAD that enables design variations to be viewed at far less cost than building physical mock-ups. As the design process proceeds towards 'cutting metal', the cockpit or flight deck can be built as a simulator in which the test pilots can become familiar with the expected flight characteristics. The simulator cockpit also provides an essential tool in the total design process because the cockpit, with its machine/human interface, is a critical element. At one time in pre-computer design days the designers of the cockpit had to depend on wooden mock-ups in which to determine the best positions for controls and instruments. They and the test pilots hoped that that they had

Modern combat simulation has come a long way from the time when model aircraft were suspended in front of the cockpit.

got everything right before the first flight took place. Today there is a near-seamless progress from CAD and to the completely simulated cockpit. Today both civil and military designers depend on simulators for achieving aircraft that will be operationally effective.

Blame the simulator

A complex legal situation might arise when the simulated performance of an aircraft does not match that of the real aircraft. For example, the certificated, placarded and demonstrated stalling speed of the real aircraft must apply also to the flight simulator, otherwise pilots might be led into a false sense of security. To avoid the latter, a flight simulator is subjected to 'flight' trials by experienced test and line pilots. These evaluation and proving 'flights' are as thorough as those to which the real aircraft is subjected. This aspect also brings in the important overall philosophy of flight simulators, which emphasises that a simulator must not just be an example of clever technology for the sake of clever technology.

Dr John Rolfe, who has made special study of the human factors involved with simulation, highlights the legal aspect. He gives as one example a pilot who attacks a bridge over which civilians are passing. The pilot is accused of committing a war crime. His defence might be that, throughout his simulator training the bridge was always occupied by military vehicles. Or, a trainee pilot fails to meet the required standards when flying the simulator. However, the standards were set by experienced pilots who had no difficulty in flying either the simulator or the real aircraft.

Flight simulators that fly

To the basic fixed-platform and moving-platform simulator types can be added those simulators which actually fly.

In 1932 the RAF's Central Flying School installed a pilot's seat, instruments and controls within the fuselage of a Vickers Victoria twin-engine biplane bomber/transport. There were no windows from which to determine aircraft attitude. The object of the exercise was to determine the effects of sensations of movement and acceleration on a pilot when attempting to maintain control in zero visibility. It was also used to teach pupil pilots the effects and dangers of 'blind' flying. To add to the control problems of the pilot under test, the 'cockpit' faced aft. Because the sensations received were completely opposite to those usually experienced in flight, this emphasised the importance of trusting the instruments and of ignoring what the aircraft seemed to be doing.

A number of aircraft types have been modified so that the stability and response to the pilot's control inputs can be varied. The aircraft is thereby endowed with the flight characteristics of a completely different type. Using a real aircraft as the motion platform has many advantages, including a natural operating environment for the pilot, and there is no need for a simulated display of the real world. In the 1970s Cranfield University operated a twin-engine Beagle Basset in this role. However, the Basset's performance and airframe limitations restricted the range of simulation. It would not have been possible to simulate a high-performance agile jet fighter.

Following on from the Beagle airborne simulator came the Astra Hawk. Again this was developed at Cranfield. The Hawk, a high-performance machine with a tough airframe, could be programmed to behave in many different ways. These ranged from a First World War fighter biplane, through most modern civil and military aircraft, to control-configured vehicles. The Astra Hawk was the solution. It provided a real pilot/aircraft environment that could be varied. It found a particular application at the Empire Test Pilots School (ETPS) at Boscombe Down. An ETPS student, already an experienced pilot, could fly a selected simulated aircraft type; the control characteristics, feel and responses of which were very different from those of the standard Hawk. The ability to vary the control characteristics and response as well as instrument display (ADI and HUD) dynamics, so as to mimic different types of aircraft, was one of the significant advantages over a ground-based simulator.

In the USA a fully equipped cockpit simulator was grafted on to the nose of a Convair C-131H. The exist-

A full-mission simulator for the Westland Sea King HAR Mk.3/3A all-weather search-and-rescue helicopter.

ing flight deck was retained for the use of the safety pilots, who at any time could 'pull the plug' on the test and experimental pilots.

Virtual reality

Present-day virtual-reality systems require the user to wear sensors and effectors so that there is an interface with the computer generated images, visual effects, sounds and other elements of the virtual environment. However, when we come to consider a virtual cockpit there are special problems. The virtual cockpit has to impart to the human body the effects of accelerations and g, and to achieve such perfection in simulation the body has to be able to receive signals that will affect the cardio-vascular system and other organs that provide the sensation of acceleration and g.

The virtual cockpit will provide images of the controls and the pilot will 'see' images of his hands touching or holding the controls. It is important that each element of the virtual cockpit interface provides 'feel'; particularly when this is essential for flying the real aircraft. The present technology available for the virtual cockpit requires equipment mounted on the pilot's body. This runs contrary to the present concept of simulation; a concept that attempts to 'deceive' the pilot into thinking that he or she is in the real thing. As a forceful advertisement for his invention, Link used to tell the story of the pilot who, during a very stressful flight on instruments, completely lost control, baled out and broke an ankle on the floor alongside the Link Trainer.

The future

In the future, greater use will be made of the concept in which real aircraft and real sea and ground vehicles are linked to simulators. Integrating real elements of a tactical force with simulated elements can enhance training. Flight simulators integrated with real aircraft and with the real environment of a tactical situation provide the important familiarisation with the way in which all tactical elements operate. By simulating all the elements of a sortie, aircrew can be trained to react correctly to all the variations and complications of real warfare.

A similar integration can be used to familiarise pilots with the operation of civil aircraft within the air traffic control environment. It can also be used to verify in detail the operation of a projected new aircraft within both the airport and airways environments. Test pilots can operate the flight simulator of the proposed aircraft from start engines, through the taxying and take-off phases and on to climb-out, noise abatement and ATC departure routeing: as was done during the development of Concorde.

The future usually has one foot in the present. Today's computers, vastly larger in memory and speed than those of the last century, along with virtual-reality head, body and seat effectors, presage a time when the massive 'six-legged' simulator is consigned to a museum. Flight simulators are already paralleled by simulators of space travel and wargames. These are proliferating to the mutual benefit of pilot training and research flight simulators and those of the entertainment world.

The help of the following with information and illustrations is acknowledged:
BAE Systems Warton; Dr M. Collier of GEC; Mike Hirst; Derek N James; Dr J M Rolfe; Moog Inc; Brain Riddle of the RAeS; Mark Rouson of Thales-tts; Jack Shlien of CAE; Wing Commander I Strachan; Vega Group.

Bibliography

Durose, C G, 'An Evaluation of some Experimental Data on Cost Effectiveness of Flight Simulators' (*Journal of the Royal Aeronautical Society*, Vol 86, 1982, pp90-93). This includes the important subject of the comparison between the costs and effectiveness of simulators and in-flight training.

Haward, D M, 'The Sanders Teacher' (*Flight*, 10 December 1910, pp1006-7). Describes one of the earlier types of simulators.

Kelly, L L, & Parke, B P, *The Pilot Maker* (Grosset & Dunlap, New York, 1970). Covers the complete history of the Link Trainer and its many developments.

Parfitt, A G, 'An Evaluation of the Nimrod Maritime Crew Trainer' (*Journal of the Royal Aeronautical Society*, Vol 86, 1982, pp86-89). Provides detailed information on a typical air force simulator.

Rolfe, J M, 'Determining the Training Effectiveness of Flight Simulators' (*Applied Ergonomics* 13, 1982, pp243-50). Discusses the important subject of the conflict between designing a 'clever' simulator and one which meets specified training needs.

Rolfe, J M, & Staples, K J, *Flight Simulation* (Cambridge University Press, Cambridge 1994). One of the 'required reading' books on the subject. Includes a comprehensive bibliography covering all aspects of the subject.

Flight Simulation Conference papers 2001 (Royal Aeronautical Society, London, 2001). These provide discussions and opinions on a wide range of simulation subjects.

12
The Future SST
John Snow

In examining the prospects for future supersonic air travel, this chapter not only looks at what is technically possible, but also considers what would be economically feasible and ecologically desirable. The technical content is broad, the detail of the technical issues is presented in those appropriate other chapters. Alternatives to the SST, such as the 'Super Jumbo' and the Sonic Cruiser, are discussed, as well as the effect of the telecommunications boom. Before we attempt to look into the future, though, we should see what can be learnt from history.

Power by nature, power by machine
The remoteness of our basic needs has been at the root of what we call the travel mechanism or 'need to travel'. Our earliest ancestors were driven to roam, hunting for food and materials for sustenance and warmth. Their effectiveness was governed by speed, range and payload-carrying capability. Clearly, early man's speed was entirely his own doing; range was related to average speed during daylight hours, while payload would have typically been (as now) a fraction of his body weight. Bipedal operating speeds were 3, 9 and 18mph (5, 15 and 30km/h) for walking, patrolling and combat respectively.

Human ingenuity then brought about advances in the science of transportation. The domestication of animals such as the horse and the invention of the wheel achieved a plateau that lasted millenia. From the 1820s, the invention of the steam engine and railways brought such increases in speed that, by the last decades of the nineteenth century, journeys were typically covered at 60mph (100km/h). Payloads were as high as 300 passengers (or 29.5 tons; 30 tonnes), and much higher for the slower freight trains. This quadrupling of average speed over the horse had been accomplished in the relatively short period of half a century, and there was little to suggest that this development should not continue. However, all technologies have their limitations, and it was not long before that of the Victorian age began to run out of steam. Then a series of developments resulted in diesel engine or (more promisingly) electric powered trains now operating at speeds of 125mph (200km/h) or more. This latter progress, though, is relatively slow. It has taken the past 100 years to achieve this doubling of speed and payload, which is minor in proportion to the gains of the nineteenth century, and so it seems that the train may be reaching the limits of its technology. Why should this be? We may identify more than one contributing factor.

The operating envelope
First, there is the 'law of diminishing returns', which means that improving on a good and acceptable level of performance requires an effort disproportionately higher than the improvement in result. Early achievements in railway development were relatively easy, as fundamental problems were overcome with contemporary technology. Once we reached the limit of this, technical development had to be driven up a steep slope against what might be called a conspiracy of the forces of nature. Faster meant more power to accelerate the train, but doubling speed requires four times the power, so sooner or later the deal turns sour. An increase in forward speed raises centrifugal force, when turning and stabilising solutions become increasingly exotic (pendulum-suspended cars, for

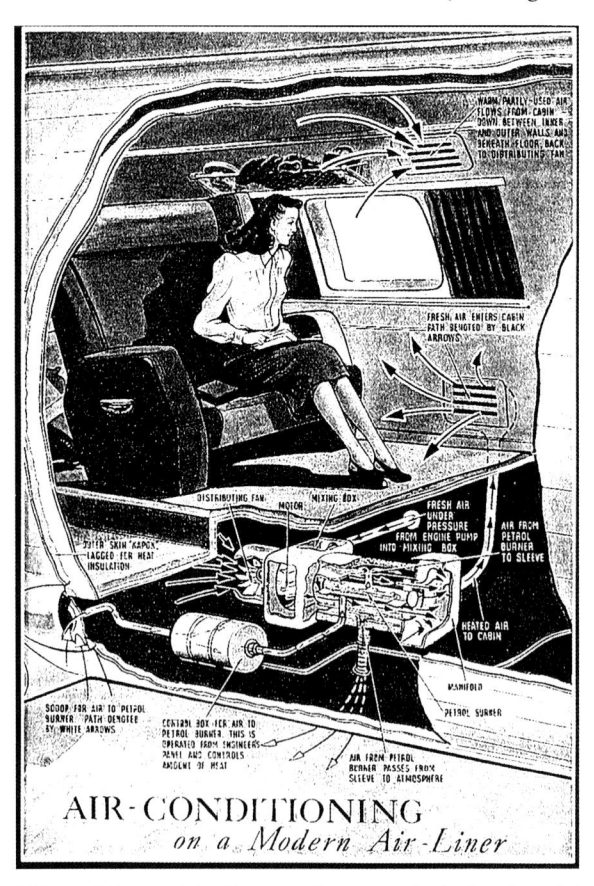

When airliner cabins began to be pressurised, the aircrafts' cruising speeds increased by 60% on account of reduced drag from the rarefied atmosphere outside.

example) requiring longer, more painstaking and expensive development effort. Thus there is a cost increase in achieving the higher speed as well as an additional cost in managing the operation at the higher speed.

The second factor is the perceived value of speed or saving of time. It seems that each means of increasing speed also requires time and effort in preparation for its use. For example, by the time we have taken the car out of the garage, driven to the newspaper shop half a mile away and then found a parking space, it might have been just as effective to have gone by bicycle or even walked, despite a ten-fold difference in cruising speed. There is then an impotence of speed over relatively short distances. Intercity trains are slower than 'sprinters' on commuter schedules, and there is a distance over which the stately DHC-6 Twin Otter will outfly Concorde, typically London to Paris rather than London to Brighton.

The third element is a corollary of the second, but considers the time-saving over greater distances where (cruising) speed is an issue. No matter how determined we are to increase the speed of the railway vehicle, it will always be easier to travel faster by air, assuming the distance is sufficient to allow the exploitation of speed.

The fourth factor comes as a result of the impact of technological change on the environment. Mechanised transport is a particular culprit, through the manufacture and ultimate disposal of vehicles, the mining and consumption of fuels, and the need to develop an infrastructure. Ecology is at last assuming its place alongside the other governing factors of transportation, even though its inclusion in the overall equation will normally temper growth and increase supply costs. Our future 250mph (400km/h) (very) high-speed trains would consume four times the fuel of current HSTs, require special tracks and generate external noise levels well in excess of current large jet airliners at take-off or approach.

So it might be said that each mode of transport has its own operating envelope where it is both effective and efficient. The envelope comprises a range of payloads and operating speeds, leading to journey times and distances. While reaching or exceeding these boundaries might be technically possible, they would tend to be tempered by economic and environmental pressures. Where operating envelopes overlap, such as between London and Paris with the train and the aeroplane, there is a choice, typically based on price, speed, comfort and convenience. If an operating envelope is to be expanded it will normally require an enabling technology. When all current technology is used up, something close to a forced invention is required, and this runs the risk of falling foul of the law of diminishing returns.

The economy of scale

Before we move fully into the realm of air transport, another economic issue needs to be discussed. If we are operating a railway where the trains have three passenger cars each, we could increase our capacity if we attached a fourth car. Clearly this would increase our operating costs, since the car would need to be purchased, cleaned and maintained. More engine power would be required, which would raise fuel consumption and wear out the engine more rapidly. However, there would probably be no need to employ additional staff, modify the signalling or strengthen the track. So the potential gain would outweigh the additional costs, leading to an economy of scale and a reduced cost of offering a seat on that journey.

This happy situation could continue on an incremental basis, but sooner or later there would be a need for a larger or additional engine, while the station platform might need extending. At that point there would be a diseconomy, since the cost of the additional car would outweigh its incremental value. We would need even more cars to recoup this cost, or we might decide that we had reached the working limit of the size of the train. Continued growth would be accommodated by increasing the number of train sets, but then we might require additional tracks on which to operate them safely. So, analogous to the operating envelope of technology, we see that there is also an economic envelope of scale. Operating at too low a scale might never enable our revenue to cover costs, while too ambitious scaling up will at a certain point impair efficiency.

Armed with the above concepts, we are in good shape to discuss the likelihood of and prospects for a future supersonic (air) transport (SST). Before we do so, however, a short review of the development of air transport technology in regard to performance might be helpful.

Taking to the skies

Air transport did not start in any measurable form until the 1920s, at the technological level of aeroplanes of the First World War. The new transport mode had to compete with railways for speed overland, and with ocean liners for range between continents. Not until air transport developed its own enabling technologies would it come of age.

The first of these was the all-metal airframe. Traditional steels would have made the basic aircraft excessively heavy, while aluminium was light but too easily damaged and deformed. The solution came with a light alloy of an aluminium base and copper particles which, while costly to produce, was virtually as strong as steel. Passengers could now be accommodated in a weatherproof cabin, heated by air ducted over the engines. Gasoline fuel was stored in integral tanks in the wing, rather than in somewhat exposed containers. We were then in the mid-1930s with the technology of the Douglas DC-3. At 185mph (300km/h), cruising speeds were around double that of the train. However, with a full cabin range of less than 620miles (1,000km), poor eco-

FASTER, FURTHER, HIGHER

nomics and inferior comfort level, surface transport did not yet suffer any serious competitive pressure.

The Second World War saw the development of larger long-range aircraft capable of delivering live payloads over oceanic distances. These were soon adapted to civil use in the ensuing peacetime. They were, however, still relatively slow and forced to fly no higher than 10,000ft (3,000m) unless passengers had auxiliary oxygen supplied. The next enabling technology was the tube-like pressurised cabin, with a sea-level temperature and pressurised to a rarefied, but easily breathable, 8000ft (2,500m) altitude atmosphere, while the aircraft could operate at 20,000ft (6,100m) in cruise. Not only was this above most of the turbulence caused by cloud formations but, in the less dense atmosphere, drag was lower. Even by using the same engine and wing technology, cruising speeds of 300mph (500km/h) could be achieved.

The limit to further speed increase was the propulsion system, and the limiting phenomenon the shock waves resulting from compressibility in air known as the sound barrier, or Mach 1. While the aircraft itself may be flying below the onset of shock waves, the propeller tips will be cutting a helix through the air where the combination of rotational and forward speed is supersonic. Thus, if the advance in speed should continue, the airliner's propulsion system would have to be a different one.

Enter the jet age

Several Second World War combat aircraft aircraft had engines generating thrust from the engine exhaust rather than from the propeller wake. These came mainly from Germany, and were powered either by rocket or gas-turbine jet engines. The safety and reliability of these early designs was inadequate for civil application, so it was not until the late 1950s that any significant number of jetliners appeared. Early examples in the UK were the Avro Tudor VIII and the better-known de Havilland Comet. Removing the propeller solved the speed limitation on the propulsion system, only to encounter the next hurdle, which was the wing. While an aircraft might be still flying well below Mach 1, the local airflow over the upper surface of the wing will tend to become supersonic, causing shock waves which are essentially an air brake. On a straight wing this limit occurs around Mach 0.6. However, the process can be fooled to an extent by sweeping the wing. Two workable solutions result. With a sweep of around 30-35 degrees, cruise speeds of around Mach 0.8-0.85 (530-590mph; 850-950km/h) can be achieved, while a sweep of 60 degrees or so allows supersonic cruising at around Mach 2, or 1,490mph (2,400km/h). Above 35 degrees sweep, take-off and landing performance is significantly impaired, while low supersonic operations (say Mach 1.2-1.4) are a challenge to flight control as well as experiencing more drag than at Mach 2.

The early successful designs of the 'high-subsonic' solution, such as the Boeing 707 and Douglas DC-8, used four of the largest civil engines available (Pratt & Whitney JT3C or JT4A) and were able to carry around 175 passengers across the North Atlantic trunk routes in seven hours or so, halving previous schedules. The jet age had arrived, and the industry set its eyes on the Mach 2 option, which had already been achieved by combat aircraft such as the Lockheed F-104 Starfighter and the English Electric Lightning.

The early jets proved very popular due to a number of favourable characteristics. The journey time was shorter, which in itself attracted passengers, but in addition their speed dramatically improved their productivity. The jet engine removed much of the noise and vibration in the cabin and proved far more reliable than the piston engine, which further improved its productivity. Although the engines consumed more kerosene than the gasoline of the reciprocating engines, fuel was inexpensive, making the overall economics of the jets so much better than that of piston-engine aircraft that, given a choice, there was no longer any reason to continue operating the latter.

It is not often that something highly desirable is improved dramatically and sold cheaper. So it is no wonder that air travel became popular on a massive scale. However, there were less-favourable aspects to this new miracle. Near the ground the jets were extremely noisy; at least 20dB or 8-10 times as loud as their predecessors. When operating at the airport, fuel was only partly burnt, resulting in large quantities of oxides of nitrogen (NOX) and an acrid tear-generating ground-level atmosphere. When cruising (at 35,000ft/11,000m) there was a risk of destabilising the mixture of gases in the upper atmosphere. At that time, though, the term ecology was not referred to very often. While even faster airliners were still under study, much technological development in air transport was directed toward the operating economics of the high-subsonic transport. If the aircraft could be made of lighter materials, it could carry more payload. If the aerodynamics could be improved, less power would be needed. If the engine fuel burn could be reduced, we could travel further. If bigger aircraft could be produced, we would benefit from the economy of scale.

Developments in the 1960s and early 1970s were most promising. The development of the bypass jet engine decreased the fuel burn of the Boeing 707 and Douglas DC-8 by 30% and increased range accordingly. In the field of aerodynamics the so-called supercritical wing, which dampened the severity of the sonic shock wave, enabled the wing to be 'unswept' by about 7-8 degrees (or 20%) without loss in cruise speed. The wing was thus able to be made smaller and lighter, significantly increasing payload. The new wing also offered the opportunity of higher cruise speed without increasing drag due to compressibility.

THE FUTURE SST

Although it was proclaimed the world's first four-engine jet airliner, the Avro Tudor VIII, a drastic modification of the second prototype piston-engined Tudor I, could hardly be said to have been purpose-built.

The sleek lines of the de Havilland Comet prototype contrast sharply with the rather inelegant form of the Tudor VIII.

FASTER, FURTHER, HIGHER

The jumbo takes wings

The demand for a massive increase in US military airlifting capability led to the Lockheed C-5 Galaxy and the Boeing 747. Here was a massive change of scale, with a doubling of passenger capacity or payload, but would there automatically be the economy? The airframe aspect was disappointing in that this weighed over 2½ times as much as the 707, implying ten engines rather than four, and probably leading to 12 with all the associated diseconomies of operation and maintenance.

Aircraft Type	OEW (Tonnes)	Pax Seats (All-Y)	OEW/Seat (Tonnes)	Max Load (Tonnes)	OEW/Load
727-200	46	150	0.31	19	2.4
707-320C	66	179	0.37	38	1.7
747-200	167	400	0.42	71	2.3

The enabling technology came in the form of the high-bypass turbofan, the first commercial example of which was the P&W JT9D, with the fan diameter approaching that of the propeller and where the 150% increase in thrust was achieved with only a 75% increase in fuel. This, then, was a true economy of scale.

In addition to increased thrust and improved fuel efficiency, the high-bypass engine was optimised to operate at high subsonic speeds and would suffer a similar fate as the propeller if it should be driven through the sound barrier. Probably more significantly, though, despite its size it was appreciably quieter (around 15dB) than the earlier jets. Just about this time, the Aérospatiale-BAC Concorde, the world's first Mach 2 airliner, was in the initial stages of a six-year flight-test programme.

Now it is Concorde's turn

The first concept of what eventually became Concorde was sketched out by the UK Supersonic Transport Committee in 1956. There was, however, a substantial shopping list of enabling technology required to allow a supersonic airliner to enter service. Some of this was available (afterburning turbojets, for example), but much required considerable development, particularly in the aerodynamics area, where the wing should be able to perform well at low speed for gentle air-to-ground manoeuvres as well as cruising at Mach 2. The basic wing was predictably of delta planform with a leading-edge sweep of around 60 degrees to avoid the 'Mach cone' of shock waves in cruise. The wing also needed to hold around 100 tonnes of fuel which, because of the movement aft of the centre of lifting force when transitioning to or from supersonic flight, needed to be pumped along the aircraft's longitudinal axis in flight, to maintain stability.

The delta wing is far from ideal for low-speed performance. Variable geometry might have been a solution, but with the engines mounted on the underside of the wing, their alignment management was perhaps a step too far. The compromise 'ogee' wing took many years of painstaking research to perfect, while another barrier to rapid development, caution, delayed matters further. The UK aviation industry was still recovering from the fatigue-related crashes of the de Havilland Comet and the loss of the prototype BAC One-Eleven from the hitherto unknown deep stall, so nobody was going to rush into something as complex as Concorde.

Much can change over a period of 20 years, including people's attitudes and expectations. When supersonic transport was first mooted, before the Suez crisis, outside North America relatively few people used private cars, let alone travelled by air. When Concorde entered service in 1976 the Vietnam war was over and the age of mass air transport was well under way in the shape of the Boeing 747. This proved to be the nemesis of Concorde, both economically and ecologically.

Concorde's main emphasis was above all on speed. To achieve its goal it had to sacrifice the fuel efficiency of the high-bypass engine. The Rolls-Royce Snecma Olympus 593 turbojets which power Concorde generate about the same 40,000lb (18,150kg) of thrust at take-off as the engines of the early 'jumbos', but because of the increased drag they have to operate close to take-off power throughout the flight, which works against fuel efficiency and engine life. Both aircraft have similar specific ranges – about 40 miles per ton of fuel – but, of course, the 747 carries up to four times as many passengers.

Other economic factors combined against Concorde. Because the aircraft was extremely complex, the cost of maintaining it was disproportionately high. The same number of pilots as the larger subsonics was required, and they had to undergo special training. Infrastructure charges (landing fees, etc.) are typically levied on maximum take-off weight (MTOW), and at 400,000lb

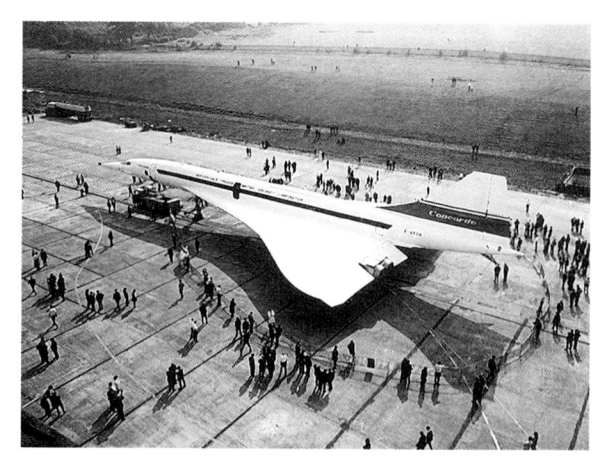

Concorde has always been an eye-catcher. In this roll-out picture, taken at Filton, Bristol, the 'ogee' compound-sweep wing planform can be clearly seen.

THE FUTURE SST

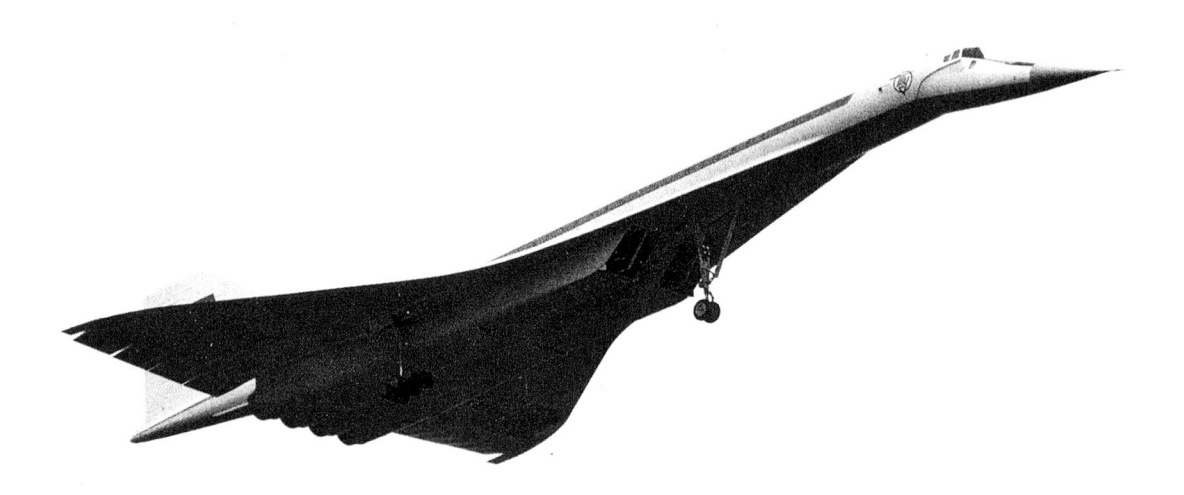

The Tupolev Tu-144 shared some design features with Concorde, and an example first appeared in public in May 1969. The Soviet airline Aeroflot's scheduled Tu-144 services only lasted between November 1977 and May 1978, when they were cancelled following an accident.

(181,500kg) Concorde has around half the MTOW of the 747.

Things were also working against the SST on the manufacturing side. Concorde was not exactly a production engineer's dream, and this was compounded by there being two production lines. The aircraft must have cost at least as much to manufacture as the Boeing 747, and then there was the twenty-year development programme to be paid for. Without labouring the point, the 747 was conceptualised in 1965 and entered service in 1970.

All things being equal, as economists say, passengers on Concorde would likely have to pay four times as much (or first class plus 50 per cent) as those in the humbler areas of the conventional aircraft, and for a seat that was no wider. Clearly there was a market, but was it large enough to support the programme?

What about environmental issues? At take-off and even more so at landing, Concorde is no louder than the early 707s and DC-8s, but by 1976 most of the latter pair had been replaced, re-engined or despatched to parts of the world less strict on noise regulations. The developed world would no longer tolerate these noise levels, particularly around the sophisticated conurbations which would provide most of Concorde's payload. Eventually exemptions were granted in what was regarded as Boeing's back yard, but the clear message was 'not a second time'.

Air displays in the 1950s often featured military aircraft breaking the sound barrier, the crowds responding to the sonic 'b-boom' with cheers. However, during sustained supersonic flight the boom is changed to something more like a roar, about which the everyday audiences on the ground were rather less than enthusiastic. Concorde, then, was essentially banned from overland supersonic flight. There were unpopulated land masses, but these had few potential first-class-plus passengers or were controlled by countries which might have mistaken Concorde for a military threat. Flying subsonic over land was not an attractive option either. Reducing speed in most powered vehicles will normally reduce fuel burn and increase range, such as the automobile at 50mph (80km/h) as opposed to 70mph (110km/h). This does not really work for an SST, since the wing is inefficient at conventional Mach 0.8-type speeds. The super- and subsonic ranges are essentially the same, so there is no benefit in sacrificing speed.

So where could Concorde operate? Ultimately this condensed to across the North Atlantic, and only between population centres within its somewhat limited range. The London/Paris to New York/Washington matrix would work, but not, say, Frankfurt to Chicago. Given that the transpacific market would demand intermediate stops and a circuitous routing, the advantage of Concorde's speed was marginalised by the time lost on the ground. The airlines which had expressed interest understandably took the less risky business path, the net result being sixteen production Concordes and well over a thousand Boeing 747s.

The USA and USSR had their own SST developments. The Tupolev Tu-144 shadowed Concorde in more senses than one, and actually operated with Aeroflot for a while. The Boeing 2707 would have been a 200-300 seat monster operating at close to Mach 3. Ironically, development costs at cancellation in 1970 had been around the same order as Concorde's in total, even though it never flew. The decision to scrap the project may well have been influenced by the results of flight testing the Mach 3

FASTER, FURTHER, HIGHER

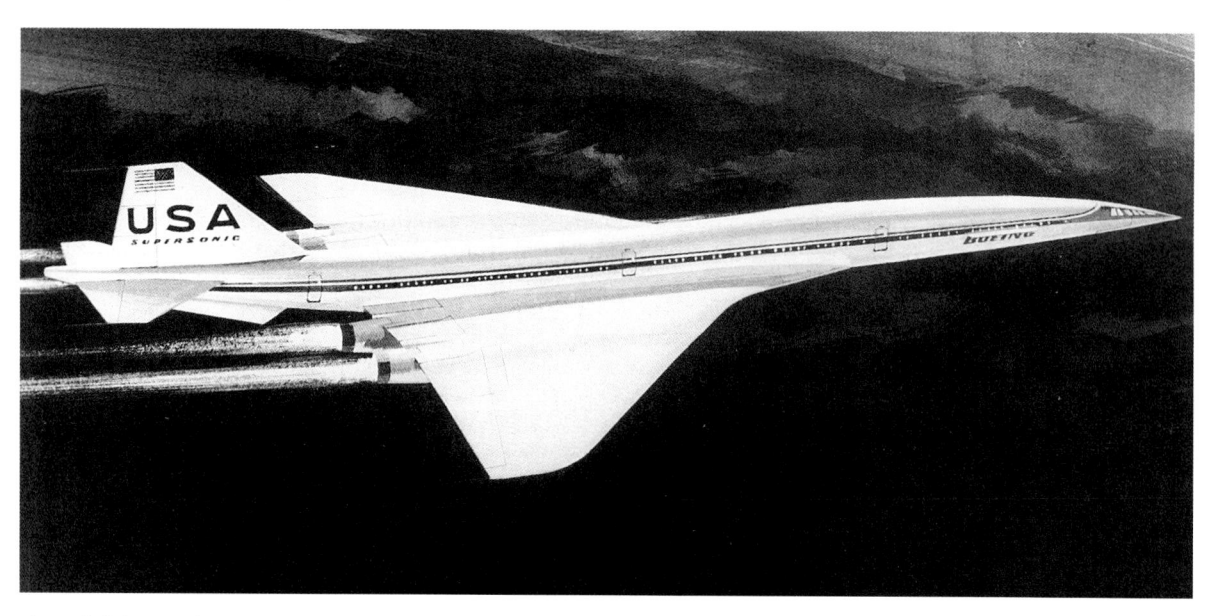

An artist's impression of the massive Boeing 2707 project of the late 1960s. The aircraft would have dwarfed Concorde and carried more than double its payload.

The mighty General Electric GE4 turbojet which would have powered the Boeing 2707. At 63,000+lb thrust it was 60% more powerful than Concorde's Olympus engine.

THE FUTURE SST

North American-Rockwell XB-70 Valkyrie strategic bomber, which never entered service either.

Whatever the commercial shortcomings of Concorde, by any standard it has been a technological success. Until Paris in July 2000, where one of its tyres encountered a piece of debris from an aircraft which had taken off earlier, it had a fatality-free safety record. Its enabling technology has been passed on to the subsequent subsonic aircraft in the form of fly-by-wire and centre-of-gravity optimisation through fuel redistribution on the current Airbus aircraft. These are produced at, among other partner factories in Europe, the same Toulouse and Bristol sites where Concorde was assembled. Although Concorde re-entered service after a period of being suspended following the Paris accident, it is clear that it is in the 'September of its years', so it is appropriate that we should turn our attention to how it might be succeeded, if not replaced.

The world of the next SST

The subject of the next-generation SST comprises interconnected issues of technology, economics and public acceptance. Since Concorde's entry into service in 1976 there has been virtually no development work performed on the aircraft, save for the recent safety-related modifications to minimise the risk of another volatile wing puncture. This is unusual for an aircraft with such a long service life and contrasts sharply with, say, the Boeing 767, which entered service five years later and which has a choice of three fuselage lengths, each with a medium and extended range version. Concorde has been neglected in this regard, but this is hardly surprising given the limited production run. When Concorde was first delivered it was at the forefront of technology, but there have been developments since. The question is, are they appropriate and adequate, or are we still an invention or two short of 'eureka'?

If a new SST is to appear, its economics have to improve. It will have to attract more passengers by flying more efficiently and effectively, not just commute across the North Atlantic. Where, then, are its markets? Concorde is clearly ecologically inappropriate in 2001, and will continue to be all the more so as it resumes and remains in service. Even assuming that the next SST's development period is halved to only ten years, it would expect to be operated for at least a further twenty, so should we be designing for the environmental standards of the years beyond 2025 rather than today?

The combinations of 'what ifs' are so great that, to make any kind of forecast, we need to fix one or two conditions. One is delivery time. If whatever new technology is required cannot be delivered within ten or fifteen years, the window of opportunity would be lost. Concorde would have been finally retired, the impatient would have adapted to this and moved on, but in another direction. We also have to assume that, in situations of increasing global environmental awareness, none of the current standards will be lowered.

Enabling technology needs and opportunities

The fundamental reason against Concorde's operational acceptance has been its sonic boom, which eliminates effective overland flying. This, then, has to be at least reduced to a level where it does not stand out among all the other noise of our daily lives. There is some cause for optimism with spin-offs from laminar-flow aerodynamics, primarily intended to reduce skin friction drag by inhibiting the formation of a turbulent boundary layer. If this or something analogous can impede or reduce the formation of shock waves, then the boom can be brought under control. Attacking the source is the better way, since it is hard to imagine any effective way of suppressing or hush-kitting the phenomenon. The research on laminar flow is promising in the laboratory, but has yet to be given any kind of operational exposure. One might suspect that the dust, grime, bumps and bruises of daily airline operation might challenge this rather clinical technology, pushing workable solutions outside our timeframe. One wonders, though, what stealth technology might have up its sleeve in that area.

If we cannot eradicate the sonic boom, can we find a workable solution to fly efficiently over land? The subsonic wing aerodynamics are the key issue here, and the most obvious solution is the swing wing. With military operating experience through the Grumman F-14, the Panavia Tornado and, probably more representatively, the Rockwell B-1, this should at least improve the subsonic range to around double that of Concorde. Obviously, such an additional complication would increase weight, development and operating cost, but

North American Rockwell's XB-70 Valkyrie strategic bomber was test flown, but that was as far as the programme went. The cost of full-scale production would have been immense.

FASTER, FURTHER, HIGHER

The Valkyrie's distant descendant, the transonic swing-wing Rockwell B-1 bomber, has gained invaluable operational exposure with the United States Air Force.

might just bring enough important markets into a potential network. The means of doing the swing-wing within the timescale are available; it comes down to cost and value. If other civil applications could be found, the costs might be shared.

But the sonic boom is not the only noise that Concorde makes. Its turbojet engines are far noisier than anything else in air transport and, as mentioned earlier, the high-bypass fans would not work in cruise. Is there a way of combining the necessary characteristics of both engines satisfactorily? The simplest in concept would be to have two different types. One turbojet or low bypass, the other high bypass. The realising of this concept soon, though, becomes a Pandora's box. We would still need four engines for supersonic operation, since there has been little in the way of thrust increase in that area since the Olympus generation of turbojets. We might get away with only an additional two high-bypass engines, but they would be very heavy, and dead weight when not in use. They would also be very draggy unless they were shielded by some of moveable fairing. Even before the safety case is made, the concept is becoming unattractive.

Research and development is under way to develop a combination engine which could change role as required. This would overcome the need to slow entire redundant engines, but the tendency towards excessive weight would remain. In addition it appears to be so complicated that civil certification in the near future is hard to imagine.

As propulsion issues are linked to the growing tendency to monitor chemical emissions, anything which appears to be moving in opposition to a cleaner, greener planet would be hard to sell. In many ways, then, the propulsion solution might be a harder nut to crack than the sonic boom. Suddenly the swing-wing appears relatively easy in comparison.

Only if the above environmental challenges can be dealt with satisfactorily could we begin to consider the economic issues. A significant factor in these would be the usual performance question: how fast, how far and how much (payload)? Concorde's specification was to fly the former ocean liner Blue Riband routes in the same number of hours as previously took days. Speed and cruise altitude were essentially set by the mechanical durability of light alloy. At Mach 2(ish) and operating in the upper stratosphere (around 60,000ft (18,300m) and −60 degrees C) the life of the leading edge of the wing would be at its effective optimum.

Operate any higher and you enter the thermosphere where temperature increases with altitude, any lower and

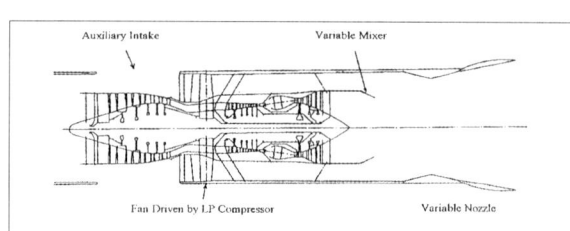

One of the less-ambitious potential solutions to ecological supersonic propulsion, the mid-tandem fan engine would need an extensive development programme before it could enter service.

THE FUTURE SST

the denser air causes more friction and heat, as would flying faster. Any slower would lower productivity without increasing range. If, then, we are to fly any faster, alternative materials have to be considered. From the military experience, titanium would be favoured. This, for example, enables the Lockheed SR-71A Blackbird to 'cruise' at over Mach 3.5, although the leading edge of the wing glows cherry red, an in-flight entertainment which might be hard for most passengers to take. Titanium, though, is pound-for-pound far more expensive than light alloy as are composite materials, which, despite being much lighter, have yet to be satisfactorily established as primary structure on conventional subsonic transports.

Concorde's range also tends to limit its application. Again, at its conception, the early Boeing 707s and Douglas DC-8s were themselves struggling to cross the Atlantic, so simply halving that journey time seemed a reasonable ambition. Now nonstop flights from Europe to the Far East and across the Pacific are commonplace, so the SST's expected range has at least doubled. Even if the 100-seat cabin were retained (see below), we are looking at a much larger and heavier aircraft than Concorde. Once the range is established, we should ensure that cruise speed enables the corresponding journeys to be effective. Concorde works well because it allows a London-New York 'day return', though it was less effective in the opposite direction. Its successor would need to pay more attention to these issues, since it would be crossing more time zones as well as datelines.

How many seats should be offered? Typically, the longer the journey, the larger the aircraft, but at this extreme end of the first-class market, is that still valid? Premium paying passengers are normally in a hurry and may not take kindly to having to schedule their travel plans around infrequent flights in a crowded cabin with little privacy. Any reduction in the cabin size would have an adverse effect on the economies of scale, automatically driving up the fare, which would consequently affect demand.

The conventional wisdom that business travellers were insensitive to airline fares has been severely challenged by the mass downgrading process which finds many former comfort-class travellers roughing it with the low-cost airline backpackers and football fans. The underlying suspicion is that, even if a new SST were produced and much improved, it would not yet be good enough economically or ecologically. The value of time saving to the passengers risks being overpriced, even with the productivity gain of speed.

Better subsonics?

Let us pause and consider if we could produce more efficient airliners flying at the same speed as today's. Is there any enabling off-the-shelf technology which can be readily applied, or do we have to invent some? We should

Because it cruised at over Mach 3, Lockheed's SR-71A reconnaissance aircraft had to have its leading edges built from titanium, rather than the aluminium alloys used in the Mach 2 Concorde.

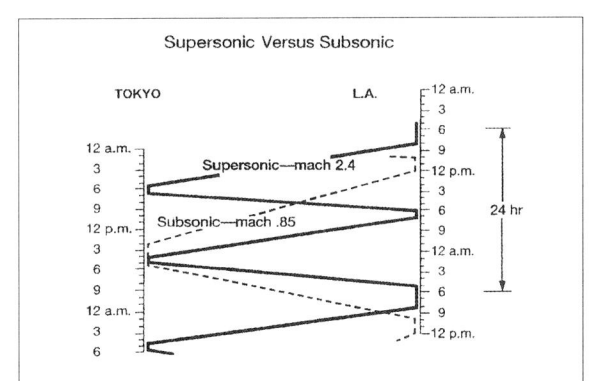

Typical scheduling between Los Angeles and Tokyo. Even if Concorde could fly the Pacific, it would be too slow to perform this schedule.

This Mach 2-2.2, 5,000-mile-plus-range supersonic business jet, seating eight to sixteen passengers, was a joint project by Gulfstream and Sukhoi. Now that low-cost travel and telecommunications dominate today's business world, does this reflect the real size of the SST?

FASTER, FURTHER, HIGHER

The Convair 990 was the first attempt to minimise the drag near the speed of sound. The area-ruled 'Coke-bottle' fuselage shape of the Rockwell B-1 proved rather more successful.

review the various parameters affecting aircraft performance and economics to see what is available. Again we come back to range, payload and speed, all of which have a price tag. As a first order approximation, range comes from leaner engines, payload comes from lighter structures and speed comes from reduced drag. If sonic shock waves hold speed constant we might use reduced drag to burn even less fuel and thus get away with an even lighter structure. All technological frontiers are being pushed back slowly, but what about breakthroughs, such as the swept wing or the high-bypass turbofan?

How far is reaching a natural limit, since most of the world's destinations can be reached nonstop, albeit in very large aircraft. Improved propulsion systems might allow smaller aircraft to increase their range to offer more frequent and direct services, saving transfer times and thus effectively speeding up the journey, which, on a smaller scale, is the *raison d'être* of the hub-avoiding regional jet. A potential breakthrough occurred in the late 1980s in the form of an unducted fan, or ultra-high-bypass engine. This offered dramatic fuel reductions, but at the expense of increased external noise and a challenging certification situation regarding blade loss management. In any case, as with the propeller, they would soon have reached the point of a thrust limit due to blade interference and ground clearance.

The same problem of scale and installation is now starting to affect the giant turbofans such as the GE-90 and Rolls-Royce Trent. Since then nothing dramatic has appeared, while alternatives, such as hydrogen-fuelled engines, are even more daunting prospects.

As mentioned earlier, lightweight composites have not been an overnight success in airline operation, mainly owing to repair difficulty and moisture absorption. Development is steady and will be of most benefit to larger aircraft, since they are generally less structurally efficient than those with single aisles. Again, breakthroughs are unlikely.

Aerodynamic development of laminar flow is moving ahead steadily, with applications on the fins of Airbus aircraft in the pipeline. This is an appropriately cautious approach, as the fin is generally accepted to be the least critical of the aerofoils. As none of the above could be called an impending dramatic jump in the state of the art, it is unlikely that a new aircraft would be launched on the strength of these improvements. Rather, they would be incorporated into aircraft currently under development, preferably after trials with aircraft currently in service. So, if an all-new-technology aircraft of conventional size is unlikely, should we expect something very much larger to reap the economies of scale? Airbus believes so, and backs this up with the launch of the A380.

Enter the 'super-jumbo'

Firstly, does the 'super jumbo' actually require any new enabling technology to make it work at all? (Recalling the development of the high-bypass engine, without which there would have been no Boeing 747.) The essential

THE FUTURE SST

answer is 'no', or at best 'not much'. This makes development easier, but by merely using technology which is also available to smaller aircraft it has to rely more and more on the economy of scale. Its engines are 'the latest' but current, its structure features extensive composites, though some of these have had little service exposure as yet.

It could, however, benefit from the laminar-flow aerofoil, but to restore aerodynamic efficiency from constraints on its plan-view configuration caused by ground handling demands. Again we are in a new age with new attitudes. When the 747 appeared, airport terminal docks and runways were customised to the aircraft, mainly because there was room to grow. Now this situation is reversed, and the new aircraft is having to be designed to fit into current airport geometry, the so-called 80m x 80m (260ft x 260ft) box.

We have already seen that economy of scale does not automatically apply when moving from single to twin aisle aircraft, and the same situation arises when moving up from single to double (passenger) decker aircraft.

Aircraft Type	OEW (Tonnes)	Pax Seats (All-Y)	OEW/Seat (Tonnes)	Max Load (Tonnes)	OEW/Load
A320-200	43	162	0.27	18	2.4
A340-300	130	349	0.37	48	2.7
A380 (est.)	275	679	0.41	83	3.3

Those elements of operating costs which will most likely benefit from scaling-up are cockpit crew, fuel and engineering. The least favourable elements would be landing charges and capital-related items which are amortised by utilisation over a fixed calendar period. The A380's cruising speed matches that of smaller aircraft, but it is almost inevitable that any larger aircraft would take longer to turn around (ground servicing between flights) than a smaller one. The proof of the pudding, however, is in the eating, and Airbus has several years to prepare the dish, so the aircraft design may well change during development. Although it is probably too late, one possible addition might have been a swing wing, this time to aid taxying (when it would be fully swept), take-off and landing (straight to shorten runway length required) and swept in flight as required. As with Concorde it would have been necessary to address the issue of engine alignment.

If we are troubled by the concept of another Mach 2(+) aircraft and fear that scaling-up alone is not enough, should we try to inch cruise speed closer to Mach 1, without entering the supersonic theatre with all its associated problems? This is the current approach by Boeing with its Sonic Cruiser.

Just a little bit faster?

Attempts to move up from the commercial 'subsonic' barrier of around Mach 0.84 are not new. Again, the value of time is under scrutiny, being measured against the almost inevitable cost increase of speeding up. A typical design approach with this type of airliner is to try to achieve the so-called 'Seers-Haack' cross-sectional profile over the longitudinal axis. Rapid changes of cross-sectional area, such as occur at the wing or tailplane, should be minimised to reduce the wave drag from shock build-up. This can be attempted either by appropriate fairings or by 'waisting' the fuselage at the wing root. The first method was evident on the Convair CV-990 Coronado, which featured fairings on the inboard trailing edge of the wing, which also served as supplementary fuel tanks. Cruise Mach number was originally planned to be at least 0.9, but results were disappointing and it took a considerable amount of sustained development to achieve even 0.87. The resulting delay inevitably handed the lion's share of the market to Boeing and Douglas.

Just about this time, the supercritical wing was being developed. As mentioned previously, this allowed the choice of reducing wing sweep and maintaining cruise Mach number or increasing this for a given sweep. Boeing investigated this latter on a 7J7-type project which also featured a waisted fuselage and was planned to fly at Mach 0.98. Again, the results did not quite match expectations, and to reach the design speed a wing sweep of over 40 degrees was required. This would have had an impact on field performance, and to match that of conventional aircraft implied extremely complex flap systems. A further restriction was the waisted fuselage. Airliner manufacturers exploit the simplicity of the tube-like fuselage to extend cabin size in successive versions of aircraft. There has been the remarkable extension of the DC-9 from a Series 10 version with eighty seats to the MD-90 with 180. The benefit of the Mach 0.98 airliners' speed increase was measurable, but was hardly dramatic even over, say, the 2,500-mile (4,000km) US coast-to-coast routes, where the time saving was around 40min. Not surprisingly, Boeing reviewed its options, and eventually produced the 757 and 767 models. These used the supercritical wing in the conventional (*à la* A300) manner and were both justifiably very successful.

Enter the Sonic Cruiser. At first glance it might appear like an attempt to revive the 7J7 of thirty years ago. The question is a two-edged one – what new enabling technology has been developed and what lessons have been learned? Rather than visibly waisting the fuselage, a compound sweep is employed, resembling that of the Tu-144, but to a lesser degree, which also had a canard (foreplane). There may be a link here. In the mid-1990s a Tu-144 was acquired by NASA as part of the Technology Concept Aircraft (TCA) project, which involved some major US aerospace companies, including Boeing. The objective was to evaluate a Mach 2.4 SST similar in capability to the earlier Boeing 2707. That project did not progress beyond the research level, but clearly there was

FASTER, FURTHER, HIGHER

a technology gain. Returning to the Sonic Cruiser, engines are mounted aft, more or less in line with the wing but relatively close to the fuselage. This is certainly far removed from the 707-type shape of the CV990, and while it certainly looks like something built for speed, will its economics justify such a revolutionary programme?

But there may be something else. As noted earlier, Boeing's response to Concorde at the time probably owed much to the Valkyrie supersonic strategic bomber project. Since then the same company has developed the B-1, which features an area-ruled fuselage and a swing wing with aft-mounted engines. It operates at Mach numbers from 0.9 to 1.25 at low altitude and its range is 6,000 miles (9,650km)-plus. Here, then, is some enabling technology for what could be an airliner which is efficient at subsonic speeds and, assuming it would reap the benefit of a 40,000ft (12,200m) cruise altitude, work effectively at Mach 1.5+. Boeing is a pragmatic organisation and does not flirt with technology for its own sake. It remains to be seen whether the Sonic Cruiser joins the ranks of the successful string of airliners produced in their many thousands or is filed away with the various 7J7s, awaiting a more acceptable relationship between technology and economics.

A practical perspective?

We have seen that each mode of transport has an effective and efficient applicable envelope of payload, range and speed. The lower limit is defined by need to compete with the next-less-ambitious form of transport, while the upper limit is set by an acceptable combination of technology, economics and now ecology. For air transport the lower limit is more or less established, but what about this upper limit?

Vehicle	Date of introduction	Speed then (kph)	Speed now (kph)	Max range (km/trip)	Max load (Tonnes)
Human	?	10	10	20	0.05
Horse & Rider	c.4000 BC	30	30	100	0.15
Horse & Carriage	c. 2000 BC	20	30	100	3.3
Train	1840	40	200	1000	1000
Airliner	1930	250	800	10 000	80

High-subsonic air transport was demonstrably effective as soon as it was introduced on a mass scale in the early 1960s. Since then it has become extremely efficient and affordable to virtually every resident of the developed world. Those passengers who took to the skies to save time now travel, albeit in greater comfort, at the same speed as everyone else. How they manage their impatience is the prime issue under discussion. For them, (super)sonic travel will need to be more effective than it

(Super) Sonic Cruiser? Boeing almost went that way thirty years ago with this 300-passenger, variable-sweep design. At the time it was stated that the company had invested more than $30 million on SST research and development.

is now, or the market will remain at the Concorde scale and scope. More effective in the first instance must mean more range and a wider choice of routes. There is little gain (probably only 30-40 minutes) for a Mach 3 London-New York journey, as opposed to several hours' saving on the current schedule between Los Angeles and Singapore by flying at Mach 1.5. This is the case for a Sonic Cruiser if it takes some spin-off from the B-1.

Assuming it would perform more or less as expected, its competition is unlikely to come from more efficient subsonics, but rather from alternatives to travel, or 'is your journey really necessary?' Concorde was launched when telecommunication was still in its steam age. There was no fax, no e-mail and telephones were anchored in a box at the end of the road. Now we are far more effective and efficient. The calamitous change in the New York skyline has brought the likelihood of a new set of values in general, and it could be that the necessity as well as just the economics of each air journey is viewed in a different light. This, of course, would also affect the likely fortunes of any project which is intended to increase the upper limit of payload, not just that of speed.

Bibliography

Airbus Industrie Web site (www.airbus.com).

Aviation Week & Space Technology, McGraw-Hill, Mid-1972 to early 1973.

Boeing Web site (www.boeing.com).

Jane's All the World's Aircraft (Jane's Information Group, various years).

Saonatsos, 'Commercial Viability of the Next Generation High-Speed Civil Transport' (Cranfield University thesis, 1997).

Snow, J, 'The Mach 0.98 Airliner' (University of Southamptom thesis, 1973).

13
Stepping into Space
Stephen Ransom

In the USA, Russia, Japan and Europe there is a desire to develop re-entry technologies and techniques which can be applied to fully-recoverable space transportation systems. During the past 40 years a number of attempts have been made to achieve this aim, but all have suffered from changes to space programme strategies and objectives and/or the effects of prevailing international and national economic depressions and crises. Consequently none of the projects has yet reached operational status.

Orbiter research programmes
Current space transport systems have been developed with the aid of small-scale unpiloted and piloted research vehicles. All of them have been produced and flown with the primary objectives of:

- measuring aerodynamic parameters during their return flight from space,
- evaluating stability, manoeuvrability and energy management techniques throughout their flight envelope,
- evaluating materials, structures and thermal protection and cooling systems, particularly during re-entry and hypersonic flight,
- developing crew-spacecraft interfaces and assess vehicle handling characteristics.

Although flight demonstration was the ultimate test to verify these objectives, all of the programmes were accompanied by numerous studies and very intensive research, and they made extensive use of ground test facilities and simulators.

NASA M2 and HL-10 (USA)
The lifting-body type of re-entry vehicle originated in work begun at the USA's National Advisory Committee for Aeronautics' (NACA, later National Aeronautics and Space Administration, or NASA) Ames Aeronautical Laboratory in 1951. This showed that a blunt object moving at very high speeds would develop a detached shock wave that would carry most of the heat load with it away from the body, and that residual heating effects could be dealt with by the application of existing materials technology. These findings were first applied to the design of the warheads of the Atlas and Titan intercontinental ballistic missiles and the Mercury space capsules.

Subsequently, NACA Ames engineers realised that the low lift/drag ratios of symmetrical blunt bodies could be improved to enable the vehicles to reach potential landing sites at considerable distances from their nominal re-entry and descent trajectory. This resulted in proposals for half-cone-shaped configurations and vehicles with elliptical cross-sections. Interest was shown by the USAF, which funded several small research programmes, including one done by The Aerospace Corporation at El Segundo, California. Three basic lifting-body configurations emerged from subsequent development work: the NASA Ames M2, the NASA Langley HL-10, and the Aerospace Corporation A3.

The A3 configuration was further developed by Martin Marietta and formed the basis of the company's proposal designated SV-5, much later to be used in the design of the X-38, a prototype of the crew rescue vehicle serving the International Space Station.

Windtunnel testing had shown that these configurations had acceptable hypersonic aerodynamic characteristics, but a number of crucial questions remained concerning their subsonic and transonic performance. These were answered by the flight testing of a number of experimental vehicles, the first of which was a small-scale model of NASA's M2 lifting body built and tested in February 1962 by Robert D Reed, a senior engineer at NASA's Flight Research Center (FRC) at Edwards Air Force Base (AFB). The model was launched from a radio-controlled mother aircraft, and the tests proved sufficiently encouraging to influence the authorisation of a six-month feasibility study of a lightweight, piloted M2 glider. By using its own personnel, the FRC managed to complete the M2-F1 glider in early 1963 for almost $30,000, and it was transferred to NASA Ames for windtunnel tests which ended in March 1963. The first towed flight of the vehicle was accomplished on 5 April 1963, and the first free flight was made on 16 August 1963, the glider, flown by NASA research pilot Milton (Milt) O Thompson, being towed aloft by a Douglas C-47 and released at an altitude of 10,000ft (3,000m). Eventually the M2-F1 completed over 100 flights and 400 ground tows.

Encouraged by the results from the M2-F1 programme, NASA began studies of rocket-propelled lifting bodies to evaluate their performance at supersonic speeds and high altitudes and their landing behaviour at realistic mission weights. In February 1964 the FRC issued Requests for Proposals to 26 companies for the construction of two mission-weight, low-speed lifting-body gliders. These were to be tested in the full-scale windtunnel at NASA Ames and also to be air-launched from NASA's Boeing NB-52. They were to be of two dif-

FASTER, FURTHER, HIGHER

Langley Research Center's HL-10, claimed to be the most successful of all the lifting-body designs, at NASA Dryden in 1966. The HL-10 had the XLR-11 rocket engine during its intial flight trials, but this was later replaced by three 500lb-thrust Bell Aerosystems rockets for the purpose of conducting controlled, powered landings. Restored after completing its test programme, the HL-10 was put on display at NASA Dryden. (NASA)

The M2-F1 in flight over NASA's Dryden Flight Research Center, which shares the same site as Edwards Air Force Base. Of steel and plywood construction, the lifting body had a length of 20ft (6.10m), spanned 14ft (4.27m) and its all-up weight was 1,138lb (517kg). The vehicle was first revealed to the news media on 3 September 1963. (NASA)

The X-24A, seen at NASA Dryden in 1968, was of similar configuration to the X-23A. Air-launched, it was flown by both USAF and NASA pilots to investigate the lifting body's static and dynamic stability and control characteristics at low-supersonic, transonic and landing approach speeds, the test results complementing those obtained in hypersonic flight with the X-23A. It had a length of 24ft 6in (7.47m), a span of 13ft 8in (4.16m) and an all-up weight of 11,000lb (5,000kg). (NASA)

ferent designs, one based on the M2 and the other on the HL-10 configuration developed by NASA Langley. Only five companies submitted proposals, and on 2 June 1964 the FRC awarded a fixed-price contract to the NorAir Division of Northrop to build the vehicles. The M2-F2 was rolled out from Northrop's Hawthorne factory on 15 June 1965, and the HL-10 on 18 January 1966. Meanwhile, it had been decided to make provisions in both vehicles for the installation of a rocket propulsion unit. The total cost of the two lifting bodies was slightly under $2.5 million.

The first gliding flight with the M2-F2 was made by Milt Thompson on 12 July 1966. During the next four months the vehicle made thirteen additional gliding flights. These confirmed that the vehicle suffered from poor lateral-directional stability, particularly at low angles of incidence and high speeds, characteristics that had been predicted from flight tests made with NASA's specially modified NT-33A variable-stability trainer. The M2-F2 was grounded on 21 November 1966 for installation of the rocket engine. On 10 May 1967, during its first powered flight, the pilot, Bruce Peterson, lost control and crash landed, seriously injuring himself and almost destroying the M2-F2. The airframe was returned to Northrop, and on 28 January 1969 NASA announced that it would be rebuilt as the M2-F3. At the same time it was modified to include a central fin as well as the two wingtip fins, to improve lateral stability.

When the M2-F3 made its first gliding flight, on 2 June 1970, its pilot, William Dana, found that its handling was much improved over that of its predecessor. On 13 December 1972 the M2-F3 attained Mach 1.6, its fastest flight, and one week later, on 21 December, its highest altitude, 71,493ft (21.8km). The vehicle was retired from its programme after completing forty-three flights (sixteen as the M2-F2 and twenty-seven as the M2-F3).

The HL-10 began its flight test programme on 22 December 1966 with a gliding flight piloted by Peterson. During the descent the pilot found that he had minimal lateral control, but was able to land it. Further windtunnel testing at NASA Langley led to small modifications to the wingtip fins. The improvement to the HL-10's handling qualities was demonstrated on its next test flight, on 15 March 1968. The first successful powered flight with the HL-10 was made by John Manke on 13 November 1968, and the lifting body was flown at supersonic speeds for the first time on 9 May 1969. The following year the HL-10 achieved the highest speed and altitude of all piloted lifting bodies: Mach 1.86, on February 18, when it was flown by Peter C Hoag, and 90,300ft (27.5km) on February 27, when piloted by Bill Dana. The HL-10 programme was terminated after thirty-seven flights (the last was made on 17 July 1970), its basic shape being considered the best of the lifting bodies and forming the basis for several early Space Shuttle concepts.

ASSET, PRIME, PILOT (USA)

The cancellation of the Boeing X-20 Dyna-Soar programme in December 1963 led to a redirection of efforts within the USAF to obtain data related to lifting re-entry vehicle technology. The redirected programme, named Spacecraft Technology and Advanced Re-entry Tests

(START), was approved in November 1964. The START programme included three specific areas of investigation, named:

- ASSET (Aerothermodynamic/Elastic Structural Systems Environmental Tests),
- PRIME (Precision Recovery Including Manoeuvring Re-entry),
- PILOT (Piloted Low-Speed Tests).

The ASSET sub-programme was conducted with small-scale, winged hypersonic vehicles built by the McDonnell Aircraft Corporation. These were boosted by Thor or Thor-Delta launchers to altitudes of 168,000 to 223,000ft (51 to 68km) and speeds of 8,900 to 13,250mph (14,250 to 21,200km/h), and were designed to be recoverable.

The lifting-body configuration SV-5, derived from a series of studies conducted by the Martin Marietta Corporation, was chosen to meet the objectives of the PRIME and PILOT sub-programmes. The hypersonic flight test programme, PRIME, was conducted with small-scale vehicles boosted to near-orbital speeds by Atlas launchers. After re-entry the test was terminated at an altitude of 98,500ft (30km), and when the vehicle's speed had decreased to approximately Mach 2. The vehicle was afterwards recovered by parachute and captured during its descent by specially equipped Lockheed Hercules. The lifting bodies used for these tests were designated SV-5D/X-23A.

The low-speed flight test programme, PILOT, was accomplished with a manned, rocket-powered, air-launched vehicle designated SV-5P/X-24A. It was used to explore the vehicle's flight envelope at speeds below Mach 2 and its horizontal landing characteristics.

ASSET (McDonnell ASV and AEV)

Although ASSET was later incorporated in the START programme, it began in April 1961 when McDonnell Aircraft signed a contract with the USAF Flight Dynamics Laboratory for the development of six experimental gliders that approximately resembled the Dyna-Soar concept. The ASSET programme included two distinct and separate research efforts: Aerothermodynamic Structural Vehicles (ASV), and Aerothermoelastic Vehicles (AEV), of which McDonnell built four of the former and two of the latter. The gliders were 5ft 9in (1.74m) long, and fully equipped weighed between 1,130 and 1,225lb (514 and 557kg). All had a low-aspect-ratio delta wing with a span of 4ft 11in (1.50m) and an area of 14ft^2 (1.3m^2). A flight control system using hydrogen-peroxide thrusters maintained the vehicle's flight attitude after separation from the booster. Measurements made during the flight were monitored by radio telemetry. The vehicle's recovery system comprised a parachute, flotation bag and locator beacon.

Although the two vehicles were superficially similar and shared common subsystems, their completely different mission and research capabilities were reflected in their totally different flight profiles. The ASVs were used to determine temperatures, heat fluxes, pressure distributions and evaluate materials and structural concepts during hypersonic gliding re-entry. They were boosted to altitudes of 190,000 to 223,000ft (58 to 68km) and velocities of 11,000 to 13,250mph (17,600 to 21,200km/h), giving them a range varying from 1,000 to 2,300 miles (1,600 to 3,700km). The AEVs, on the other hand, were boosted to 167,300 to 187,000ft (51 to 57km) at a velocity of 8,900mph (14,250km/h), achieving ranges of 620 to 830 miles (990 to 1,330km). The AEVs were used to study panel flutter phenomena and the effect of hypersonic flight on aerodynamic control surfaces.

The first ASSET flight took place on 18 September 1963, with a launch from Cape Canaveral Air Force Station towards Ascension Island. Telemetry data obtained from the flight confirmed that the vehicle, an ASV, performed a successful re-entry, but it was lost when its flotation bag burst on splashdown and it sank. The second ASV, launched on 24 March 1964, was destroyed when its booster failed. The booster was modified, and the third ASSET mission, launched on 22 July 1964, was successful, the vehicle being recovered and post-flight examination indicating that it could be flown again. The fourth and fifth vehicles, both AEVs, were launched on 27 October and 8 December 1964 respectively. Although telemetry data was obtained, both were lost after impacting with the sea. The last flight, made on 23 February 1965, was deemed successful even though the vehicle was not recovered.

The five vehicles provided data on the flight environment at a wide range of speeds and angles of incidence. The ASSET programme studied structural responses, system and subsystem performance, various testing techniques, and correlated in-flight data with theory and ground tests. The programme also proved conclusively that a winged vehicle could be built to withstand the demands of re-entry. The total cost of ASSET, including boosters, was $41 million.

PRIME (Martin Marietta X-23A)

The X-23A programme was initiated in 1964. Martin Marietta built four X-23As specifically to test configurations, control systems and ablative materials for hypersonic lifting-body-type re-entry vehicles. The X-23A was the product of some two million hours of development work by Martin Marietta in the field of hypersonic re-entry vehicles and research initiated by NASA. The work that culminated in the X-23 began in 1961, when the company received a USAF contract to investigate the aerodynamic characteristics of manoeuvrable-lifting-body configurations, and the first windtunnel tests were conducted in May 1962. Subsequently, about fifty low-speed gliding flights using a model X-23A were successfully completed before the actual re-entry prototypes were built. They were 6ft 8in (2.03m) long, spanned 3ft 10in (1.16m) and had a fully-equipped weight of 894lb (406kg).

Three successful X-23A missions were eventually flown. On the first flight, on 21 December 1966, only pitch manoeuvres were conducted. This vehicle was lost over the Pacific when aerial recovery efforts failed. On the second flight, on 5 March 1967, a differential movement of the flaps was used to bank the vehicle at hypersonic speeds. Distances of up to 660 miles (1,050km) away from the nominal flight path, with successful returns, were accomplished. This vehicle was also lost when its flotation gear separated and it sank. The third flight was made on 19 April 1967, the vehicle being launched from Vandenberg Air Force Base and recovered intact off Kwajalen Island in the Pacific. All of these missions were judged successful and led to the decision to cancel the fourth planned flight. The total cost of the PRIME flight tests, including the Atlas boosters, was $70.5 million. The information derived from the X-23A programme qualified lifting-body designs for space flight. Technologies developed for and verified in this programme included manoeuvring re-entry techniques, ablative materials and steam-cooled equipment.

PILOT (Martin Marietta X-24A and X-24B)

The X-24 lifting-body vehicles were developed under a USAF contract awarded to Martin Marietta. The initial X-24A configuration was similar to that of the X-23A. The PRIME series of tests had demonstrated the lifting body's hypersonic stability, but the USAF also wished to obtain data concerning its low-speed handling and landing qualities. The objectives of the PILOT programme were to investigate the static and dynamic stability characteristics and control surface control laws for the vehicle during the low supersonic, transonic and landing approach speed regimes. Additional data was to be collected on various aerodynamic forces and pressures, hinge moments, pilot-machine interfaces and the energy management requirements for lifting-body designs. Post-flight analysis was to look at the correlation of flight data with windtunnel predictions.

The programme began slowly, and by December 1964 had still not received official approval. The following year was spent resolving various managerial, procurement, funding and technical problems. In February 1966 the PILOT project office received proposals from Northrop and Martin Marietta for the manufacture of one rocket-powered vehicle. The proposals were evaluated quickly and Martin Marietta was awarded the contract to build the X-24A, which was rolled out at the company's

The X-24B during its descent to land at Edwards. Built using components from the X-24A, the vehicle was fitted with a slender delta wing representative of those being considered during Phase B of the development of the Shuttle. (NASA)

Baltimore factory on 11 July 1967. It was not immediately handed over for flight testing, but was sent to NASA Ames for windtunnel testing to establish a database for the post-flight analyses. It was then delivered to Edwards, where it underwent various static ground tests. A year later it was cleared for flight, the first unpowered gliding flight being made on 17 April 1969 by USAF test pilot Jerauld Gentry. The first powered flight was made on 19 March 1970, also by Gentry, after a further eight unpowered flights. On 14 October 1970 NASA test pilot John Manke flew the X-24A for the first time at supersonic speeds. It accumulated a total of almost three hours' flight time during 28 missions, achieving a maximum altitude of 71,407ft (21.8km) on 27 October 1970, and a maximum speed of Mach 1.6 on 27 March 1971. On both occasions it was flown by Manke. The last portion of the vehicle's test programme was dedicated to simulating Space Shuttle approaches and landings, and the X-24A made its last flight on 4 June 1971.

Martin Marietta also completed two jet-powered low-speed lifting bodies based on the SV-5 configuration as a company-funded venture. These were to be used as astronaut trainers, but they were considered significantly underpowered for this purpose and were never used. In 1968 the USAF System Command solicited proposals for their use, and the USAF Flight Dynamics Laboratory's suggestion that one of them should be modified to have a highly-swept, low-aspect-ratio delta planform was accepted. (The wing geometry was similar to that proposed for the manned hypersonic research vehicle that would have been designated X-24C.) However, further studies concerning the choice of propulsion led to the selection of a rocket engine. This choice, and the fact that the X-24A was nearing completion of its test programme at that time, resulted in the decision to modify the X-24A rather than one of the jet-powered airframes. On 1 January 1972 the USAF awarded Martin Marietta a $1.1 million contract to modify the X-24A (half of the cost was met by NASA, which had a vested interest in obtaining data from the test programme). The modified airframe, designated X-24B, was rolled out on 11 October 1972. After several months of ground testing, Manke made the first gliding flight on 1 August 1973, and also the first powered flight of the X-24B, on 15 November 1973. The X-24B was retired from the test programme on completion of its last flight by NASA test pilot Thomas McMurtry on 26 November 1975. Thirty-six flights were accomplished, the accumulated flight time amounting to 3hr 47min. The vehicle reached a maximum speed of Mach 1.76 (Lieutenant Colonel Michael Love on 25 October 1974) and an altitude of 74,130ft (22.6km) (John Manke on 22 May 1975).

Both the X-24A and X-24B made significant contributions to various design aspects of the Space Shuttle programme.

MUSTARD (United Kingdom)

MUSTARD, an acronym of multi-unit space transport and recovery device, was the result of a feasibility study initiated partly by a British Ministry of Aviation contract awarded to English Electric Aviation early in 1963, and partly by English Electric's own project research group at Warton Aerodrome, dating from about mid-1955. The study embraced long-range, high-speed cruise aircraft, rocket-boosted vehicles then named spaceplanes, recoverable launchers and boost glide vehicles. Its object was to highlight a profitable line of further research. Proposals from the company, by then renamed the British Aircraft Corporation Preston Division following mergers within the industry, were submitted to the Ministry in January 1964, and there followed two contracts for further research to be conducted over a 15-month period. The MUSTARD concept was formulated in February 1964, soon after the placing of the second contract. Work on the project continued after the expiry of the Ministry contracts, but ended in January 1968.

English Electric's own research was a logical development of work done at Warton in connection with the P.10 high-altitude Mach 3 reconnaissance aircraft project. The extension of the P.10 study resulted in a series of air-breathing hypersonic and cruise booster aircraft capable of launching rocket-powered orbital spacecraft. These aircraft were grouped under the project designation P.42.

At the beginning of 1963 a new department, the Aero Research Group, was formed to undertake research into hypersonic vehicles. Later in the year another department, the Aero Space Group, was created to take over the P.42 project and continue hypersonic research. Further work on air-breathing vehicles led to the conclusion that the system's high research and development costs greatly outweighed its operational cost advantages. The air-breathing booster concept was subsequently shelved and attention given to recoverable, rocket-powered vertical-take-off lifting bodies, with the aim of substantially reducing research and development costs.

In the USA, Douglas Aircraft published details of its Astro project, which had a first-stage booster launcher and a second-stage payload in the form of a lifting body.

Both the British Aircraft Corporation and Hawker Siddeley Aviation, the two major groups to emerge after the consolidation of the British aircraft industry, investigated recoverable space launchers during the early 1960s. BAC's proposal, MUSTARD, was launched vertically and, after separation, each of the three components landed horizontally. (BAe Systems)

Both stages were rocket-powered and geometrically similar. The latter feature was claimed to be an advantage, since considerable effort and cost would be saved in developing only one configuration.

The Aero Space Group at Warton realised that this advantage of the Astro concept offered new possibilities if both stages were made geometrically identical and there was a very high degree of commonality of the stages' components. Further, since both stages had to operate in similar environments it became a prerequisite that each unit was made from the same materials. These three factors meant that only one unit need be produced and developed, with consequent reduction in cost. An investigation followed into the most efficient method of assembling the units, or modules, as they became known. This work showed that at least three would have to be combined to meet launching requirements. At first sight, the most attractive method of assembly appeared to be a triangular cluster, but this arrangement suffered from the effects of a venturi formed at its centre. The modules were subsequently arranged in a stack, which eliminated the aerodynamic problem, occupied a smaller launching base area, and also lent itself to the use of additional modules if required. At this juncture the acronym MUSTARD came into being to describe the modular concept. As envisaged, MUSTARD was made up of two units acting as first-stage boosters, with the third unit as the spacecraft.

To gain a more realistic assessment of the research and development costs, the group investigated in detail the various programmes required to produce and develop a MUSTARD module. The main programme was visualised in the form of major events supported by the results from specific areas of research. To this end the group suggested the construction of a simple lifting-body piloted glider and a more sophisticated model to be launched by the first stage of the proposed Black Arrow satellite launcher. The model was to be used to confirm the methods of calculating the heat input due to hypersonic re-entry, to confirm the design and manufacturing processes for thin-gauge structures capable of withstanding the heat loads, to provide information on heat transfer throughout the structure, and to determine hypersonic aerodynamic characteristics. Launches were to be made at Woomera, Australia. Work on the metals and form of structure needed for this research was well advanced before the MUSTARD project was discontinued.

SPIRAL and BOR (USSR)

The following account of the development of reusable space transportation systems in the USSR is based mainly on published reports. Although it is claimed that some of this information was obtained through exclusive interviews with persons directly involved in the design and testing of these systems, contradictions and inconsistencies are to be found in all of the reports. This is not surprising, since the distribution of data within the USSR was made on a 'need-to-know' basis, security measures and the style of management of such programmes as practised in the USSR prohibiting or restricting an exchange of information, even among companies and institutes directly involved in the same programme. Consequently, very few people had, or were permitted to have, complete access to all information related to these programmes, and what follows should be treated with circumspection.

The first serious attempt by the USSR to develop a reusable space transportation system began in the mid-1960s at the Mikoyan OKB (Experimental Design Bureau). This was a two-stage-to-orbit system based on a horizontal-take-off, hypersonic, winged first stage propelled by hydrogen-fuelled, air-breathing engines, and a rocket-boosted orbital upper stage. The latter combined the features of a lifting body with those of a winged vehicle, the wings being positioned at 45 degrees and 95 degrees to the vertical during re-entry and transonic/subsonic flight respectively.

A full-scale piloted prototype of the orbiter, named SPIRAL, was completed and flown successfully. About ten piloted subsonic flights were made, which included at least three air launches from a Tupolev Tu-95, the first being made on 27 October 1977 at an altitude of 9km (3,000ft), and a number of conventional take-offs, for which SPIRAL was fitted with a small jet engine.

One unconfirmed report suggests that a rocket launch of SPIRAL and its pilot was also made to obtain data at much higher speeds and altitudes than could be obtained from previous flights. That such a flight could have taken place, or was intended, is indicated by the arrangement of the four-legged, deployable undercarriage, which was installed without disrupting the heatshield, and by the fact that the air intake for the jet engine could be sealed by a clamshell-type door. The forward undercarriage could be equipped with wheels or skids and the rear undercarriage with skids.

Further development of the project was continued under the leadership of NPO Molniya at Tushino with small-scale, unmanned versions of the SPIRAL re-entry configuration, named BOR. Also associated with the design of the BOR vehicle were: TsAGI, Zhukovsky, (aerodynamics and flight dynamics); TsNIIMash, Kaliningrad on the outskirts of Moscow, (aerothermodynamics and thermal protection materials); LII, Zhukovsky, (flight testing).

No official information has been released regarding the flight testing of BOR. It is believed that its testing took place between 1976 and 1979, and comprised dual launches designated Cosmos 881/882 (15 December 1976), Cosmos 997/998 (30 March 1978) and Cosmos 1100/1101 (22 May 1979). In all cases the launch vehicle was an SL-13 Proton, the launch site was Baikonur, the launches took place before dawn, and the vehicles were recovered by parachute in daylight at a site in eastern Kazakhstan after completing one or two almost circular orbits at an altitude of approximately 125 miles (200km). More recent information suggests that the vehicles used in these tests were named BOR 1, BOR 2 and BOR 3 respectively. At least one of the dual missions failed and the Cosmos spacecraft were lost (this launch took place on 4 August 1977). The pairs of BOR vehicles used for these tests had different aerodynamic configurations. The test programme was halted soon after the launch of Cosmos 1100/1101 but revived again in 1982 to provide data for the development of the Soviet space shuttle, Buran.

The initial series of space vehicles used for the development of Buran were designated BOR 4 and had a span of 2.6m (8ft 6in), a length of 3.4m (11ft 2in) and a mass

Spiral was a manned vehicle which would have been launched by a rocket booster at high speed from the upper surface of a winged aircraft propelled by air-breathing engines. Work on the project began in 1962 but stopped seven years later. A prototype of Spiral, which was the third stage of the overall launcher concept, was tested at subsonic and supersonic speeds. The outer wing panels were to be folded upwards to create wingtip fins for flight at very high Mach numbers. The vehicle is now one of a large collection of exhibits at the Russian aerospace museum at Monino. (Mikoyan)

Several models designated BOR 4 were launched into orbit from Kapustin Yar to obtain heatshield performance data for the design of Buran. The BOR 4 configuration was based on the Spiral. It was publicly displayed at the International Luftfahrtausstellung (ILA) held in Berlin in 1992 at a time when Russia was promoting the vehicle's use for the orbiting and recovery of scientific payloads, which would have been accommodated in its voluminous fuselage. The launch weight of BOR 4 was 1,000kg (2,200lb). (Original source of photograph not identified)

of about 1,500kg (680lb). They were used to obtain aerodynamic data and measure the characteristics of the thermal protection materials to be used on the Russian space shuttle, BOR-4's re-entry trajectory being identical to that proposed for Buran. The first launch of BOR 4 (Cosmos 1374) was made with an SL-8 from Kapustin Yar on 3 June 1982, the flight being monitored by the Soviet tracking ships *Patsayev*, *Dobrovolsky* and *Chumikan*. Re-entering after one complete orbit, the vehicle then made a cross-range manoeuvre of about 600km (375 miles) and was recovered 560km (350 miles) south of the Cocos Islands in the Indian Ocean. A Royal Australian Air Force (RAAF) patrol aircraft photographed the recovery. The second flight of BOR 4 (Cosmos 1445) took place on 15 March 1983, the tracking ships *Volkov* and *Beljajev* monitoring the flight. Again the vehicle was recovered south of the Cocos Islands and photographed by the RAAF. The third flight was made nine months later (Cosmos 1517, 27 December 1983), but the craft was recovered from the Black Sea to avoid further publicity. The fourth and last flight took place on 19 December 1984 (Cosmos 1614), the recovery again being made from the Black Sea.

The aerodynamic dissimilarity of the BOR 4 and Buran configurations led to the design and testing of BOR 5, a one-eighth-scale version of the space shuttle. The first flight of BOR 5 was made on 4 July 1983, the spacecraft probably being launched by an SL-8 from Kapustin Yar. The recovery was made by parachute over land. At least three more flights were made.

LB-21 and ART-24 (Germany)

Concept studies of reusable space transport systems began in Germany in 1962. Some of these were conducted by the Entwicklungsring-Nord (ERNO) and were to lead, some ten years later, to the development and partial realisation of a lifting-body project designated LB-21 Bumerang, which, according to then current planning, would have provided data to support the development of a European orbiter and crew rescue system that was expected to become operational in 1983.

ERNO's preliminary work in 1962-63 concentrated on vertical-launch, horizontal-landing, two-stage fully-reusable winged launchers, both stages being propelled by liquid-fuelled rocket motors. Interestingly, the second stage, designated Research Project 623, then bore a remarkable resemblance to the much later and last configuration of Hermes, designed to support European missions to the International Space Station. In 1963-64 ERNO, together with the French companies SNECMA and Nord-Aviation, investigated horizontal-take-off-and-landing, two-stage launchers. The delta-winged first stage was ramjet powered, and the rocket-propelled lifting-body second stage was carried underneath the first stage's fuselage.

These studies showed that a much greater knowledge of the flight-mechanic and aerodynamic characteristics of these vehicles was needed for the realisation of such projects. ERNO therefore proposed a programme of development work which could be advanced in incremental steps. Begun in 1966, it was almost entirely concerned with the theoretical and experimental determination of the aerodynamic and aerothermodynamic characteristics of lifting bodies, which were chosen for these studies because it was thought they would provide data applicable to other types of re-entry vehicle. This choice was influenced by work then being done in the USA.

The programme was funded initially by the Bundesministerium für wissenschaftliche Forschung (BWF), later by the Bundesministerium für Bildung und Wissenschaft (BMBW) and afterwards by the Bundes-

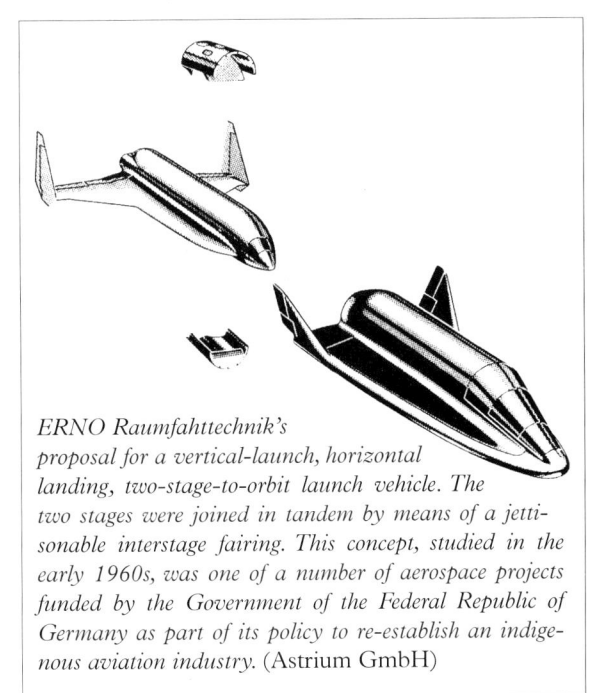

ERNO Raumfahttechnik's proposal for a vertical-launch, horizontal landing, two-stage-to-orbit launch vehicle. The two stages were joined in tandem by means of a jettisonable interstage fairing. This concept, studied in the early 1960s, was one of a number of aerospace projects funded by the Government of the Federal Republic of Germany as part of its policy to re-establish an indigenous aviation industry. (Astrium GmbH)

ministerium für Forschung und Technologie (BMFT). Annual funding of the initial phase of the programme was DM 800,000; by 1970 it had increased to about DM 2 million.

Low-speed windtunnel testing of the first of the lifting-body configurations, LB-1, began in the spring of 1968 and was done by the Institut für Angewandte Gasdynamik of the Deutsche Forschungs- und Versuchsanstalt für Luft- und Raumfahrt (DFVLR) at Porz-Wahn. Supersonic windtunnel tests at speeds up to Mach 4.5 of an aerodynamically-improved version of the lifting body, designated LB-10, were carried out in 1969. These were followed in 1970 by more extensive tests at speeds up to Mach 25 of another variant, LB-21, incorporating additional refinements.

The first phase of free-flight testing of a model of LB-21, named Bumerang I, began on 12 August 1971 with an uncontrolled gliding flight to verify the launch of the vehicle from the cargo hold over the loading ramp of a Transall transport aircraft, to verify the vehicle's parachute recovery system and obtain initial aerodynamic data. The test was conducted over the North Sea northwest of Helgoland, with the support of the Luftwaffe Test Centres E 61 in Manching and E 71 in Eckenförde and the Transport Squadron LTG 63 based at Hohn. Unfortunately, during the launch sequence the parachute used to withdraw the palletised Bumerang from the cargo hold opened too soon and the model, still fastened to its pallet, was pitched violently nose upwards by the turbulent airflow around the rear fuselage of the Transall. On separation from its pallet Bumerang I performed a 180-degree turn manoeuvre, went into a dive, thereby gaining speed, pulled up and nosed over into another dive. It flew in this fashion until the retrieval parachute opened automatically at the desired altitude. The model was recovered intact. Subsequent test flights were conducted in better weather conditions over the Mediterranean from the Italian test centre at Salto di Quirra, Sardinia. The second flight of Bumerang I, on 21 October 1971, even though on this occasion it was successfully air-launched, exhibited the flight characteristics observed during its first test. The model was again recovered and subsequently equipped for radio-controlled flights, for which it was renamed Bumerang II. The next three test flights, the last of which took place on 27 October 1971, were accomplished successfully. Further tests of Bumerang II were scheduled for 1972 but, because of other factors which were to influence the programme, they did not take place until 1973.

Meanwhile, NASA had invited the European industry to participate in its Post-Apollo Programme and, in particular, in the development of the Space Shuttle. The experience and knowledge gained by ERNO Raumfahrttechnik with its lifting-body programme allowed the company to respond quickly and, in 1970, it became a member of the Space Shuttle Phase B development team led by McDonnell Douglas. ERNO, together with the DFVLR, conducted low-speed windtunnel testing and system studies related to the Shuttle MDC 050 configuration. This work culminated in 1973 in free-flight trials of a model, designated Orbiter I, using techniques developed during the testing of the LB-21.

Parallel studies undertaken in the period 1962-71 by both Junkers and Messerschmitt-Bölkow-Blohm (MBB) in Munich also investigated two-stage-to-orbit concepts. In 1972 the results of these studies and the work done by ERNO Raumfahrttechnik were combined and further developed within the framework of an industrial group funded by the BMBW/BMFT and named, ART – Arbeitsgemeinschaft (later Arbeitsprogramm) Rückkehrtechnologie. This group, which included various institutes of the DFVLR and, soon after its formation, Dornier System in Friedrichshafen, proposed and initiated a programme for the development of re-entry technologies that could be applied to a 'second generation' European orbiter. This work culminated in 1983 in the flight demonstration of a test vehicle which could be deployed in orbit by the US Shuttle or launched by the Europa III. (The US Shuttle completed its first orbital mission in April 1982. The development of the Europa III launcher was discontinued in June 1972, when the Europa I/II programme was cancelled.)

The ART programme was subdivided into three relatively distinct phases.

Phase 1 (1972): This began on January 28 with preliminary discussions at Trauen between ERNO Raumfahrttechnik and MBB. General agreement on various aspects was reached at a subsequent meeting in Bremen on March 24. This was followed by a symposium in Munich on July 14, during which lifting-body, winged and 'wave-rider' configuration re-entry test vehicles were

Primarily of wooden construction, the LB-21 Bumerang lifting body was dropped from an underwing pylon on the Transall transport aircraft for radio-controlled, free-flight tests conducted over the western Mediterranean. (Astrium GmbH)

Both MBB's ART-24A (left) and ERNO's ART-24B proposals were investigated in considerable depth and were extensively tested in windtunnels at speeds up to Mach 20. Each configuration had its merits, but the latter was selected for further development. ERNO's proposal was based on work it had done in connection with the MDC 050 Orbiter. (Astrium GmbH)

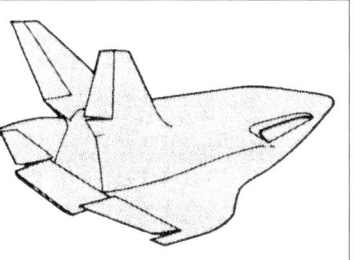

considered. At the next symposium, in Bremen on August 31, it was decided to select two baseline designs for comparison. Both were winged configurations. They were designated ART-24A (MBB proposal) and ART-24B (ERNO proposal). During a succeeding series of monthly meetings the content and schedule of the re-entry technology programme were defined and detailed in considerable depth.

Subsonic windtunnel testing of both configurations was undertaken in this period, and by the end of the year was 50% complete, and the calibration of the models for supersonic windtunnel testing had begun. A preliminary assessment of the vehicles' aerodynamic characteristics was attempted, with the object of selecting the most promising configuration as the baseline for future work. The choice, however, was deferred until more conclusive results from transonic and supersonic windtunnel tests, and from analyses of the vehicles' dynamic flight characteristics, became available.

Phase 2 (1973-76): This phase encompassed the development and definition of the final configuration, and included planning and preparations for subsonic free-flight gliding tests, which were to begin in September 1974. Transonic and supersonic free-flight tests were to be conducted in 1976 with rocket-powered models. Only the final definition of ART-24 was to be flight tested.

By November 1973 supersonic and hypersonic wind-tunnel testing of both configurations up to speeds of Mach 20 had been completed. An evaluation of the aerodynamic characteristics revealed, however, that both configurations needed to be altered slightly to achieve the required overall performance. In general, ART-24A exhibited the more desirable characteristics at hypersonic speeds, and ART-24B better characteristics at lower speeds. These results, combined with those from system analyses, led at the end of the year to the selection of the ART-24B configuration, subsequently redesignated LK 74. Subsonic free-flight testing was to begin in 1974, but the schedule was modified by the decision to postpone further free-flight testing of LB-21 until August 1973 and, consequently, to include LB-21 in the ART programme.

Phase 3 (planned 1977-83): Planning for this phase included the implementation of all ground facilities required for the hypersonic flight testing of small and full scale vehicles and their recovery. The small models were to be flown at speeds between Mach 15 and 25, and the full-scale vehicles were to be used to obtain data at speeds between Mach 7 and 15. Re-entry tests of the full-scale spacecraft were also proposed, these being scheduled to begin in 1982. The wingspan of the re-entry vehicles was limited to 14ft 9in (4.5m) so that they could be accommodated, if required, in the cargo bay of the US Shuttle. (This idea has also been more recently reconsidered for in-orbit and re-entry tests of both the X-38 and X-40A.)

In 1974 all work on these projects was cancelled in favour of Europe's participation in the NASA/European Space Agency (ESA) Spacelab programme, following a decision made in the USA to build the Shuttle without a European contribution to its development and manufacture.

Maia, Falke and Hermes X2000 (Europe)

The Hermes spaceplane was part of ESA's plans to develop an European orbital infrastructure complementary to its contribution to the International Space Station programme. ESA planned to deliver, among other infrastructure elements, a pressurised laboratory attached to the space station and a man-tended, free-flying laboratory (MTFF), which could be docked periodically to the space station to exchange its scientific payload. Hermes was intended to service the MTFF. During the early stages of the development of what became, and is still, known as the Columbus programme, the MTFF was cancelled and the need for Hermes diminished until it too was cancelled in November 1992 and replaced by attempts to develop an unmanned demonstrator, Hermes X2000, and a crew return capsule. Hermes was cancelled for both budgetary and technical reasons.

Within the framework of the Hermes programme, Avions Marcel Dassault-Breguet Aviation Espace and MBB conducted a feasibility study in 1986 for a one-third-scale prototype, intended to verify hypersonic technologies needed for the development of Hermes. Known as Maia, the spacecraft was to be launched on Ariane 4.

Also proposed to support Hermes development was a high-altitude balloon-launched vehicle for verifying the supersonic and subsonic flight characteristics of Hermes. To verify the feasibility of the proposal, designated Falke, a scale model of the Shuttle was launched in this fashion

Work on a European shuttle recommenced in the 1980s with the Hermes project, which would have provided the Europeans with a piloted launch capability. The Hermes programme was closely coupled to ESA's contribution to the International Space Station, at that time named Freedom. *ESA's participation, known as the Columbus programme, envisaged the development of a number of elements attached to and co-orbiting with the space station. Two of the free-flying elements, an unpressurised platform and a pressurised laboratory, both to be used for scientific research purposes, were to be serviced and re-equipped by Hermes, or at the space station itself. Both free-flying elements were dropped from the Columbus programme for budgetary reasons, and with their cancellation the need for Hermes diminished.* (CNES)

for calibration purposes, the telemetry data obtained during the flight afterwards being compared with the Shuttle's performance. One flight was made. The flight test programme with Falke was not continued because of the cancellation of Hermes.

HIMES, OREX, HYFLEX and ALFEX (Japan)

As part of Japan's long-term space plans, the Space Engineering Committee of the Institute of Space and Astronautical Science (ISAS) formed a dedicated working group in 1981 to outline a research programme for the development of winged space vehicles. In 1983 ISAS announced Japan's intention to develop a space shuttle. By 1986 the group was advocating a phased approach to the development of such a spacecraft, the first phase to end with a launch of a small research vehicle on the H-II launcher in 1991. The second phase envisaged the launching of a larger vehicle by the H-II fitted with booster rockets, and the third phase proposed developing a horizontal-take-off variant of the spaceplane. In 1987 Japan revealed its plans, which were expected to culminate in the unmanned H-II Orbiting Plane, HOPE, and which would provide logistic support for Japan's contribution to the International Space Station.

The vehicle considered in the first two phases of the programme was named HIMES (an acronym derived from highly-manoeuvrable experimental spacecraft). This winged vehicle was expected to conduct short-duration scientific experiments in orbit under weightless conditions and to evaluate technologies for future vehicles. Later versions of HIMES were to be launched horizontally from an electromagnetically levitated and propelled sled. The overall length of HIMES was 13.7m (45ft), its span was 9.25m (30ft) and its launch mass was 14,100kg (31,020lb). Five one-seventh-scale models of this spacecraft were tested in 1986 and 1987 by launching them from a helicopter at altitudes of between 1,000 and 2,000m (3,300 and 6,550ft) and a speed of 55m/s (125mph). The tests were conducted at ISAS's Noshiro Test Centre over the Sea of Japan. All models were recovered; the last two flights being regarded as successful. A final test of a slightly larger, rocket-boosted model of HIMES, which was intended to obtain aerothermodynamic data at speeds of between Mach 3.5 and 4.0, conducted on 21 September 1988 at the Kagoshima Space Observation Centre on Kyushu Island, was not successful. It had been intended to lift the model to an altitude of 20km (65,500ft) by balloon before igniting the rocket booster, but the balloon failed to reach the required height and the model was lost at sea.

In 1989 Japan's spaceplane programme was redefined and the first flight of HOPE was postponed until 1999. Its development was to be supported by flight tests of orbital spacecraft and air-launched models. These included:

- OREX   Orbital Re-Entry Experiment
- ALFLEX   Automatic Landing Flight Experiment
- HYFLEX   Hypersonic Flight Experiment

OREX was designed to measure the performance of thermal protection systems during re-entry. Initially scheduled to be orbited in 1992, it was finally launched from Tanegashima Space Centre on 4 February 1994. It completed one orbit and splashed down 460km (300 miles) south of Christmas Island. Although telemetry data was received throughout the flight, OREX was not recovered.

Preparatory flight tests of ALFLEX began in 1992 with trials conducted over a period of 18 months with a

ALFLEX is seen shortly after it was released from a helicopter over Woomera, before making an unmanned, automatic gliding flight and landing. (NASDA)

vehicle resembling the HOPE configuration suspended below a helicopter. ALFLEX was successfully tested at Woomera between 6 July and 15 August 1996, automatic landings following ten flights made during this period.

Meanwhile, HYFLEX had been prepared for launch on the then new Japanese J-1 launcher. It was successfully orbited on 12 February 1996 from Tanegashima Space Centre, and although telemetry data was obtained during its short flight, the vehicle sank soon after splashdown when its flotation gear failed to function. Attempts to locate the vehicle using a remote-controlled deep-sea surveyor on the seabed of the Pacific had to be abandoned.

In July 1997 the overall HOPE development programme was slowed down considerably when Japan decided to make drastic cuts in its space budgets. However, HOPE-X, an unmanned, high-speed flight demonstrator, is scheduled to fly in 2003.

PLATO

PLATO (an acronym derived from platform orbiter), was a small, unmanned, winged re-entry vehicle designed for launch by Ariane and retrieval by conventional landing at a site in Europe. The vehicle was intended to provide resources for conducting material and life science experiments in space and for developing materials, structures, navigation and control techniques, and other technologies required for future recoverable spacecraft. It could also have been used for Earth observation and environmental monitoring missions in polar orbits.

At the time the PLATO project was proposed (April 1987), NASA's Space Transportation System was undergoing extensive modification and testing following the fatal accident to the Shuttle *Challenger* in January 1986, and there was much speculation as to when, or indeed if, the Shuttle would again become operational. This situation led ESA to place contracts to examine the feasibility of launching the European Retrievable Carrier (EURECA) by Ariane 4, or Ariane 5, instead of the Shuttle, with the expectation that the spacecraft would eventually be recovered by the Shuttle. The EURECA free-flying platform, then approaching final completion, was due to be orbited in the summer of 1987. Continuing uncertainty regarding the recommencement of Shuttle operations led to consideration of the possibility of retrieving EURECA using a European system based on the Hermes winged re-entry vehicle, or even a large re-entry capsule. An alternative approach proposed by MBB/ERNO Raumfahrttechnik in Bremen considered

An artist's impression of Kawasaki's proposal for HOPE, one of three submitted by the Japanese industry to NASDA. The unmanned vehicle was designed to provide logistic support for Japan's contribution to the International Space Station and for scientific experiment purposes. The programme was scaled down for budgetary reasons, but NASDA still envisages testing a small-scale vehicle designated HOPE-X. (Kawasaki Heavy Industries)

An artist's impression of PLATO in orbit. The scientific experiments could be accommodated in a large, enclosed bay behind the forward rudder, and observation sensors or a conformal synthetic aperture radar could be mounted flush with the surface on the opposite side to the heatshield. Power was provided by a deployable, gimballed solar array. PLATO was designed to have a mission duration of up to 18 months. During descending flight the spacecraft was controlled by trailing-edge flaps and elevons and by the forward rudder. PLATO was required to perform a 180-degree roll manoeuvre at subsonic speeds before landing, for which skids were fitted to the tips of the winglets and forward rudder. A ¹⁄₁₀-scale model of PLATO was tested in the large, open-section, low-speed windtunnel at Dresden-Klotzsche. The vehicle was designed to be artificially stabilised. (Astrium GmbH)

integrating EURECA's functions and scientific mission profile in the design of a re-entry vehicle. The whole mission, including the landing, was to be performed automatically. A number of design iterations eventually led to a proposal for a very-low-aspect-ratio flying wing with low wing loading. PLATO had a launch mass of about 4,500kg (9,900lb), an overall length of 12m (39ft 4in) and a span of 6.4m (21ft). Subsequent comparative low-speed windtunnel testing of delta and rectangular planforms showed the latter to have the more desirable aerodynamic characteristics, and this planform was consequently used for captive free-flight windtunnel testing of a one-tenth-scale model. One of the more unusual features of the vehicle was that it was required to perform a roll manoeuvre before landing, a measure adopted to simplify the design of its heatshield. This idea was considered during the initial design phases of the later X-34 project. Proposals to conduct scientific and technology development missions are being considered for the X-37 and X-40A programmes.

Work on the PLATO project stopped in 1993 through lack of funding. Although Shuttle missions were suspended following the explosion that destroyed the *Challenger* in January 1986, missions resumed in the autumn of 1988; EURECA was launched by the Shuttle in July 1992 and successfully retrieved after a mission lasting 11 months.

Future perspective

The current worldwide recession has meant that space agency budgets for a number of future retrievable re-entry vehicles have been drastically reduced or cancelled. Work in the USA on the X-33, a demonstrator for the Venture Star, a vertical-launch, horizontal landing, single-stage-to-orbit vehicle envisaged as a successor to the Shuttle, and on the X-34, was stopped in March 2001 for technical and budgetary reasons. Further development of the X-38, which was being undertaken with ESA as a rescue vehicle for the crew of the International Space Station, was stopped later that year in attempts to reduce

NASA's programme costs. The development of NASA's X-37 and the flight testing of the USAF's X-40A are continuing. Japan is developing its own shuttle concept, but at a very much slower pace. In Europe, plans are being made for Hopper, a sub-orbital, reusable, horizontal-take-off-and-landing launcher, for which a sub-scale prototype named Phoenix is in an advanced stage of design.

Bibliography

Berkes, U L, 'Hermes Technology – Basis for Future European Hypersonic Developments', DGLR-89-143, Annual Conference, DGLR, Hamburg, 2-4 October 1989.

Courtois, M, 'The Hermes program – Hermes X2000 project', IAF-92-0830, 43rd Congress, IAF, Washington DC, 28 August - 5 September 1992.

Draper, A C and Sieron, T R, 'Evolution and development of hypersonic configurations 1958-1990. Final Report, July 1990-March 1991', Wright Laboratories, Wright-Patterson AFB, Ohio, September 1991.

Fischenberg, D, et al, 'FALKE - Ein Beitrag zur Flugerprobung von Hermes-Konfigurationen', DLR-Nachrichten, No.61, November 1990, pp.53-57.

Föhlisch, P, 'Entwicklung, Bau und Erprobung des Lifting-Body Freiflugkörpers Bumerang', Astronautik, 1971, Vol.8, No.3/4, pp.118-125.

Hallion, R P, 'The path to Space Shuttle: The evolution of lifting re-entry technology', Journal of the British Interplanetary Society, December 1983, pp.523-541.

Inatani, Y, et al, 'Status of HIMES re-entry flight test project', AIAA-90-5230, 2nd International Conference, Aerospace Planes, AIAA, Orlando, Florida, 29-31 October 1990.

Izumi, T, et al, 'The development status of OREX (Orbital Re-entry Experiment) project in Japan', IAF-91-174, 42th Congress, IAF, Montreal, 5-11 October 1991.

Jenkins, D R, Space Shuttle – The History of Developing the National Space Transportation System (Motorbooks International, Osceola, USA, 1992).

Kidger, N, 'The Soviet Shuttle Story', Spaceflight, January 1990, pp.4-6.

Ladnorg, U, 'Bumerang Lifting Body von ERNO', Flug Revue + Flugwelt International, 6/1972, pp.70-74.

Maita, M, 'Space plane program in Japan', Proceedings, 10th International Symposium, Air Breathing Engines, AIAA, Nottingham, GB, 1-6 September 1991, Vol.1, pp.62-70.

Miller, J, The X-Planes X-1 to X-31 (Aerofax Inc, Arlington, Texas, 2nd Edition, 1988).

Nagatomo, M, et al, 'Low-Speed Glide Characteristics of a Winged Space Vehicle', Journal of Space Technology and Science, Japanese Rocket Society, Vol.4, No.2, 1988.

NASA, 'Flight test results pertaining to the Space Shuttlecraft', NASA TM X-2101, Proceedings, Symposium, NASA Flight Research Center, Edwards, California, 30 June 1970.

Newkirk, D, 'Soviet Space Planes', Spaceflight, October 1990, pp.350-355.

Peebles, C, 'The Origins of the US Space Shuttle', Spaceflight, Part 1, 11 November 1979, pp.435-442; Part 2, 12 December 1979, pp.487-492.

Ransom, S, and Hoffmann, R, 'PLATO – A European Platform Orbiter Concept', Paper, AeroTech 89, Birmingham, GB, 31 October - 3 November 1989.

Sasaki, H, et al, 'Concepts of flight experiments for HOPE development', Proceedings, 17th International Symposium on Space Technology and Science, Tokyo, 20-25 May 1990, Vol.2, pp.1319-1324.

Smith, T W, 'An Approach to Economic Space Transportation', Journal of the Royal Aeronautical Society, August 1966, pp.802-810.

Tolle, H, et al, 'Entwicklung der Konfiguration eines aerodynamischen Wiedereintrittsflugversuchskörpers unter besonderer Berücksichtigung der Stabilität und der Manövrier-barkeit', Paper, 7th International Congress of the Aeronautical Sciences, Rome, 14-18 September 1970.

Williams, T, 'Soviet re-entry tests: a winged vehicle', Spaceflight, May 1980, pp.213-214.

Yamanaka, T, 'Overview of Japanese Aerospace Plane', AIAA-92-5005, AIAA 4th International Aerospace Planes Conference, Orlando, 1-4 December 1992.

Zagainov, G I and Plokhikh, V P, 'USSR aerospace plane program', AIAA-91-5103, 3rd International Conference, Aerospace Planes, AIAA, Orlando,

Index

Page references in *italics* refer to illustrations. Aircraft types and engine types are grouped under their manufacturers. Colons[:] separate aircraft type numbers from page numbers.

A4 rocket 33, *33*
ablative materials 194-195
accelerometers 139
advanced supersonic transport research 19-21, 20, 228-240
Aerial Refuelling Systems Advisory Group (ARSAG) 69
aerodynamic compressibility 29
aerodynamics 10-28
aeroelasticity 96-99
AeroVironment Inc Centurion *108*
aileron reversal speed 98
air data 137-138, 139
air traffic control transponder 145-146
air-to-air refuelling (AAR)
  crossover method 55-56
  flying-boom method 59, 62, 63, 72
  hose-based systems 61
  looped-hose system 56, 61, 71
  military use 62-69
  probe-and-drogue method 59-63, 71-72
  trailing-hose system 57
Airbus Industrie
  A300-600: *18*, *221*
  A300-600ST *18*
  A310: *18*, *170*, *197*, *201*
  A320: *18*, *208*, *219*, *223*
  A321: *18*, *203*
  A330: *18*, *205*, 220
  A340: 10, *10*, *18*, *130*
  A380: 17-18, *17*, *210*, 239
  A400M: 15, *15*
  Corporate Jetliner *18*
  family 17, *17*-*18*, *18*
airflow modelling 204
Airspeed Courier 55, *56*
ALFLEX 250-251, *250*
alternative fuels 132-134
aluminium alloys 100-101, 190-191, 195
aluminium-lithium alloys 107, 108-110
Antonov An-70: *16*
Arado Ar 234B Blitz 115
Armstrong Whitworth
  A.W.23: 56, *57*

Siskin 75
ART programme 248-249, *249*
ASSET 242-243
Atcherley, Richard 55
Aurora project 87, 89
autoclaves 201
Automatic Player Piano Company 215
autopilot 161, 163-164
Avro
  707: 180-182, *181*
  Lancaster *60*, 186
  Tudor VIII: 230, *231*
  Vulcan *66*

BAe EAP demonstrator *108*
British Aircraft Corporation (BAC)
  One-Eleven 16, 232
  221: 180, *180*
  TSR.2: *95*
  VC10: *68*
balloons 73, 74
Beamont, Roland 177
Bell
  X-1: 35, *35*, 36, *36*, 38, *38*, 39, 80, *81*, 100, *100*, 173
  X-2: 80, *105*, 173
  X-5: *39*, 47, *47*, 173
  BMW 003: 115, *115*
Boeing
  707: 83, *127*
  737: 218
  747: 84, 216, *218*, 222
  777: *222*
  2707: 19, 84, 233, *234*, 235, 239
  B-29 Superfortress 36, 78, 79, *79*
  B-47 Stratojet *title page*, 98, *98*, *204*
  B-50: 59
  C-135F *67*
  KB-29: *62*, *63*
  KC-135: 63, *70*
  KC-97: *63*
  Sonic Cruiser 18-19, *19*, 239-240
  V-22 Osprey 22-23, *23*
  X-32: 21-22, *22*, *109*
  X-43: 25
BOR 246-247, *247*
Boulton Paul P111: 185-186, *186*
Bristol
  Centaurus *111*, 111-112
  Olympus 120-121, *121*
  Type 138A: *75*, 76

Type 188: 104, 179, *192*, 193-194, 195
Bruce, Hon Mrs Victor 54
Buran Shuttle 90-91
bypass ratio 128, 129-131

CAD systems 11, 209-211
carbon fibre composite (CFC) 106-107, 191-193, 199-200
Celestial Navigation Trainer (CNT) 213-214
chemical power 134
choking 116
clear-air turbulence (CAT) 83-84
Cobham, Sir Alan 55, 56, *57*, 58-61
Cold War 91
combat training 223-225
combination engine 236
communications aids 167
communications systems 142-146
composite materials 105-108
compressibility 31, 116, 171
computational fluid dynamic evaluation 210
computer-aided design (CAD) 11, 209-211
Concorde 20, 29, 84-85, 191, *191*, 195, 232, *232*-233, 235-237
Convair
  990: *238*, 239
  B-58 Hustler 101, *102*
  F-102A 97
  NB-36H 133, *133*
  R3Y-2 Tradewind *65*
  XF-92A: *39*, *45*, 46, 173-174
  YF-102 Delta Dagger 46, *46*

Dassault Mirage *67*, 123, 174
De Havilland
  DH-4B 53, *54*
  D.H. 9: 54, *56*
  D.H. 60: 56
  D.H.98 Mosquito 189-190, *190*
  D.H.106 Comet 15, 62, *63*, 83, *214*, *215*, 216, *231*, 232
  D.H. 108 Swallow 36, *37*, 173, 177-178
deep-stall phenomenon 16
delta wing design 45-47, 232
Derry, John 177
design 203-211
  CAD systems 209-211
  concepts and process 203-207

details 205-206
electronics 207-211
diffusion bonding 103-104
dimensional tolerance 211
direct-lift concept 21-22
Doppler radar 141, 142
Douglas
  D-558-1 Skystreak *34*, 35, 37, *39*, 173
  D-558-2 Skyrocket *title page*, 34, 35, *37*, *39*
  X-3 Stiletto: 39, 48, 123, *124*, 173-174
Dowding, Sir Hugh 55
drag divergence Mach number 16-17, 31
drag reduction 40
drag rise 30, 31, 32
drive-by-wire technique 26
Dutch-roll phenomenon 16
Eardley Billing Oscillator *212*
economy of scale 229
ejection seats 81-82
endurance 52-72
English Electric
  Canberra *81*
  P.1: 123, *123*, 178
Entwicklungsring Nord
  ART-24B *249*
  LB-21 247-249, *248*
Entwicklungsring Sud VJ 101C 185, *185*
Eurofighter Typhoon 13, *13*, 106-107, *107*, 109, 201, *201*
Eurojet EJ200 engine *128*
European Supersonic Research Programme project 20
Fairey
  F.D.1: 183-184
  F.D.2: 8, 49, *49*, 82, 118-119 (cutaway drawing), 122-123, 178-179
Falklands Conflict 66, *67*
fighter design, latest 13-15
flight deck displays 154-159
flight management system 161, 164, 166
Flight Refuelling Limited 55-57, 58-60
flight training 212-227
fly-by-wire systems 161-163
flying simulators 226
Fokker C-2A 54, *54*
Formula One (F1) motorsport 25-26
France, supersonic development 49

Index

fuel economy 121-122
fuselage stretching process 10

g simulation 223-224
General Aircraft GAL 41: 77, *77*
General Atomics RQ-1A 23, *24*
General Dynamics RB-57F 85
General Electric
  GE4 *234*
  GE90 238
  J47 121-122
  J79 *121*
  J93 124, *125*
Germany
  ballistic missiles 33, *33*
  high-speed aerodynamics 32-33
Gloster
  Javelin 62
  Meteor *60*, 114
graceful design 189-190
Griffiths, A A 126
Grumman
  F9F-5 Panther *64*
  XF-10F-1 Jaguar 173, *174*
Gulfstream and Sukhoi supersonic project *237*
gyroscopes 138, 139

Handley Page
  H.P.88: *182*, 182-183
  H.P.115: *44*, 179-180, *179*
  Harrow 56, *58*
  Victor 64-65, *65*, 69, *69*
  W.10 55, *56*
Hawker
  Harrier *68*, 82-83, 107, 200
  P.1127: 184-185, *185*
head-up display 224
heat exchanger exhaust ducting 197-198
Hermes project 249-250, *250*
hiduminium 84, 101
high-altitude civil airline flying 83-85
high-altitude flying 73-91
high-speed flight 29-52
high-speed flight test vehicles
  France 42, 44-45
  Great Britain 42, 44-45
  Soviet Union 42, 44-45
  United States 41, 44-45
Hiller Ryan XV-142: 23
HIMES 250-251
HOPE 250-251, *251*
Houston-Westland PV3: *74*
HUD techniques 158-159
Hunting H.126 187, *187*
hydrogen power 134, 135
Hyfil 131
HYFLEX 250-251
hypersonic flight 24-25, 89
hypoxia 73

image generating systems 220-221
Imperial Airways C class flying boat 56, *58*
in-flight refuelling 52-72
inertial coupling 48
inertial navigation 139-141
inertial sensing 138-139
insect flight 23-24
integral construction 95-96
Ivchenko PROGRESS/Zaporozhye D-27 engines 16

jet engines, early 113-116
jet training 215-217
Junkers
  Ju 86: *78*, 79
  Jumo 004 engine 115

Kawasaki HOPE proposal *251*
Kevlar 105
Korean War 80
Kotcher, Major E 33-34
  design by *33*

Leduc ramjets 31, 117
0.21 No 02: *31*
LeMay, Curtis Emerson 68
Lifting-body aircraft 90, 247-248
Link Trainers 212-215, *213*
Lippisch, Alexander 32
Lockheed and Lockheed Martin
  A-12/YF-12A/SR-71 Blackbird 29
  C-5A *129*
  Constellation *112*
  F-22A Raptor 29
  F-35: 20-21, *21*
  P-80: 30, *31*, 63
  F-104 Starfighter *82*, 93, *94*, 96, *96*, 196-197
  F-117 Nighthawk 7, 13, *198*, 199
  SR-71 Blackbird 29, 85, *86*, 89, 195, *196*, 197, *237*
  TR-1: 85
  TriStar K.1: *69*
  XC-35: 76, 77
  YP-38 Lightning *29*, 30
lofting 206

magnetic heading reference system 139, 140
magnetic sensing 138
Martin RB-57F: *85*
Martin Marietta
  X-23A: 242, 243-244
  X-24: 90, *90*
  X-24A and B: 90, *90*, 242, *242*, 244, 244-2455
Martindale, A F (Tony) 33
materials 99-110
May, Wesley 53, *53*
MBB ART-24A *249*
McDonnell Douglas
  AV-8B Harrier 107, 200, *200*
  F-15: *88*
  F/A-18 Hornet 99, *205*, 206

F2H-3 Banshee *65*
Phantom II *121*, 188, McFadden six-axis motion platform *220*
Messerschmitt
  Me 163: 30, *30*
  Me 262: 15, 80
  P.1101: 47, 173
Mikoyan
  MiG-15 92
  MiG-23-01: 176, *176*
  MiG-25 Foxbat 85, 195-196
  MiG-I-270: *43*
  Ye-152A: *43*
Miles M.52: 33-34, *34*, 48-49, *116*, 116-117, 122, 171, 177
Miles, F G 116
Mitchell, General 'Billy' 53
Mojave Desert 37, 170
Moog electric motion platform *220*
MUSTARD 245-246, *245*

NASA
  HL-10: 241-242, *242*
  M2: 241-242, *242*
  NASA/Boeing unmanned combat air vehicle *25*
  Space Shuttle: 88, 89-91, *208*
navigation 150-154
navigation aids 146-149, 167-168
navigation, future systems 166-167
nickel-based alloys 132
Nord 1500 Griffon *42*
North American
  AJ-1 Savage 64, *64*
  F-86 Sabre 80, 92, *92*
  F-100 Super Sabre 93, *93*
  P-51 Mustang 31, *31*, 171, *172*
  X-15: 86-87, *87*, *103*, *106*, 134-135, 173, *193*, 194-195
  XB-70 Valkyrie: 104-105, *106*, 124, *125*, 194, 195, 235, *235*
North Atlantic trials 59, 62
Northrop Grumman
  B-2: *14*, 15, 201-202, *202*, 211
  RG-4A Global Hawk 23, *24*
  X-4 Bantam: *39*, 48, 173
nuclear power 132-134

Ohain, Hans Pabst von 115
operating envelope 228-229
operational ceiling in WWI 73-74
Operations *Desert Shield* and *Desert Storm* 67
Orbiter research programmes 241
OREX 250-251
outside loop 171
oxygen 74-75

Panavia Tornado *12*, 70, 103,

104, 197-198, *211*
Parker, Billy 76
performance development 229-232
Piaggio P.111: *78*, 79
PILOT 242, 244, 244-245
piston engines 111-113
PLATO 251-252, *252*
Post, Wiley 76
Powers, Gary 196-197
Pratt & Whitney
  F119 engine *128*
  J58 engine *124*
  JT3 engine 121
  JT11/J58 engine 125
  TF33/JT3D engine 126, *127*
  Wasp Major engine 111, *112*
pressurisation experiments 75-76
pressurised cabins 77-80
PRIME 242, 243-244
production 189-202
propulsion 111-135

quiet supersonic platform (QSP) 19-20

radar altimeter 141-142
radar cross-section (RCS) 87, 198-199
radar sensors 141-142
radar signature 10, 87
railway development and comparisons 228-229
ramjet engines 117, 135
Reid Reaction testing device 214, 217
Republic
  F-84F Thunderstreak 62
  F-105 Thunderchief 66
  XF-91 Thunderceptor *175*
research aeroplanes 29-52
research flying 170-188
Rockwell B-1: 103, *104*, *236*
Rolfe, John 226
Rolls-Royce
  Avon 120, *120*, 122-123
  Conway 126, *126*
  Olympus 232
  RB.163 Spey 128
  RB.211 130, 130-131
  Trent *131*, *132*, 238
Rumpler Taube 4C: *73*
Rutan, Burt, design *209*

SAAB 210: *44*
safety aids 159-160
Saunders-Roe SR.53: 80, *80*
Scaled Composites 151: *209*
scramjet 91
sensors 136-144
SFECMAS Gerfaut *41*
Short
  C-class flying boat 56, *58*
  S.B.4 Sherpa *186*, 186-187
  S.B.5: 183, *183*
  S.C.1: 184

255

short take-off and vertical landing (STOVL) 20-24, 82-83
Shute, Nevil 210
Sikorsky
  HH-3E helicopters 65, 67
  MH-53 Pave Low IV helicopter 72
simulation demand 218-220
simulator costs 217
simulator platforms
  two-axis 216
  four-axis 216
  six-axis 220, 221, 223
simulator testing 226
simulators 212-227
Smith, George F 81-82
sonic boom 235-236
'sound barrier' 80
South Atlantic trials 59
Soviet Union experimental vehicles 175-177
spaceflight 89-91, 241-253
Space Shuttle model 208
Space Shuttle Orbiter 88, 89-91
Spatz, Carl 54
SPIRAL 246-247, 246
spoilers 98-99
Stack, John 32
stainless steel 193-194, 195
steels 104-105, 193-194
Stinson SM-1B Detroiter 55
structural techniques 92-110
subassemblies 189-190, 190
subsonic advantages 237-240
Sud-Oeust SO.9050 Trident 43
Sukhoi Su-47 Berkut 12, 13
Sullivan, James V 86
super-jumbo 238-239
super-plastic forming 103-104
supersonic flight pioneers 36

supersonic transports 228-240
supersonic turbojet engines 122-125, 128
surveillance aids 168
Swain, F R D 75
Sweden, supersonic development 49
sweepback 92-93
swept wing design 45-46
swing wings 235-236
systems 136-169

T-tail 16, 43
test pilots 7
thermosphere 236-237
thin wings 92-93
'Tiger Force' 58, 59
tilt-rotor concept 22-23
titanium 101-103, 124, 125
torsional stiffness 97-98
training 212-227
transonics problems 29-30
transport aircraft development 15-16
transport modes comparison 240
Tupolev
  Tu-16 Badger 61, 61
  Tu-144 85, 233, 233
turbofan engines 84, 125-129, 129-131, 238
turbojet engines 117, 120-122, 122-125, 128
turning performance of fighters 50
twin-dome simulator 224-225
Twiss, Peter 82, 122

unmanned combat air vehicle (UCAV) 11, 25

V1 flying bomb 58
V2 rocket (Vergeltungswaffe Zwei) 33, 33
Variable-geometry wing shape 11-15
variable wing sweep 47-48
vertical/short take-off and landing (VSTOL) 20-24, 82-83
Vickers Valiant 62, 62
Vietnam War 65
Virden, Ralph 30
vortex ring state (VRS) 22

Wallis, Barnes 47
weather radar 142
Weems, P V H 213
Westland
  PV.3: 74, 76
  Sea King HAR helicopter 226
White, Robert M 87
Whittle WU engine 113-114, 113, 114
Whittle, Sir Frank 113-115, 116, 126
Widdifield, Noel F 86
Wide Screen Visual Simulator 224-225, 224, 225
windtunnels 29, 204
wings
  delta 45-47, 232
  high-speed airliner design 16-17
  ogee 232
  skins 94-95, 105-106
  structure 94-95
  swept 45-46
  swing 235-236
  thin 92-93

variable-geometry 11-15
variable-sweep 47-48
Wright turbo-compound engine 111, 112, 112

X-series high-speed flight research aircraft
  contribution to development 174
  contribution to research 38-41, 39
  programme 117
  X-1 (Bell) 35, 35, 36, 38, 39, 80, 81, 100, 173
  X-2 (Bell) 80, 105, 173
  X-3 Stiletto (Douglas) 38, 39, 48, 123, 124, 173-174
  X-4 Bantam (Northrop) 39, 48, 173
  X-5 (Bell) 39, 47, 47, 173
  X-6: 133
  X-15 (North American) 86-87, 87, 103, 106, 134-135, 173, 193, 194-195
  X-23A (Martin Marietta) 242, 243-244
  X-24 (Martin Marietta) 90, 90
  X-24A and B (Martin Marietta) 242, 242, 244, 244-245
  X-32 (Boeing) 21-22, 22, 109
  X-33, X-34, X-35: 252-253
  X-43 (Boeing) 25

Yakovlev Yak-36 177, 177
Yeager, Charles E (Chuck) 36, 37-38, 80, 713